SMALL-TELESCOPE ASTRONOMY ON GLOBAL SCALES
IAU Colloquium 183

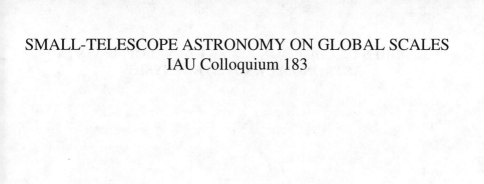

COVER ILLUSTRATION:

Hα survey of the Southern sky (see Gaustad et al., p.75).

A SERIES OF BOOKS ON RECENT DEVELOPMENTS IN ASTRONOMY AND ASTROPHYSICS

Publisher

THE ASTRONOMICAL SOCIETY OF THE PACIFIC
390 Ashton Avenue, San Francisco, California, USA 94112-1722
Phone: (415) 337-1100   Fax: (415) 337-5205
E-Mail: catalog@aspsky.org   Web Site: www.aspsky.org

A listing of all other ASP Conference Series Volumes and IAU Volumes
published by the ASP is cited at the back of this volume

ASTRONOMICAL SOCIETY OF THE PACIFIC
CONFERENCE SERIES

Volume 246

SMALL-TELESCOPE ASTRONOMY ON GLOBAL SCALES
IAU Colloquium 183

Proceedings of a Colloquium held in
Kenting, Taiwan
4-8 January 2001

Edited by

Wen-Ping Chen
Graduate Institute of Astronomy, National Central University
Chung-Li, Taiwan, R.O.C.

Claudia Lemme
Graduate Institute of Astronomy, National Central University
Chung-Li, Taiwan, R.O.C.

and

Bohdan Paczyński
Princeton University, Princeton, New Jersey, USA

SEP/AE
MATH

Library of Congress Cataloging in Publication Data
Main entry under title

Card Number: 2001093955
ISBN: 1-58381-084-6

ASP Conference Series - First Edition

Printed in United States of America by Sheridan Books, Chelsea, Michigan

Conference Poster xii
Preface xiii
List of Participants xiv
Conference Photograph xvii

Opening Remarks
Bohdan Paczyński 1

I. Telescope Arrays and Networking

The Whole Earth Telescope: An International Adventure in
Asteroseismology
Steven D. Kawaler 3

MONET: a MOnitoring NEtwork of Telescopes
Frederic V. Hessman 13

The STARBASE Network of Telescopes and the Detection of Extrasolar
Planets
Charles H. McGruder, III, Mark E. Everett and Steve B. Howell 23

STARBASE: A Network of Fully Autonomous Telescopes for Hands-on
Science Education
Richard Gelderman, David Barnaby, Michael Carini et al. 31

The NORT: Network of Oriental Robotic Telescopes
Roger Hajjar, François R. Querci and Monique Querci 33

Taiwan Oscillation Network and Small-Telescope Research at
Tsing Hua University
Dean-Yi Chou 39

II. Monitoring and Surveys

Monitoring Variability of the Sky
Bohdan Paczyński 45

The All Sky Automated Survey (ASAS-3) System – Its Operation and
Preliminary Data
G. Pojmański 53

RTLinux Driven Hungarian Automated Telescope for All Sky Monitoring
Gáspár Á. Bakos 59

The Grid Giant Star Survey for the Space Interferometry Mission
Richard J. Patterson, Steven R. Majewski, Catherine L. Slesnick et al. ... 65

A Robotic Wide-Angle Hα Survey of the Southern Sky
John E. Gaustad, Wayne Rosing, Peter McCullough et al. 75

MOA Extra-Solar Planet Research via Cluster Supercomputing
Nicholas Rattenbury, Ian Bond, Phil Yock et al. 77

YSTAR :
Yonsei Survey Telescopes for Astronomical Research
Yong-Ik Byun, Won-Yong Han, Yong-Woo Kang et al. 83

III. New Trends in Small-Telescope Technology

The 1.3-meter Robotically Controlled Telescope: Developing a Fully
Autonomous Observatory
Richard Gelderman 89

The Development of Advanced-Technology Automated/Robotic Telescope
Systems and the Future of Small-Telescope Astronomy
Richard J. Williams and James Mulherin 95

Progress Report for the KAO 1.0 meter Robotic Telescope
Peter Mack, Wonyong Han, Matthew Bradstreet et al. 103

ACE FlexGrid Telescope Flexure and Pointing Software
Peter Mack, John Stein, and Wonyong Han 111

Development of the Far-ultraviolet Imaging Spectrograph on
KAISTSAT-4
Wonyong Han, Kyoung Wook Min, Jerry Edelstein et al. 113

Design of a Multi-CCD Controller
Yong-Woo Kang, Yong-Ik Byun, Sung-Yeol Yu et al. 115

The Multiple Telescope Telescope, an Inexpensive Fiber Fed
Spectroscopic Facility
Reed L. Riddle and William G. Bagnuolo, Jr. 117

IV. Transient Events

The Lick Observatory Supernova Search with the Katzman Automatic
Imaging Telescope
Alexei V. Filippenko, W.D. Li, R.R. Treffers et al. 121

The Beijing Astronomical Observatory Supernova Survey
Yulei Qiu, Jingyao Hu, and Weidong Li 131

Optical Identification of Gamma-Ray Bursts at Kenting Observatory
Wei-Hsin Sun and Shun-Tang Tseng 143

Early Results from HETE-2
N. Kawai, A. Yoshida, T. Tamagawa et al. 149

RIBOTS: An Automatic Telescope System for Gamma-Ray Burst
Follow-Up Observations
Yuji Urata, Nobuyuki Kawai, Atsumasa Yoshida et al. 155

A Spectrograph for Prompt Observations of Gamma-Ray Bursts with a
1-m Telescope
Tetsuya Kawabata, Kazuya Ayani, Mitsugu Fujii et al. 157

V. Variability Study

Cataclysmic Variables: A 'SWOT' Analysis
Brian Warner 159

Optical and Near-IR Monitoring of Symbiotic Binary Systems
Joanna Mikołajewska 167

The Rotation and Variability of T Tauri Stars: Results of Two Decades
of Monitoring at Van Vleck Observatory
W. Herbst 177

High-Speed Photometry of Bright roAp Stars With Small Telescopes
D. W. Kurtz 187

The Moscow Long-Term Program of Cepheid Radial Velocities
Nikolai Samus and Natalia Gorynya 197

High Speed CCD Photometry of Flare Stars
Sun-Youp Park and Yong-Ik Byun 203

Search for δ Scuti Type Pulsating Components in Eclipsing Binary
Systems
S.-L. Kim, J.W. Lee, J.-H. Youn et al. 205

Long-period Red Variables in the Large Magellanic Cloud from the MOA
Database
Mine Takeuti, S. Noda, F. Abe et al. 207

Spectroscopic Detection of an Extraordinary Flaring-Event on DF Tau
J.Z. Li, W.P. Chen, and W.H. Ip 209

Search for Variable Stars in the Open Cluster NGC 2539
K.J. Choo, S.-L. Kim, T.S. Yoon et al. 211

The Ongoing Search for Variables in Young Clusters: Up-to-Date Results
and Perspectives
A. Pigulski, G. Kopacki, Z. Kołaczkowski et al. 213

Variable Stars in the Globular Cluster M92
Grzegorz Kopacki 216

Blazhko Cycles of ω Centauri RRab Stars
Johanna Jurcsik 217

A Near Infrared Camera Refrigerated by Two Stirling Machines –
an Alternative to Robotic Telescopes
José K. Ishitsuka I., Takehiko Wada, Fumihiko Ieda et al. 219

Observations of Variable Stars by the 76-cm SuperLight Telescope of NCU
J.Z. Li, C.H. Wu, Z.W. Zhang et al. 221

Observations of Variable Stars With a Small Telescope at Tabriz
University
D.M.Z. Jassur, F. Adabi, and N.A. Cham 223

VI. Solar System Studies

The Role of Small Telescopes in the Discovery and Follow-up of
Near Earth Objects
Andrea Boattini 233

NEOPAT: Near-Earth Object PATrol program
Hong-Kyu Moon, Moo-Young Chun, Yong-Ik Byun et al. 240

An Education Program Using Tera-Byte NEA Observation Data
A. Asami, D.J. Asher, T. Hashimoto et al. 245

CCD Photometry of Two Asteroids (895) Helio and (165) Loreley
H.-S. Woo, S.-L. Kim, M.-Y. Chun et al. 251

The Taiwan-America Occultation Survey for Kuiper Belt Objects
Sun-Kun King 253

Distinguishing KBO from Asteroid Occultations in TAOS
Claudia Lemme and Chyng-Lan Liang 259

The Humps of KBO's Size Distribution
Cheng-Pin Chen and Ing-Guey Jiang 261

The True Colors of KBOs
Hui-Chun Hsu and Wing-Huen Ip 263

Revealing Variety of Comets by Long-Term Monitoring Observation
with a 50-cm Telescope
Jun-ichi Watanabe and Hideo Fukushima 265

A Simulation of Shell Structures of Comet Hale-Bopp in February 1997
J. Tao and B.C. Qian 271

Cometary Polarimetry
Asoke K. Sen 275

Observation of Comet C/1999 S4 (LINEAR)
C.Y. Lin and W.H. Ip 277

High Spatial Resolution Observations of the 1998 and 1999 Leonid Meteors
X.J. Jiang and J.Y. Hu 279

Project MONICA for the Study of Time-Variable Phenomena of the
Jovian Sodium Cloud and the Io Plasma Torus
Chien-Pang Chang and Wing Ip 281

VII. Science With Small Telescopes

Long-Term Coudé Radial-Velocity Studies With a 1.2-m Telescope
C.D. Scarfe 283

INTEGRAL and Small Telescopes
Nami Mowlavi, Peter Kretschmar, Marc Türler et al. 289

The Carl Sagan Observatory: A Telescope for Everyone
J. Saucedo-Morales, A. Sánchez-Ibarra, and D. Lunt 295

The NCU Lu-Lin Observatory
Wean-Shun Tsay, Alfred Bing-Chih Chen, Kuang-Hsiang Chang et al. ... 299

Monitoring of AGNs at the Shanghai Astronomical Observatory
B.C. Qian and J.Tao 305

Fast Drift-Scan CCD Imaging and Photometry with Small Telescopes:
Lunar Occultations and Speckle Interferometry
Jorge Núñez and Octavi Fors 309

The Automated Telescope of Novosibirsk State University
A. Nesterenko, M. Nikulin, D. Vyprentsev et al. 315

Some Aspects of Astronomy at Maidanak Observatory
Alisher S. Hojaev 317

Observational Results with the 1 Meter Telescope at Yunnan Observatory
During 1990-2000
P.S. Chen, and W.Y. Zhang 324

Systematic Spectroscopic Observations on Small Telescopes:
Past and Future Research of Stellar Kinematics
M.E. Sachkov, E.V.Glushkova, and A.S.Rastorguev 327

Advantages and Drawbacks of the ISM Method in Globular Cluster
Photometry
József M. Benkő 329

Spatial Structure of Star Clusters by the 2MASS Database
J.W. Chen and W.P. Chen 331

The Kinematics of Globular Cluster NGC 288
Chan-Kao Chang, Alfred B. Chen, Wean-Shun Tsay et al. 333

Bright Young Star Candidates in the Rosette Nebula
P.S. Chiang, W.P. Chen, J.Z. Li et al. 335

Deprojection of Planetary Nebulae
Z.W. Zhang and W.H. Ip 337

Chemical Abundances of the Planetary Nebulae NGC 2392 and
NGC 3242
C.H. Wu, J.Z. Li, Z.W. Chang et al. 339

Former Soviet Union / West Europe Consortium for AGN Monitoring
N.G. Bochkarev, A.I. Shapovalova and A.N. Burenkov 341

Russian/Former Soviet Union Experience in Small Telescope Usage for
Investigation of Interstellar Matter (ISM) and Nebulae
N.G. Bochkarev 343

Russian/Former Soviet Union Experience in Professional Small Telescope
Usage
N.G. Bochkarev 345

Concluding Remarks
Brian Warner 353

Appendix: a List of Robotic Telescopes
Frederic V. Hessman 357

Small IAU Colloquium 183
Telescope Astronomy on Global Scales
小型望遠鏡 ⋯⋯

January 4 (JD2451914.0) to 8, 2001
Kenting National Park, Taiwan (E120°47', N21°57')

Scientific Programs
Telescope Networking and Array
Sky- and Time-Coveraged Surveys by Small Telescopes
Observations of Transient Events
Challenges in Data Archiving, Processing, and Communications
Future of Small-Telescope Astronomy
New Trends in Hardware Design and Manufacturing

To Contact
Ms. Kelly Chen, c/o IAUC183
Graduate Institute of Astronomy, National Central University,Chung-Li, 32054 Taiwan
Phone: +886-3-426-2302
Facsimile: +886-3-426-2304
Email: iauc183@joule.phy.ncu.edu.tw
www.astro.ncu.edu.tw/iauc183

SOC
Charles Alcock (USA)
Yong-Ik Byun (Korea)
Wen-Ping Chen (Taipei, China; Co-Chair)
Alex Fillipenko (USA)
Syuzo Isobe (Japan)
Don Kurtz (South Africa)
Joanna Mikolajewska (Poland)
Bohdan Paczynski (USA; Co-Chair)
John Percey (Canada)
Hui-Song Tan (Nanjing, China)
Shyam N. Tandon (India)
Andrzej Udalski (Poland)

LOC
Hsiang-Kuang Chang (NTHU)
Wen-Ping Chen (NCU; Chair)
Claudia Lemme (NCU)
Chih-Kang Chou (NCU)
Rue-Ron Hsu (NCKU)
Wing Ip (NCU)
Chorng-Yuan Hwang (NCU)
Sun-Kun King (IAA)
Wei-Hsin Sun (NCU)

Sponsored by
International Astronomical Union(IAU)
National Science Council (ROC)
Ministry of Education (ROC)
Chinese Astronomical Society (Taipei)
Chinese Physics Society (Taipei)
NCU Foundation

Preface

One particular advantage of editing something is that one has to read it from cover to cover. For the Proceedings of the International Astronomical Union Colloquium 183 on Small-Telescope Astronomy on Global Scales, this advantage has been well rewarding. The meeting took place during January 4 to 8, 2001 at the very southern tip of Taiwan, in the Kenting National Park. Some one-hundred and sixty people, from more than 20 countries, attended the meeting in which, beautiful beaches notwithstanding, the science by, and emerging technology on, small telescopes—with emphasis on global networking and time/sky monitoring—were reported and discussed.

We would like to express our gratitude for the authors who turned in their manuscripts in a timely fashion, in response to our policy to expedite the publication process, and for their patience in our pestering about deadlines and editing changes. These Proceedings collect 75 papers presented during the meeting, the abstracts of which, along with the conference group photo, can be found on the conference website www.astro.ncu.edu.tw/iauc183/.

We thank the faculty and the graduate students of the National Central University for their enduring help before and during the meeting. Without them, the meeting would not have been successful, or even possible. We are also grateful to Debbie Nester for proofreading the manuscripts, Peggie Cheng, Kelly Chen, and Jessica Chen for their clerical assistance, and Chan-Kao Chang for designing the conference poster which, instead of the stereotype dark or deep blue backgound color, carries a delight tone of sky and ocean blue.

On behalf of the Science Organizing Committee and the Local Organizing Committee, we thank the International Astronomical Union, the National Science Council, the Ministry of Education, the Chinese Physics Society (ROC), the Chinese Astronomical Society (ROC), the NCU Foundation, and the Chung-Hua Foundation for financial supports.

Wen Ping Chen
Claudia Lemme
Bohdan Paczyński

June 2001

List of Participants

Name	Affiliation	Email
Charles Alcock	U. of Pennsylvania	alcock@hep.upenn.edu
Gaspar Bakos	Konkoly Obs.	bakos@konkoly.hu
Jozsef Benko	Konkoly Obs.	benko@konkoly.hu
Andrea Boattini	Ast Obs. of Rome	boattini@ias.rm.cnr.it
Nikolai Bochkarev	Sternberg Ast. Inst.	boch@sai.msu.ru
Yong-Ik Byun	Yonsei U.	byun@darksky.yonsei.ac.kr
Chien-Pang Chang	Nat. Central U.	m889001@astro.ncu.edu.tw
Hsiang-Kuang Chang	Nat. Tsing Hua U.	hkchang@phys.nthu.edu.tw
Ming-Hsin Chang	Nat. Central U.	changmh@venus.seed.net.tw
Ya-chun Chang	Nat. Central U.	rex@astro.ncu.edu.tw
Yung-Hsin Chang	Nat. Central U.	lidodo@astro.ncu.edu.tw
Chan-Kao Chang	Nat. Central U.	rex@astro.ncu.edu.tw
Alfred Bing-Chih Chen	Nat. Central U.	alfred@astro.ncu.edu.tw
An-Le Chen	Taipei Ast. Museum	alchen@tam.gov.tw
Chin-Ming Chen	Nat. Taiwan Normal U.	jimmy@sgrb2.geos.ntnu.edu.tw
Hui-Chen Chen	Nat. Central U.	m889003@astro.ncu.edu.tw
Jing-Wei Chen	Nat. Central U.	awei@outflows.astro.ncu.edu.tw
Jui-Fu Chen	Nat. Central U.	m899012@astro.ncu.edu.tw
Kelly Ke-Li Chen	Nat. Central U.	kelichen@astro.ncu.edu.tw
Pei-Sheng Chen	Yunnan Obs.	iras@public.km.yn.cn
Wen-Ping Chen	Nat. Central U.	wchen@astro.ncu.edu.tw
Po-Shih Chiang	Nat. Central U.	skylion@outflows.astro.ncu.edu.tw
Huey-Ling Chiou	Nat. Taiwan Normal U.	sibyl@sgrb2.geos.ntnu.edu.tw
Shwu-Huey Chiou	Nat. Central U.	m899003@astro.ncu.edu.tw
Kyoung Ja Choo	Kyungpook Nat. U.	kjchoo@sirius.knu.ac.kr
Din-Yi Chou	Nat. Tsing Hua U.	chou@phys.nthu.edu.tw
Mei-Yin Chou	Nat. Central U.	m889010@astro.ncu.edu.tw
Chih-Kang Chou	Nat. Central U.	chou@astro.ncu.edu.tw
Min-Shu Chu	Nat. Central U.	m889007@astro.ncu.edu.tw
Moo-Young Chun	Korea Ast. Obs.	mychun@boao.re.kr
Wan-Ling Chung	Nat. Taiwan Normal U.	hitoshi@ms18.hinet.net
Kem Cook	LLNL	kcook@llnl.gov
Der-Chang Dai		
Zi-Gao Dai	Nanjing U.	daizigao@public1.ptt.js.cn
Alex Filippenko	UC, Berkeley	alex@astro.berkeley.edu
Hsieh-Hai Fu	Nat. Taiwan Normal U.	geofv027@scc.ntnu.edu.tw
John E. Gaustad	Swarthmore College	jgausta1@swarthmore.edu
Richard Gelderman	Western Kentucky U.	gelderman@wku.edu
John Gon	Nat. Central U.	johngon@astro.ncu.edu.tw
Roger Hajjar	Notre Dame U.	rhajjar@ndu.edu.lb
Wonyong Han	Korea Ast. Obs.	whan@kao.re.kr
Debra K. Herbst	Guest of W. Herbst	wherbst@wesleyan.edu
William Herbst	Wesleyan U.	wherbst@wesleyan.edu
Frederic V. Hessman	Goettingen U.	hessman@uni-sw.gwdg.de
Pei-Li Ho	Central Weather Bureau	hopeili@cwb.gov.tw
Alisher Hojaev	Mirzo Ulugh Beg Ast. Inst.	ash@astrin.uzsci.net
Jin-Liang Hou	Shanghai Ast. Obs.	hjlyx@center.shao.ac.cn
Shu-Huei Hou	Nat. Central U.	m899009@astro.ncu.edu.tw
Hui-Chun Hsu	Nat. Central U.	coronaa@pchome.com.tw
Rue-Ron Hsu	Nat. Cheng Kung U.	rrhsu@mail.ncku.edu.tw
Jin-Yao Hu	Beijing Ast. Obs.	hjy@class1.bao.ac.cn

Name	Affiliation	Email
Hui-Tsuan Huang	Nat. Taiwan Normal U.	hspring@sgrb2.geos.ntnu.edu.tw
Kui-Yun Huang	Nat. Central U.	m899002@astro.ncu.edu.tw
Mei-Ting Huang	Nat. Central U.	m899013@astro.ncu.edu.tw
Chorng-Yuan Hwang	Nat. Central U.	hwangcy@astro.ncu.edu.tw
Wing-Huen Ip	Nat. Central U.	wingip@astro.ncu.edu.tw
Jose Kaname Ishitsuka	U. of Tokyo	pepe@chianti.c.u-tokyo.ac.jp
Syuzo Isobe	Nat. Ast Obs.	isobesz@cc.nao.ac.jp
D.M.Z. Jassur	Tabriz U.	jassur@ark.tabrizu.ac.ir
Ing-Guey Jiang	Academia Sinica	jiang@asiaa.sinica.edu.tw
Xiao-Jun Jiang	Beijing Ast. Obs.	xjjiang@class1.bao.ac.cn
Ho Jin	Korea Ast. Obs.	jinho@kao.re.kr
Shiaw-Eel Juang	Nat. Taiwan Normal U.	s44082@cc.ntnu.edu.tw
Johanna Jurcsik	Konkoly Obs.	jurcsik@konkoly.hu
Yong-Woo Kang	Yonsei U.	ywkang@galaxy.yonsei.ac.kr
Tetsuya Kawabata	Bisei Astronomical Obs.	kawabata@bao.go.jp
Nobuyuki Kawai	RIKEN	nkawai@postman.riken.go.jp
Steve Kawaler	Iowa State U.	sdk@iastate.edu
Jonathan Kemp	Ctr for Backyard Astrophysics	j.kemp@jach.hawaii.edu
Seung-Lee Kim	Korea Ast. Obs.	slkim@kao.re.kr
Sun-Kun King	Academia Sinica	skking@asiaa.sinica.edu.tw
Daisuke Kinoshita	Grad U. for Adv. Studies	daisuke@pub.mtk.nao.ac.jp
Grzegorz Kopacki	Ast. Inst. of Wroclaw U.	kopacki@astro.uni.wroc.pl
Yi-Jehng Kuan	Nat. Taiwan Normal U.	kuan@sgrb2.geos.ntnu.edu.tw
Hui-Jean Kuo	Precision Inst. Development Ctr	hjkuo@pidc.gov.tw
Donald W. Kurtz	U. of Central Lancashire	dwkurtz@uclan.ac.uk
Wei-Chien Lai	Nat. Central U.	m889006@astro.ncu.edu.tw
Megan E. Lavery	Guest of R. Gelderman	
Hsu-Tai Lee	Nat. Central U.	eridan@astro.ncu.edu.tw
Tse-Lin Lee	Nat. Central U.	m899001@astro.ncu.edu.tw
Typhoon Lee	Academia Sinica	typhoon@earth.sinica.edu.tw
Claudia Lemme	Nat. Central U.	lemme@astro.ncu.edu.tw
Jin-Zeng Li	Nat. Central U.	ljz@astro.ncu.edu.tw
Yang-Shyang Li	Nat. Central U.	u860232@phy.ncu.edu.tw
Huan-Hsin Li	Nat. Central U.	huanhsin@ms5.hinet.net
Chung-Yi Lin	Nat. Central U.	m889008@astro.ncu.edu.tw
Shih-I Lin	Nat. Taiwan Normal U.	kiwi519@cm1.hinet.net
Yi-Hui Lin	Nat. Central U.	d862001@joule.phy.ncu.edu.tw
Zue-Yuan Lin	Nat. Central U.	m899010@astro.ncu.edu.tw
Ming-Hsiung Liu	Nat. Central U.	m899011@astro.ncu.edu.tw
Xue-Fu Liu	Beijing Normal U.	Lxf@bnu.edu.cn
Kowk-Yung Lo	Academia Sinica	kyl@asiaa.sinica.edu.tw
Phillip Lu	Western Connecticut U.	PhillipKLu@aol.com
Peter Mack	ACE, Inc.	mack@astronomical.com
Charles McGruder	Western Kentucky U.	charles.mcgruder@wku.edu
Krisztina Meiszel	Guest of G. Bakos	
Joanna Mikolajewska	N. Copernicus Ast. Ctr	mikolaj@camk.edu.pl
Nami Mowlavi	INTEGRAL Sci Data Ctr	nami.mowlavi@obs.unige.ch
Igor Nesterenko	Inst. of Nuclear Physics	nesterenko@inp.nsk.su
Jorge Nunez	Univ. de Barcelona	jorge@am.ub.es
Bohdan Paczynski	Princeton U.	bp@astro.princeton.edu
Danny Pan	U. of Maryland	pan@wam.umd.edu
Jang-Hyun Park	Korea Ast. Obs.	jhpark@kao.re.kr
Sun-youp Park	Yonsei U.	sunyoup@galaxy.yonsei.ac.kr

Name	Affiliation	Email
Joe Patterson	Columbia U.	jop@astro.columbia.edu
Richard Patterson	U. of Virginia	ricky@virginia.edu
Miao-Ling Peng	Nat. Taiwan Normal U.	mlpen@sgrb2.geos.ntnu.edu.tw
Andrzej Pigulski	Ast. Inst. of Wroclaw U.	pigulski@astro.uni.wroc.pl
Grzegorz Pojmanski	Warsaw U. Obs.	gp@sirius.astrouw.edu.pl
Bo-Chen Qian	Shanghai Ast. Obs.	qbc@center.shao.ac.cn
Yu-Lei Qiu	Beijing Ast. Obs.	qiuyl@nova.bao.ac.cn
Nicholas Rattenbury	U. of Auckland	nrat001@phy.auckland.ac.nz
Reed Riddle	Iowa State U.	riddle@iastate.edu
Mikhail Sachkov	Russian Acad. of Science	msachkov@inasan.rssi.ru
Nikolai Samus	Russian Acad. of Science	samus@sai.msu.ru
Julio Saucedo	Univ. de Sonora	jsaucedo@cosmos.cifus.uson.mx
Ann Scarfe	Guest of C. Scarfe	SCARFE@phys.UVic.CA
Colin Scarfe	U. of Victoria	scarfe@uvic.ca
Asoke Kumar Sen	Assam U.	aksen@dte.vsnl.net.in
Ching-Yu Shao	Chung-Ping Senior High	shu94549@ms9.hinet.net
Guo-Xuan Song	Shanghai Ast. Obs.	gxsong@center.shao.ac.cn
Han-Tzong Su	Nat. Cheng Kung U.	htsu@mail.ncku.edu.tw
Wei-Hsin Sun	Nat. Central U.	sun@astro.ncu.edu.tw
Mine Takeuti	Tohoku U.	takeuti@astr.tohoku.ac.jp
Sadako Takeuti	Guest of M. Takeuti	takeuti@astr.tohoku.ac.jp
Jun Tao	Shanghai Ast. Obs.	taojun@center.shao.ac.cn
Yi Tong	Beijing Normal U.	yitong@bnu.edu.cn
An-Le Tsai	Nat. Central U.	
Min-Yan Tsai	Nat. Central U.	m889009@astro.ncu.edu.tw
Wean-Shun Tsay	Nat. Central U.	tsay@astro.ncu.edu.tw
Shun-Tang Tseng	Nat. Central U.	m889005@astro.ncu.edu.tw
Wei-Ling Tseng	Nat. Taiwan Normal U.	wltseng@sgrb2.geos.ntnu.edu.tw
Yuji Urata	RIKEN	urata@crab.riken.go.jp
Jing-Hua Wang	Nan-Kang High School	shu94549@ms9.hinet.net
Shiang-Yu Wang	Academia Sinica	sywang@asiaa.sinica.edu.tw
Jen-Hung Wang	Academia Sinica	andrew@asiaa.sinica.edu.tw
Brian Warner	U. of Cape Town	warner@physci.uct.ac.za
Junichi Watanabe	Nat. Astronomical Obs.	jun.watanabe@nao.ac.jp
Marina Watanabe	Guest of J. Watanabe	
Naoki Watanabe	Guest of J. Watanabe	
Yoshie Watanabe	Guest of J. Watanabe	jun.watanabe@nao.ac.jp
Yumi Watanabe	Guest of J. Watanabe	jun.watanabe@nao.ac.jp
Jian-Yan Wei	Beijing Ast. Obs.	wjy@yac.bao.ac.cn
Douglas Welch	LLNL/McMaster U.	welch@physics.mcmaster.ca
Richard Williams	Torus Technologies	richjwilliams@msn.com
Hwa Sung Woo	Kyungpook Nat. U.	hswoo@sirius.knu.ac.kr
Chang-Jen Wu	Nat. Taiwan Normal U.	bronto@cm1.hinet.net
Chi-Hsiung Wu	Nat. Central U.	
Chi-Hung Yan	Nat. Taiwan Normal U.	tseng@sgrb2.geos.ntnu.edu.tw
Tae Seog Yoon	Kyungpook Nat. U.	yoonts@knu.ac.kr
Dah-Lih You	Nat. Central U.	m899005@astro.ncu.edu.tw
Sung-Yeol Yu	Yonsei U.	astro96@nownuri.net
Wen-Yuan Zhang	Yunnan Obs.	iras@public.km.yn.cn
Zhi-Wei Zhang	Nat. Central U.	m889002@astro.ncu.edu.tw
Jun-liang Zhao	Shanghai Ast. Obs.	jlzhao@center.shao.ac.cn

International Astronomical Union Colloquium 183

第 一 八 三 屆 國 際 天 文 聯 合 會 學 術 研 討 會

Small-Telescope Astronomy on Global Scales 大規模的小型望遠鏡天文學

Small-Telescope Astronomy on Global Scales
ASP Conference Series, Vol. 246, 2001
W.P. Chen, C. Lemme, B. Paczyński

Opening Remarks

It is a great pleasure to see so many astronomers at a meeting devoted to small telescopes. It's a clear indication that while a lot of effort, and even more funds, are directed to build ever bigger and more expensive telescopes, there is plenty of most interesting science to be done with small and low cost instruments. It is not clear what is the upper size limit for a telescope to be classified as small: 1 meter? 2 meters? perhaps even larger. And what is the lower size limit? Note that among the most spectacular discoveries made in 1999, two were done with instruments of only 10 cm in diameter. These were, the first (and so far the only) optical flash associated with gamma ray burst GRB 990123, and the first (and so far the only) planetary transit in front of a star (HD 209458). Perhaps during this meeting we shall learn about interesting science done with even smaller instruments.

I am convinced that while there are very important astrophysical goals that can be achieved only with the largest telescopes, there are other important problems for which small instruments are far more efficient. We shall certainly hear about many of these during this meeting.

Only small instruments, because of their low cost, allow for a great luxury for their users: there is no need to fight with Time Allocation Committees. To be truly efficient, small instruments have to be made as fully robotic as possible. I am tempted to use these two properties: no TAC and being fully robotic, as the definition of a modern small instrument. Certainly, these are among the most important properties that make small instruments so attractive to me, and perhaps to many of you as well.

A very special application of small telescopes are global networks. The Internet will certainly lead to the expansion of networks, perhaps well beyond our current imagination, certainly well beyond the current systems which will be presented during this meeting. It is possible, even likely, that monitoring all sky for any unusual events will be done from many sites, to provide instant alerts to be used for follow-up observations by a variety of large telescopes.

Another very useful feature of small telescopes is their usefulness for education and public outreach, made possible by their low cost. This is very important, not only in our search for funds, but also in justifying what we do. In my view astronomy, just as any science, is an entertainment, certainly for the scientists, but also for the general public, which, one way or the other, is the source of our funds. We are lucky that astronomy is popular and that very many people find it attractive. While big telescopes are attractive because of their size, small instruments allow a broad range of amateurs and students to actually have a "hands-on" experience. And even more: non-professionals can do genuine science with small telescopes. A very well known example is provided by the observations of selected variable stars. It is less well known that thousands of new variable stars can be discovered by amateurs using very small instruments.

As enthusiastic as I am about small telescopes, and as much as I am looking forward to the next few days, I am only responding to the initiative of professor Wen Ping Chen. It was his idea to have this meeting, and it was his effort and a huge amount of work that made it happen. I would like to use this opportunity to thank him for all his efforts, and let us all enjoy the meeting.

Bohdan Paczyński

Small–Telescope Astronomy on Global Scales
ASP Conference Series, Vol. 246, 2001
W.P. Chen, C. Lemme, B. Paczyński

The Whole Earth Telescope: An International Adventure in Asteroseismology

Steven D. Kawaler

Department of Physics and Astronomy, Iowa State University, Ames, IA 50014 USA

Abstract. Today, we are beginning to probe the interior of stars through the new science of stellar seismology. Certain stars, ranging from our own Sun to white dwarfs, undergo natural vibrations that can be detected with sensitive time-series photometry and/or spectroscopy. Since the signal we seek is an unbroken time-series to allow determination of the vibration frequencies, data from a single-site is usually incapable of uniquely identifying the pulsation modes, no matter how large the telescope being used. In many cases, the observational goals can be achieved using small–ish telescopes in well-coordinated global networks. Here, I briefly describe the work of one such international network of observatories and scientists known as the Whole Earth Telescope (WET). With the WET, we have sounded out the interiors of a large number of nonradially pulsating stars. Over the past 14 years, WET has observed dozens of stars in 20 separate observing campaigns. Our team has wide span of interests, and has observed several other classes of objects such as delta Scuti stars, CV stars, pulsating sdB stars, and rapidly oscillating Ap stars.

1. Introduction - a Little WET History

In 1986, astronomers from the University of Texas established a world–wide network of cooperating astronomical observatories to obtain uninterrupted time–series measurements of variable stars. The technological goal was to resolve the multi-periodic oscillations observed in these objects into their individual components; the scientific goal was to construct accurate theoretical models of the target objects, constrained by their observed behavior, from which their fundamental astrophysical parameters could be derived (Nather et al. 1990). This approach has been extremely successful, and has placed the fledgling science of stellar seismology at the forefront of stellar astrophysics.

This network, now known as the Whole Earth Telescope (WET), is run as a single astronomical instrument with many operators. The collaboration includes scientists from around the globe in data acquisition, reduction, analysis, and theoretical interpretation. For the first decade of its existence, the WET was headquartered at the University of Texas in Austin. After 1994, with Dr. J. Christopher Clemens as a Hubble Fellow at Iowa State University, and with support from the International Institute of Theoretical and Applied Physics

(IITAP), headquarters for WET runs began to be held at Iowa State. Tables 1 and 2 list all WET observing runs through 2000.

Table 1. WET runs 1988 – 1994				
Run / Date	Target	Type	PI	Status
Xcov 1	PG 1346	CV	Winget, Provencal	Published
Mar 88	V803 Cen	O'Donoghue	Published	
Xcov 2	G29-38	ZZ Ceti	Winget	Published
Nov 88	V471 Tau	CV	Clemens	Published
Xcov 3	PG 1159	GW Vir	Winget	Published
Mar 89				
Xcov 4	AM CVn	CV	Solheim, Provencal	Published
Mar 90	G117-B15A	ZZ Ceti	Kepler	Published
Xcov 5	GD 358	DB	Winget	Published
May 90	GD 165	ZZ Ceti	Bergeron	Published
Xcov 6	PG 1707	GW Vir	Clemens	In analysis
May 91	GD 154	ZZ Ceti	Vauclair	Published
Xcov 7	PG 1115	DB	Barstow	in analysis
Feb 92	G226-29	ZZ Ceti	Kepler	Published
	WET-0856	δ Scuti	Handler	Published
Xcov 8	PG 2131	GW Vir	Kawaler, Nather	Published
Sep 92	G185-32	ZZ Ceti	Moskalik	in analysis
Xcov 9	PG 1159	GW Vir	Winget	Published
Mar 93	FG Vir	δ Scuti	Breger	Published
Xcov 10	GD 358	DB	Nather, Bradley	Published
May 94				
Xcov 11	RX J2117	PNN	Vauclair, Moskalik	in prep
Aug 94				

WET co–founder Dr. Edward Nather retired as director in 1997. The directorship is now shared by Darragh O'Donoghue (at SAAO) and the author; responsibility for coordinating WET runs moved to Iowa State University. Fortunately, the founders, Drs. Ed Nather and Don Winget, continue active involvement in WET science.

Through December 2000, the WET has managed 20 observing runs (called "Xcovs"). The principal targets have been pulsating white dwarfs, ranging from the coolest (ZZ Ceti stars) through the pulsating central stars of planetary nebulae. WET has also observed a pulsating sdB star (PG 1336). In November 2000, the WET observed a rapidly oscillating Ap star - the pre-est of pre-white dwarfs to be observed by this collaboration.

WET observations have played a central role in over 12 Ph.D. dissertations, generated over 30 refereed publications, and been highlighted in 5 dedicated international workshops. Proceedings from the past four have been published by the journal *Baltic Astronomy*. Active WET members number approximately 50, with home institutions in 16 countries.

Table 2. WET runs 1994 – 2000

Run / Date	Target	Type	PI	Status
Xcov 12	PG 1351	DB	Hansen	In analysis
Apr 95	L19-2	ZZ Ceti	Sullivan	in prep
Xcov 13	RE 0571+14	CV	Marar, Seetha	in prep
Feb 96	CD-24 7599	δ Scuti	Handler	published
Xcov 14	PG 0122	GW Vir	O'Brien	published
Sep 96	WZ Sge	CV	Nather	in clouds
Xcov 15	DQ Her	CV	Nather	In analysis
Jul 97	EC 20058	DB	O'Donoghue	in analysis
Xcov 16	BPM 37093	ZZ Ceti	Kanaan, Nitta	in analysis
May 98				
Xcov 17	PG 1336 (N)	sdB	Kilkenny, Reed	in prep
Apr 99	BPM 37093 (S)	ZZ Ceti	Nitta	in analysis
Xcov 18	HL Tau 76	ZZ Ceti	Dolez	in analysis
Nov 99	PG 0122	GW Vir	O'Brien	in analysis
Xcov 19	GD 358	DB	Kepler	in analysis
Apr 00	PG 1159	GW Vir	Kepler	in analysis
Xcov 20	HR 1217	roAp	Kurtz	in analysis
Nov 00				

2. WET Science Goals and Technical Challenges

The science goals of WET revolve around fully resolving the pulsation spectra of multiperiodic nonradially pulsating stars. Once fully resolved, we attempt to match the observed pulsation periods with those of stellar models. Success in doing so provides a determination of global properties of the pulsating stars (such as mass, rotation rate, luminosity, and distance) and, more importantly, gives us a window into the stellar interior structure.

For us to succeed in meeting these goals for the stars we are interested in, several technical requirements need to be met. To avoid the unresolvable confusion caused by 1 cycle/day aliases in data from a single terrestrial site, we must obtain uninterrupted time-series photometry of these rapidly variable stars. Our "instrument" must be sensitive in the temporal frequency range between 700 and 50,000 μHz (i.e. to periods between 20 and 1400 seconds). To obtain a frequency resolution of 1 μHz (though we sometimes wish for even better) a run must last at least 1 week. To see the pulsations, which are small amplitude, we require an amplitude sensitivity of less than 1 millimagnitude for $V < 17$ - that is, we need telescopes, at good sites, of 1 meter aperture or more. Finally, we try to keep the maximum 1 cycle/day alias amplitudes below 20%

We meet these technical challenges by attempting to obtain global coverage through coordinated observations at up to 15 observatories. If the weather cooperates, we can obtain 24 hour/day coverage and squash the 1 cycle/day alias. The weather rarely cooperates to this degree. However, the weather is a random variable - and with persistence, we can obtain data from all sites around the globe in a typical 2–week WET run. This means that while the coverage may not be continuous, we still can get a high duty cycle (frequently better than 70%) and, most importantly, cover all phases of the 24-hour rotation period of our Earth. Nather et al. (1990) illustrate the significant reduction of these 1 cycle/day alias structures that we can achieve with WET.

Success also requires using detectors of uniform wavelength sensitivity, and care in combining data from different telescopes. These chores are minimized with a "standard" photometer design (Kleinman et al. 1996) and a standard set of software tools. By working with astronomers who are interested in the science of the WET and who are also those at the telescopes during the observations, we are able to meet the technical challenges. Because of the sensitive nature of the signal and instrumentation needed to measure it, we have found that automating the data collection is not practical or wise. We really need to have observers – observers who care about the data – at the telescopes.

3.   WET Operations

The WET is an organization that allows astronomers from around the world work together — first in planning an observing run, next in obtaining the data during the actual observing campaign, and finally in analyzing the data and publishing the results. With over 50 astronomers from more than a dozen countries, this mode of operation would be nearly impossible without the use of the Internet to link observers and scientists. By using computers linked through the Internet, such "virtual collaborations" are viable. In the case of the WET, we have shown that not only can we have a virtual collaboration, we can actually do the science in a networked fashion - thus the term "virtual collaboratory."

3.1.   Target Selection and Observing Time

The WET can only obtain high–quality data on one (or maybe two) targets at a time. The resources (in both time and money) needed for a WET run are considerable. Given these facts, we can only observe a limited number of targets, and selection of which stars to observe is an important and difficult assignment.

Approximately 8 WET members serve on a "Council of the Wise" or simply COW. Once or twice a year, proposals for WET targets are solicited from the entire community by the COW. The COW then decides, based on the science case as well as the case for the need of the peculiar strengths of the WET - which targets are to be observed. The successful proposer becomes the PI for that target, and prepares a generic scientific justification for use in applying for telescope time. The PI works with the Associate Director for WET Operations (or ADWO) in coordinating proposals for the various WET sites.

The WET has no dedicated telescopes. When the target is approved, all members of the WET collaboration prepare and submit proposals to their local observing facility for time to participate in the run. These proposals are coordinated so that the separate time allocation committees are aware of the collaborative nature of each proposal. Observing sites range from smaller instruments that are local to the home research institutions, through national facilities such as KPNO, CTIO, OHP, and SAAO.

3.2. The WET Control Center

With up to 15 observers around the world observing the same target, there is ample room for confusion and waste of resources. To maximize the effectiveness of our collaborative observations, during a WET run (which typically lasts 2-3 weeks) we maintain a continually staffed control center. At the control center (or HQ) are WET members with significant experience in observing, data reduction and analysis. HQ personnel usually include the PI for the target of the run.

The HQ staff communicate with all observers daily to update them on the progress of the run, and to help solve any problems that may arise at a site. They also provide an independent test of the observatory and data acquisition clocks. Throughout the run, as an observer completes his/her night of data collection, s/he sends the data via e-mail or ftp to the HQ. HQ staff then reduce the data and examine it for quality, timing errors, etc.

With a fully staffed HQ, we can maintain an up-do-date light curve and compute current pulsation spectra for analysis. The reductions and analysis are made available to all members of the WET team, including those not directly involved in the run, via the WWW. The WWW site for the run–in–progress, linked to the main WET website (currently http://wet.iitap.iastate.edu), shows the log of each individual run, the combined lightcurve, and amplitude spectrum. To help plan the immediate future, and assess the run, the website also has weather updates and other status variables from each site. An open discussion page allows collaborators to discuss the results so far, speculate on new discoveries, etc.

Another important role of the HQ staff is allocating observatories to secondary targets. When overlapping sites are both active, the HQ staff can move one of the observers to a secondary target, allowing the WET to obtain data on more than one target at a time. Occasionally, we have been able to get nearly 48 hours of data in one 24 hour period!

At the end of a run, the HQ computers have accumulated all of the raw data. In addition to the raw data, we archive the first–look reductions by the HQ staff, plots of light curves, spectra, and other reductions, e-mail messages from all sites pertaining to the run, and the current version of all reduction and

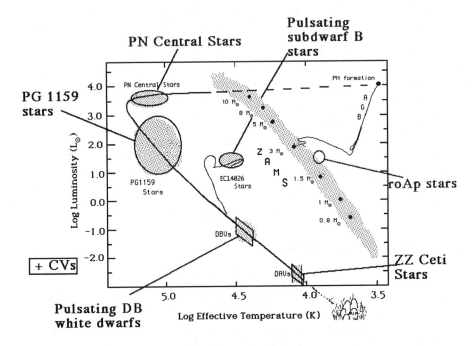

Figure 1.   A schematic Hertzsprung-Russell diagram including representative main sequence and evolutionary tracks. Indicated on this diagram are the classes of variable stars that have been examined as targets of WET observations (after Kawaler 1986).

analysis software. A copy of this archive, which fits easily on a standard CD, is taken home by the PI for a more leisurely and thorough analysis.

3.3.   Data and Publication

Despite the apparent simplicity of time–series photometry (i.e. low data volume compared to imaging or spectroscopy) correct reduction and analysis of WET data is very difficult. The PI has the responsibility to handle the data and draft a paper (or papers) for publication, but this can take years. To try to stimulate our members, the WET has adopted an 18 month proprietary period, wherein the PI has all rights to publishing the data. Typically the first paper based on a WET run will have as authors all observers and HQ staff that participated in the run. Author order questions are solved by the nature of the collaboration - first–author honors go to the PI and the person or people who did the analysis after the run was over. Authorship continues starting at the geographic location of the HQ, and working West. Subsequent papers that use the data have an author list that includes only those who participated in the additional analysis.

After the proprietary period is over, or the first paper is published, the rest of the collaboration is free to publish independent analysis of the data in whatever form they choose. Of course, all WET collaborators have free access to all the data from the time of the run forward. The raw data, and all reduction software are publicly available after the proprietary period is over.

4. WET Science

In this brief review, there is no chance to fully describe the many exciting science results of the WET, or to adequately describe the wide range of impact of WET (and WET-style) data. This section simply gives a few "headlines" about these sample targets, with references to papers in the literature. Reviews of many WET results can be found in the proceedings of the last three WET workshops (Meistas & Vauclair 2000, Meistas & Moskalik 1998, Meistas & Solheim 1995). See also Kawaler & Dahlstrom (2000) for a broad overview of white dwarf stars and the impact of WET.

Figure 1 illustrates the range of variable stars that the WET has investigated in terms of the H–R diagram. Clearly, we have covered a large portion of this map, with the inclusion of white dwarf stars over 6 orders of magnitude in luminosity, roAp and δ Scuti stars, and the pulsating sdB variables.

Sample amplitude spectra of WET observations are shown in Figure 2. This figure shows the remarkable uniformity in the pulsation spectra of a wide variety of stars. It is this similarity in photometric behavior that gives WET such a wide range of targets. It is an instrument designed with white dwarfs in mind, but applicable to a much larger range of targets.

4.1. PN Central Stars and Pre-White Dwarfs

WET has observed one central star in detail: RX J2117. This star, first identified as an evolved star by its identification as a ROSAT X-ray source, was found to be variable in 1993. This star is a complex pulsator - a paper with analysis of the WET data is in preparation (Moskalik et al. 2001).

Arguably the most productive WET target has been PG 1159, a pulsating hydrogen–deficient pre-white dwarf. Observations in this star from 1991 have yielded accurate determinations of its mass, distance, and subsurface composition profile (Winget et al. 1991, Kawaler & Bradley 1994). These data have been combined to reveal a secular period change that challenges theoretical models of these stars (Costa et al. 1999) We continue monitoring PG 1159 - and it continues to bear fruit.

Another pre-white dwarf, PG 2131, has provided an important test of asteroseismological analysis method (Reed et al. 2000). Perhaps the acme of asteroseismological inference is the distance to the pulsating star. All of the many analysis steps must be correct to derive an accurate distance. PG 2131 has a close companion that is a low-mass main sequence star. The companion has been resolved (with a separation of 0.3 arc seconds) in a pair of images obtained by the Hubble Space Telescope. Reed et al. (2000) analyze these images and obtain a spectroscopic parallax distance to the system that they compare with the seismological distance determined through a reanalysis of the WET data of Kawaler et al. (1995). Reed et al. (2000) find that the two distance agree within the 1 σ level, with the seismological distance having a small formal error of about 10 %.

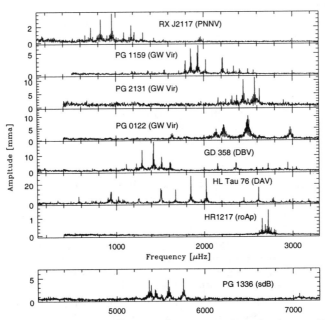

Figure 2. Pulsation spectra from WET. Panels show a PN central star (PNNV), pre-white dwarfs (GW Vir stars), pulsating DB white dwarf (DBV), and ZZ Ceti star (DAV). Bottom two panels show spectra for a rapidly oscillating Ap star and subdwarf B star.

Stellar probes of other areas of physics now include the coolest of the pre-white dwarfs, PG 0122. The evolution of this star should be dominated by energy loss through neutrino emission - the neutrino luminosity exceeds its photon luminosity by a factor of about three. O'Brien & Kawaler (2000) show how to use WET observations of this star, coupled with single–site monitoring, to measure the influence of neutrinos on the change of the pulsation periods.

4.2. Cooler White Dwarfs: the DB and DA (ZZ Ceti) Pulsators

For the pulsating DB white dwarfs, the principal object has been GD 358, first observed with the WET in 1990 (Winget et al. 1994). Results for GD 358 show it to be the most prolific of the pulsators, with over 185 separate modes visible. This high density of modes provides a good probe of the outer layers (Bradley & Winget 1994, Metcalfe et al. 2000), and constraints on models of white dwarf chemical evolution (Dehner & Kawaler 1995) Its pulsation patterns change dramatically from year to year, providing grist for the nonlinear pulsation theorist's mill (Vuille et al. 2000)

ZZ Ceti stars have been observed throughout the history of the WET. Results for ZZ Ceti stars at the hot end of the instability strip are collected in Clemens (1994), while Kleinman (1995) did a comprehensive study of those at the cool end of the strip. One of the most exciting individual targets has been BPM 37095, a ZZ Ceti star with a mass high enough for it to have a crystalline C/O core (Montgomery & Winget 1999); observations by WET are included in the Ph.D. dissertation of A. Nitta (2000). Finally, WET observations have played a key role in the determination of the cooling rate of the ZZ Ceti star G117-B15A by Kepler et al. (2000b).

4.3. Cataclysmic Variables, Subdwarf B Stars, and Main Sequence Stars

One of the early class of targets were CVs that showed short-period phenomena. See, for example, the results for PG 1346 (Provencal et al. 1997), and AM CVn (Solheim et al. 1998).

WET has been used to look at short–period δ Scuti stars - because we do relative photometry, our pulsation spectra are unreliable for periods longer than 20 minutes or so. Table 1 lists those stars that have had successful observations, including one ("WET-0856") that was discovered to be a variable during a WET run because it was used as a comparison star (Handler et al. 1996). In 1999, we observed a pulsating sdB star within a short-period eclipsing binary. The most recent WET run had as its target the rapidly oscillating Ap star HR 1217.

5. Conclusions

The WET, a network of "small" telescopes for global photometry of variable stars, has enabled a broad range of inquiry in stellar astronomy. The first 14 years of the WET have been extremely fruitful. This "new" way of observing produces front-line data using modest instrumentation . . . employed in creative ways. WET results have shaped our view of white dwarf evolution ranging from the formative stages down through the ZZ Ceti instability strip.

At the recent European Workshop on White Dwarfs, for example, most of a day contains papers directly related to WET observations and analysis, as well as results from WET that are directly incorporated into atmospheric and evolution studies. Many current white dwarf researchers earned their Ph.D.s working on WET or using WET data and results - and have gone on to expand their careers well beyond pulsating white dwarfs.

Sites on the WET network are testing CCD photometry systems that can provide rapid photometry (cycle times of 10 seconds or less) of much fainter stars for a given aperture. Additional "expansion" of the WET network has already included simultaneous observations with time-resolve Hubble Space Telescope spectroscopy (Kepler et al. 2000a). With new classes of objects, of which the WET has only made some tentative observations, there is great promise for it to continue its legacy of providing entirely new views of the interiors and environments of interesting stars.

Acknowledgments. The author expresses thanks to the National Science Foundation for support of Whole Earth Telescope science under Grant No. AST-9876655 to Iowa State University, and support from Iowa State University, through the International Institute for Theoretical and Applied Physics.

References

Bradley, P.A. & Winget, D.E. 1994, ApJ, 430, 850

Clemens, J.C. 1994, Ph.D. dissertation, University of Texas

Costa, J.E S., Kepler, S.O., & Winget, D. E. 1999, ApJ, 522, 973

Dehner, B.T. & Kawaler, S.D. 1995, ApJ, 445, L141

Dreizler, S. & Heber, U. 1998, å, 334, 618

Handler, G. et al. (the WET collaboration) 1996, å, 307, 329

Kawaler, S.D. 1986, Ph. D. dissertation, University of Texas

Kawaler, S.D. & Bradley, P.A. 1994, ApJ, 427, 415

Kawaler, S.D. et al. (The WET Collaboration) 1995, ApJ, 450, 350

Kawaler, S.D. & Dahlstrom, M. 2000, American Scientist, 88, 498

Kleinman, S.J., Nather, & T. Phillips, 1996, PASP, 108, 356

Kleinman, S.J. 1995, Ph.D. dissertation, University of Texas at Austin

Kepler, S.O., Robinson, E.L., Koester, D., Clemens, J.C., Nather, R.E., & Jiang, X.J. 2000a, ApJ, 539, 379

Kepler, S.O., Mukadam, A., Winget, D.E., Nather, R.E., Metcalfe, T.S., Reed, M.D., Kawaler, S.D., & Bradley, P.A. 2000b, ApJ, 543, 185

Meistas, E. & Solheim, J.-E. 1995, *The Third WET Workshop Proceedings,* Baltic Astron., 4, 104

Meistas, E. & Moskalik, P. 1998, *The Fourth WET Workshop Proceedings,* Baltic Astron., 7, 1

Meistas, E. & Vauclair, G. 2000, *The Fifth WET Workshop Proceedings,* Baltic Astron., 9, 1

Metcalfe, T.S., Nather, R.E., & Winget, D.E. 2000, ApJ, 545, 974

Montgomery, M.H. Winget, D.E. 1999, ApJ, 526, 976

Moskalik, P. et al. (the WET Collaboration) 2001, in preparation

Nather, R.E., Winget, D.E., Clemens, J.C., Hansen, C.J., & Hine, B.P. 1990, ApJ, 361, 309

Nitta, A. 2000, Ph.D. dissertation, University of Texas at Austin

O'Brien, M.S. & Kawaler, S.D. 2000, ApJ, 539, 372

Provencal, J. et al. (the WET Collaboration) 1997, ApJ, 480, 383

Reed, M.D., Kawaler, S.D., & O'Brien, M.S. 2000, ApJ, 545, 429

Solheim, J.-E. et al. (the WET Collaboration) 1998, å, 332, 939

Vuille, F. et al. (the WET Collaboration) 2000, MNRAS, 314, 689

Winget, D.E. et al. (the WET Collaboration) 1991, ApJ, 378, 326

Winget, D.E. et al. (the WET Collaboration) 1994, ApJ, 430, 839

Small-Telescope Astronomy on Global Scales
ASP Conference Series, Vol. 246, 2001
W.P. Chen, C. Lemme, B. Paczyński

MONET: a MOnitoring NEtwork of Telescopes

Frederic V. Hessman

Universitäts-Sternwarte, Geismarlandstr. 11, 37083 Göttingen, Germany

Abstract. In an age of 8m-class telescopes, we should rethink the way we use small telescopes. While 2-4m telescopes are still needed and will continue to be operated largely in the traditional fashion, new 1m-class telescopes operated robotically in global networks will enable a wide range of new and exciting scientific and educational projects, both by themselves and in conjunction with much larger telescopes. I describe our plans for such a global network of two 1m-class robotic telescopes.

1. Introduction

The research tools of an astrophysicist at the beginning of the 21st century are very different from those of his/her colleague a decade or two earlier. In Göttingen, we now routinely use the 8m ESO *Very Large Telescopes* in Chile and the 9.2m *Hobby-Eberly-Telescope* in Texas and will soon have access to the *Southern African Large Telescope* and the *Large Binocular Telescope* in Arizona. A wide range of large-aperture ground and space telescopes covering almost the entire range of wavelengths is available or on the horizon. The present 2-4m-class ("medium-sized") telescopes are still needed, both for their own sakes and for supporting observations for the larger telescopes. And, of course, many institutes still have working 0.5-1.5m (a modern definition of "small") telescopes.

The dramatic increase in the available telescope area per working astronomer has not been accompanied by an equal increase in our ability to carry out more projects or sub-projects simultaneously. Even though computers have made it possible to process Terrabytes of data almost routinely, our mental capacity to plan, organize and carry out observations is limited (a simple "zero-sum" mental model of an astrophysicist). Thus, given the finite number of things we can do in a typical working day, is there any place for general-purpose small telescopes in an age of 8m giants?

Most of the participants at this colloquium wouldn't be here if they thought the answer was "no", but it is instructive to consider what service small telescopes can and should still provide and what constraints on their operation are now relevant.

Support observations for 4-10m telescopes:
> if such observations are not to be a burden on the preparations for and analysis of larger-telescope observations, they must be easy to obtain. Since the large telescopes are based on both hemispheres, simple and effective access to telescopes on both hemispheres becomes very important.

Projects which require large numbers of observations:
since our access to 4-10m telescopes is limited, this is a natural field for small telescopes, but such observations must be easy to manage and have low personnel/travel costs. A classic example of such a project and the inherent costs involved with performing them using traditional observational methods is the COYOTES collaboration (e.g. Bouvier et al. 1997): while their monitoring of the photometric modulations of pre-main sequence stars defined our knowledge of the rotational evolution of such stars, that knowledge came at a considerable personnel and financial cost.

Projects which require all-sky or extended time coverage:
large-scale surveys, simultaneous observations with spacecraft, and observations of phenomena on timescales greater than a night require access to the whole sky. (the WET Consortium is a good example of the need for both sky and time coverage: see the article by Kawaller). The organization of loose consortia for this type of work (e.g. AGN monitoring) has traditionally been painful and often unsuccessful. One needs automatic access to several telescopes spread out in latitude and longitude.

Projects which require extreme flexibility:
an immediate access to the telescope is required, e.g. to detect the optical counterparts of γ-ray bursts or to follow the lightcurve of a supernova or a micro-lensing event. Immediate access is usually only possible when one has guaranteed access – usually only available on small telescopes.

Limited access to larger telescopes:
ideally, this should not be a major constraint. In practice, our colleagues from less-developed countries have a difficult time finding travel funds. As more telescopes become remotely and/or robotically operated, the opportunities to do interesting science will hopefully increase significantly. Thus, we can hope that this reason to use small telescopes will diminish with time.

Training:
the best way to produce future generations of instrument builders is to let undergraduate and graduate students have direct access to working telescopes and the possibility to construct and manipulate low-cost but relatively "state-of-the-art" instrumentation.

Education:
even when cutting-edge science is mostly done with larger telescopes, small telescopes can be used very effectively by high school and even elementary school students in programs which enable them to learn about the scientific process "hands-on".

While *Training* requires direct access to telescope hardware, all of the other small telescope uses do not or indeed are only possible when several telescopes distributed over the globe are at least remotely steerable and even fully robotic and autonomous. Thus, the future of small telescope astronomy lies in the use of global networks of robotic telescopes.

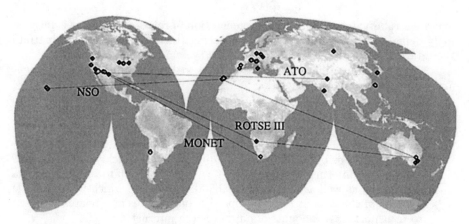

Figure 1.   A map showing the global distribution of robotic telescopes. The networks are indicated by the connected sites: *MONET* is our own project; *ROTSE III* is the planned extension of the *ROTSE* experiment, *ATO* is the *Antipodal Telescope Observatory*, and *NSO* is the *National Schools Observatory* (see the table of robotic telescopes in the appendix).

2.   Robotic Telescopes

While there might be slight nuances in the definition of what constitutes a "robotic" telescope, the basic definition is that it performs the observations given to it without the need for any local or remote operator.   This means that the system must respond (quickly) to changes in weather and must gather it's own calibration observations.   Modern robotic telescopes should be able to handle their own scheduling dynamically – e.g.   to respond to a *Target-of-Opportunity* (ToO) request without losing track of what it should be doing otherwise or to respond to changing seeing and weather conditions and not just blindly go through a list of proposed targets. Robotic telescope systems should perform some level of data-pipelining (dark, bias, flatfield, flux and astrometric calibration) and analysis (photometry of target objects, search for variability or motion across the sky).

There are nearly 60 robotic (or "autonomous") telescopes around the world either in operation, under construction, financed, or planned (see Fig. 1, derived from the table of robotic telescopes in the Appendix) and the number is increasing rapidly. This does not include a number of telescopes which are nearly robotic – e.g. which are capable of performing a long list of observations once a local operator has opened the dome and prepared everything.

The first generations of robotic telescopes – e.g. the *Automatic Photometric Telescopes* (APT) at Fairborn Observatory developed by Lou Boyd and others (Boyd, Genet & Hall 1986; Genet, Boyd & Hall 1986), the Berkeley supernova-search telescopes (Richmond, Treffers & Filippenko 1993), the Univ. of Indiana's *RoboScope* (Honeycutt 1994; Honeycutt et al. 1994ab), and the *Iowa Robotic Telescope Facility* (Deleo & Mutel 1992) to mention just a few – were largely constructed from scratch or used existing hardware and required the writing of the full range of telescope, dome, weather-station, and scheduling software needed to operate a robotic telescope. The new generation of robotic telescopes is increasingly being purchased "off the shelf", reducing (but certainly not eliminating) the total effort necessary to bring a working system on-line.

There are many difficult issues associated with the operation of an efficient and simple-to-use robotic network:

The Data Avalanche:

Robotic telescopes are capable of amassing astounding volumes of data which either must be passed on to a well-organized easily accessible data-bank (via ftp or on some storage medium) and/or quickly processed into higher level information. The data-flow must be kept to a minimum via a standard as well as a user-extensible data-pipeline which permits some analysis of the data at the telescope site and reduces or obviates the need to transport large amounts of data over the internet.

Flexibility:

in order to perform whole classes of interesting science, the network must be able to respond rapidly to user-defined events, Targets-of-Opportunity, and be useful as a trigger for large telescope observations.

Robotic Software:

while it is no longer necessary to write *all* of the software from scratch for each project – practically all telescope venders now offer some form of robotic capability – there is (still) no turnkey system and the scientific constraints on the software are bound to be different for each telescope and network.

Administration & Maintenance:

the administration of the network and projects carried out by the network must be simple. A robotic network with heterogeneous scientific and educational users must be able to handle the different users needs and constraints.

Ease of Use:

the users must have fast and transparent access to the network and have to deal with a minimum of red tape and overhead. Ideally, a creative user should be able to pose a question, submit the observations needed to answer that question, and receive not the raw data but the analysis of the data if not the answer itself – and all within a minimum amount of waiting time.

Communication within the Network and between Networks:

while it may be simple to organize the transfer of information between the users, administrators and telescope within a simple robotic system, a complex network of users and multiple telescopes and certainly the communication between different networks of different telescopes using different telescope control software requires an adequate means of transfering information about the observations.

The latter challenge is being met within an expanding fraction of the robotic telescope community via an *eXtensible Markup Language* (XML) dialect called *Robotic Telescope Markup Language*[1] (RTML). Actually, RTML should be called "**Remote** (rather than Robotic) Telescope Markup Language", since its true function is not to steer robotic telescopes but to permit the seamless transfer of information about an astronomical observation between telescopes and between networks of telescopes. This protocol looks somewhat like the *HyperText Markup Language* (HTML) used by internet browsers, is self-defining, easily extensible, and is easily parsed both by computers and by humans. Here is an example of a normal RTML request:

```
<?xml version="1.0" encoding="UTF-8" standalone="no"?>
<!DOCTYPE rtml SYSTEM "http://hou.lbl.gov/rtml/rtml.dtd">
<RTML version="1.0" mode="obslist" >
    <CONTACT email="aeinstein@princeton.edu"
             school="Princeton University"
             teacher="Professor"/>
    <REQUEST request_id="1852753"
             username="aeinstein"
             observers="N/A"
             reason="Astronomy Research"
             timestamp="Mar 29, 2000">
        <SCHEDULE priority="5"> <MOON phase="1" /> </SCHEDULE>
        <TARGET id="12345678" name="NGC 2266" numobs="1">
            <COORD ra="06:41:42.00"
                   dec="+27.00.00.00"
                   equinox="unknown" epoch="2000" hamax="4"
                   raFormat="hh:mm:ss.dd"
                   decFormat="deg.mm.ss.dd"
                   dformat="mm.dd.yyyy" />
            <EXPOSE exptime="60" tunits="sec" filter="Clear" />
            <TRACK rate="sidereal" />
        </TARGET>
        <CORR zero="true" dark="true" fixpix="true"
              crpix="true" flat="true" dome="true"
              fringe="true" />
    </REQUEST>
</RTML>
```

RTML promises to become the universal standard in telescope control and observation specification just as FITS is our standard data format, enabling telescopes and networks using totally different hardware and software to communicate seamlessly.

---

[1] http://hou.lbl.gov/rtml/intro.html

3.   Educational Use of Small Telescopes

While the participants at the meeting and the readers of the proceedings are undoubtably aware of the many scientific uses of small and particularly robotic telescopes (see, e.g., the many contributions in this volume), they may not be aware of how much activity there has been in the educational use of small telescopes (mostly high school and undergraduate but increasingly even down to elementary school). There are a wide variety of remote and robotic telescopes which are now or will soon be accessible by schools, ranging from quite small "department store"-sized telescopes to 2m telescopes.

The most successful remote telescope project is probably "Telescopes In Education" of the Mt. Wilson Observatory. Using standard (commercial) astronomy software for PC's, registered classrooms all over the globe can have access to a 24-inch. While there is no formal curriculum, the project has a well-documented list of interesting projects which can be performed using the data.

The use of remote telescopes in classrooms has the advantage of being immediate and actively usable, but has two substantial problems: (1) the available bandwidth and informal and interactive choice of objects means that the telescope is not used efficiently; and (2) there is the danger of using the telescopes for more astronomical "site-seeing" rather than truly scientific (if small and simple) projects. The attractiveness of astronomy as a natural science with high public awareness and interest should be used to teach how science is done by letting the students *do* scientific projects. This means placing the goal – answering interesting scientific questions by posing and carrying out a well though-out program of observations and reductions – above the means.

This is the approach taken, e.g. by the *Iowa Robotic Telescope Facility*[2] (IRTF) and the *Hands-On Universe Project*[3] (HOU). HOU is a privately and federally-funded project complete with a curriculum designed and tested by educators and comes with enough data to carry out astronomical exercises without any telescope at all. HOU is presently expanding its global capacity for providing new data via a global queue system. R. Mutel of the IRTF has written an undergraduate curriculum based on the availability of a robotic queue-driven telescope and is even developing a complete hard- plus software plus curriculum solution (including dome, CCD and spectrograph!) with *Torus* called *Rigel* for interested schools, colleges and universities.

4.   The MONET Project

The reasons why we became interested in robotic telescopes are very simple and probably very typical of many other institutions. While we work in a lovely old observatory and have local access to a old 34cm astrograph and a 50cm solar telescope, the weather in Göttingen is bad even by German standards. Research groups within the observatory have long needed the monitoring of variable stars

---

[2]http://denali.uiowa.edu

[3]http:///hou.lbl.gov, http://www.uni-sw.gwdg.de/~hou

and AGNs, but the increasing use of 8m-class telescopes has made it essential to obtain timely and continual information on the photometric state of large-telescope targets. Finally, a drop in the number of incoming physics students has put us under pressure to help improve the quality and hence attractiveness of the physics curriculum in local high schools: we have entered the *Hands-On Universe* collaboration and have conducted several teacher workshops but can't offer our participating classes any useful access to new data. Thus, we finally decided that we needed our own robotic telescopes for research, university, and school purposes. Since our programs need observations of objects all over the sky, we need at least 2 telescopes stationed on both the northern and southern hemispheres. The observatory already maintains a major solar observatory on Teneriffe, so we are unable to cover the local operating costs of additional telescopes and need partners willing to exchange such support for observing time on both telescopes. Since such partners are primarily interested in the scientific use of the network, the telescopes have to be large enough, e.g., to reach the same faint objects which should be observed spectroscopically with large telescopes.

Given the substantial costs of two telescopes and our intent to use a large fraction of our time for purely educational purposes, Prof. Klaus Beuermann and I approached a private charitable foundation with the idea. The *Alfried Krupp von Bohlen und Halbach-Foundation*, one of the largest charitable foundations in Germany, generously agreed to be the sole capital contributor – mainly because of the educational benefits promised by the network.

Due to our close contacts with the McDonald and South African Astronomical Observatories via our participation in the construction of the 9.2m Hobby-Eberly-Telescope and with the 9.5m Southern African Large Telescope, we have natural northern and southern partners who are very interested in the science which can be done both with the network alone and in conjunction with large telescopes as well as in the educational use.

MONET[4] will be different from other robotic telescope projects less in the sense that we will do any particular thing differently from any one else but rather because we will combine many of the individual highlights of other projects into a single network:

Sky Coverage:
> we will have complete northern and southern sky coverage with a fully integrated network;

Time Coverage:
> our two sites in the USA and Africa should give us very good longitudinal coverage for extended monitoring programs on timescales longer than a night at a single site;

Direct Connection to 8m-class Telescopes:
> we plan on directly connecting the network to the operation of our two 9m telescopes having the same all-sky coverage, HET and SALT, permitting the automatic triggering of large telescope observations via the regular monitoring of our robotic telescopes;

---

[4] www.astro.physik.uni-goettingen.de/~hessman/MONET

Aperture:
> we have enough funds for 80-120cm telescopes which can easily reach the faint objects which we want to observe spectroscopically with 8m-class telescopes or provide high S/N observations of variables on short timescales - a network of smaller telescopes would make things much more difficult;

International:
> the project will be run as a fully international collaboration;

Education:
> roughly as much time will be given to educational as to scientific projects (a full 50% of the Göttingen time!); and

Remote & Robotic:
> our participating schools and our students will have access to the telescopes both remotely and via a robotic queue (with emphasis on the latter, however).

MONET will also help to redefine the use of educational telescope networks: not only will it be available to participating classrooms either remotely or via a queue, we will also provide access to users in external queues like that being set up for the global HOU in order to use up any left-over educational time; we would like to provide a virtual reality access to the telescope in the Göttingen XLAB science center; we plan on targeting young women (the fraction of woman in German physics & astronomy is one of the lowest in the world); and we want our university students to administer their own fraction of time on the network with their own TACs and projects.

5. Conclusions

I have argued that the most effective use of small telescopes in the next decade will come from the operation of networks of robotic, autonomous telescopes. While a decade ago, the construction, programming and operation of such telescopes was a major undertaking made by individual pioneers, the state-of-the-art has progressed far enough to make it nearly (but not quite) possible to buy robotic telescopes "off-the-shelf". This permits the creation of homogeneous robotic telescope networks which, thanks to the efforts and good will of many users and venders, will also be connected into very inhomogeneous but powerful networks. The efficient operation and use of these telescopes will make it possible to conduct significant and unique scientific investigations both alone and in combination with large ground- and space-based telescopes. Just as important is the use of the telescopes by schools: we can make it possible for schoolchildren all over the world to learn about science by performing their own and collaborative astrophysical experiments. We hope that MONET – our "MOnitoring NEtwork of Telescopes" which should become operational starting in 2002 – will live up to these high expectations.

I would like to acknowledge the travel support provided by the Deutsche Forschungsgemeinschaft and particularly the generous funding of our project by the Alfried Krupp von Bohlen und Halbach-Foundation.

References

Bouvier, J., et al. 1997, A&A 318, 495

Boyd, L.J, Genet, R.M., Hall, D.S. 1986, PASP 98, 618

Deleo, D.V., Mutel, R.L. 1992, in Robotic Telescopes in the 1990s, ASP Conf. Ser. Vol. 34, e.d. A.V. Filippenko (San Francisco, ASP), 97

Genet, R.M., Boyd, L.J., Hall, D.S. 1986, in Instrumentation and research programmes for small telescopes, Proc. IAU Symp. 118 (Dordrecht, Reidel), 47

Honeycutt, K. 1994, in Optical Astronomy from the Earth and Moon, ASP Conf Series Vol 55, ed. D.M. Pyper, J. Angione (San Francisco, ASP), 103

Honeycutt, R.K., Adams, B.R., Swearingen, D.J., Kopp, W.R. 1994a, PASP, 106, 670

Honeycutt, R.K., Robertson, J.W., Turner, G.W., Vesper, D.N. 1994b, in Interacting Binary Stars, ASP Conf Series Vol 56, ed. A.W. Shafter (San Francisco, ASP), 277

Richmond, M, Treffers, R.R., Filippenko, A.V. 1993, PASP 105, 1164

Strassmeier, K.G., Boyd, L.J., Epand, D.H., Granzer, T. 1997, PASP 109, 697

Kawaler, Welch and J. Patterson

Small–Telescope Astronomy on Global Scales
ASP Conference Series, Vol. 246, 2001
W.P. Chen, C. Lemme, B. Paczyński

The STARBASE Network of Telescopes and the Detection of Extrasolar Planets

Charles H. McGruder,III

Department of Physics and Astronomy, Western Kentucky University, Bowling Green KY 42101, USA

Mark E. Everett and Steve B. Howell

Astrophysics Group, Planetary Science Institute, 620 N. 6th Avenue, Tucson, AZ 85705-8331, USA

Abstract. A network of longitudinally spaced imaging telescopes is described. Due to the limitations of the radial velocity method extrasolar planets have only been found around bright stars (less than 10 mag). Employment of the network and the photometric method to detect extrasolar planets will lead to the discovery of extrasolar planets at much fainter magnitudes (less than 19 mag).

1. Introduction

In the past few decades, significant advances have been made in our understanding of how stars and planetary systems are formed. In the last few years both extrasolar planets and brown dwarfs, whose masses are between those of planets and stars $(10^{-3}M_{sun} < M < 0.1M_{sun})$ have been found. Most stars (60 to 95%) are not single stars like the sun, but are found in binary systems. The theory of binary star formation is too rudimentary to make conclusions about possible associated planetary systems. However, our theoretical understanding leads to the profound conclusion that planetary systems should be found around virtually every single star. This connection is so strong that if planetary systems around single stars are not found it means that the basic tenets of the theory of star formation need to be revisited. Thus, the existence of planetary systems represents a test of the theory.

A number of different techniques to detect extrasolar planets are currently being employed or in the planning stages. They are: astrometry, spectroscopy (two approaches - radial velocity and line identification), photometry, the monitoring of pulsar periods, and high spatial resolution imaging (both optical and infrared). All extrasolar planets discovered to date (~ 55) have been found with the radial velocity method.

The radial velocity approach involves the measurement of the back and forth motion of the parent star in the radial (line of sight) direction due to the gravitational attraction of the planet on the parent star. It yields only minimum masses and orbital information. In order to determine the chances of extraterrestrial life however, we need to know the physical properties of an extrasolar planet - radius, density, surface gravity, temperature, etc. With a network of robotic imaging telescopes, Western Kentucky University (WKU) and Planetary Science Institute (PSI) astronomers will work to detect extrasolar planets using a different approach - the photometric method, which can lead to the determination of the above mentioned physical quantities.

2. The Photometric (Transit) Method

The photometric or transit method involves the measurement of the decrease in stellar brightness due to the transit of a planet in front of the parent star. The decrease, ΔI, in brightness, I, of the star is given simply by the ratio of the cross-sectional areas of the planetary, R_p^2, and stellar disks, R_s^2:

$$\frac{\Delta I}{I} = \frac{R_p^2}{R_s^2} \tag{1}$$

The shape of the transit curve, that is $\Delta I(time)$, is determined by five quantities: Planetary radius, stellar radius, stellar mass, orbital inclination and limb darkening. Normally one can assume that the stellar quantities (radius, mass and limb darkening) - are relatively well known. The unknown quantities of planetary radius and orbital inclination are then derived from the best fit to the transit curve.

3. History of Extrasolar Planet Searches

The basic idea that a planet transiting a parent star could be a cause of a decrease in the brightness of a star goes back to at least the middle of the nineteenth century (Lardner 1858). Last century Sturve (1952) revived this idea and it was studied by Huang (1963). But it was first Rosenblatt (1971) who developed a viable approach. He suggested a system of three wide-field, widely-spaced, robotically-controlled, photometric telescopes. Our approach (some thirty years later) is basically a realization of the Rosenblatt method. It was kept alive by Borucki and Summers (1984) and others in the intervening years.

It was however Mayor and Queloz (1995) using the radial velocity approach who first succeeded in detecting planets around ordinary stars. They found most unexpectedly a probable Jupiter-mass planet in a 4.2 day orbital period around the G2V star 51 Pegasi. The distance of the planet from the parent star is only .05 AU, far closer than Mercury is from the sun (.39 AU). Rosenblatt expected to find only $\sim$ 1 planet/year. However, the presence of large planets ("hot Jupiters") in short period orbits means that the likelihood of detecting planets is significantly higher.

Because the radial velocity method can only derive minimum masses from observations, it was not completely clear that the tens of probable Jupiter-mass bodies discovered where actually planets until 1999. Employing the photometric method Henry et al. (2000) and Charbonneau et al. (2000) found a transiting planet. Charbonneau derived the radius, mass, density, surface gravity, and orbital inclination for the planet orbiting the parent star HD209458, which had been previously discovered by astronomers using the radial velocity method.

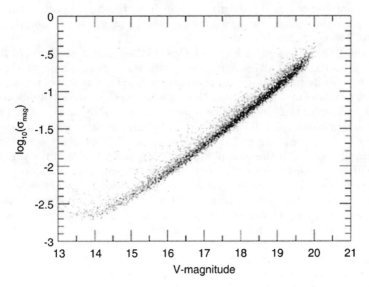

Figure 1.   The log of the calculated uncertainty in each star's light curve vs. V magnitude of the star. The data comes from one of the CCDs of the KPNO 0.9 m mosaic camera. The best precisions are 0.0019^m (V∼ $14.2 - 14.8$). The variable stars are easy to pick out as they lay above the main concentration of the plot. The best theoretical precision for this CCD and telescope combination is 0.002^m.

Now that Henry and Charbonneau have successfully applied the transit method, the PSI and WKU will employ it to discover previously unknown extrasolar planets.

4.   Photometric Precision

According to Equation (1) a transit of a Jupiter-like planet in front of a sun-like star produces a 1% (0.01 mag) decrease in the stellar brightness. Ground-based stellar photometry can easily achieve this precision. Therefore, such a transit is detectable using ground-based telescopes. In fact, PSI and WKU will undertake a program of photometrically monitoring field stars in search of these amplitude variations, as described in the following sections. We will show below that we expect to achieve an order of magnitude higher precision (∼ .1%, 0.001 mag) with the 1.3 m telescope at the Kitt Peak National Observatory (KPNO) for our brighter stars ($V \sim 14$ mag).

Figure 1 presents the photometric precision of observations from Everett & Howell (2001). The observations were made at KPNO on the NOAO .9 m telescope using the wide-field MOSAIC CCD Camera on 16-20 March 2000. They determined their photometric precision by calculating the standard deviation, σ, of each star's light curve over the five nights of observations. In Figure 1, $\log_{10} \sigma$ is plotted vs. the mean V magnitude of the light curve. Each point in the figure represents the standard deviation from the light curve mean based on all their single 3-minute integrations over the five nights of observations. No light curve averaging has been done.

For the brightest stars in Figure 1 (V~ 14.2 − 14.8) a precision of 0.0019 mag (0.17%) is achieved. This agrees well with the best precision (0.002 mag) that is theoretically possible (from Poisson statistics) for this telescope, exposure time and CCD detector. Everett and Howell achieved this through application of differential ensemble photometric techniques as pioneered by Howell et al. (1988).

The attainment of the theoretical limit of photometric precision means that all sources of systematic errors have been eliminated. This is achieved through careful application of differential photometry. Specifically, careful data reduction, use of high-quality hardware, good observing techniques, and proper error assignment to each data point (not an approximate value assigned to some mean datum) essentially eliminates instrumental errors and random effects. Now that we have demonstrated we can obtain the limit of theoretical precision as given by Poisson statistics, we consider the 1.3 m telescope located at KPNO, which we will refer to as the RCT.

For a 2-minute integration using the RCT, saturation will occur at $V \sim 14.3$ mag. Compared to the mosaic data (Figure 1) the much deeper CCD well depths of the RCT will allow not only a higher precision but also a larger overall dynamic range in which Jupiter-sized planets can be detected. For the RCT this range is 5 magnitudes as opposed to only 2.5 magnitudes for the mosaic. This larger magnitude range greatly increases the total number of stars we can search for transits (McGruder et al. 2000).

5.   Types of Detectable Planets

Now that we understand the precision we can obtain in CCD photometric measurements we turn to the question of what types of planets can we detect with this precision. We will show that not only Jupiter-size (both hot and cold), but also Neptune-size and even earth-size planets can be detected via the transit method.

In Figure 2 the relationship between main sequence spectral type and relative transit depth, $\Delta I/I$ (as given by equation 1) for different planetary radii is shown. We consider four planetary radii - hot Jupiter ($1.3R_J$), cold Jupiter (R_J), Neptune ($.34R_J$) and Earth ($.09R_J$). We have chosen to use $1.3R_J$ for hot Jupiters because this value agrees with the only radius of a hot Jupiter that has been measured (Charbonneau et al. 2000). However, theory (Guillot et al. 1996) indicates that R for hot Jupiters can go up to just shy of $3R_J$. Thus, our predictions may be conservative.

So far no extrasolar planets have been found around stars with $R_s \geq 1.2R_{sun}$. This is a limitation of the radial velocity method. Figure 2 makes clear that the transit method can detect "hot Jupiters" revolving around stars up to $R_s \approx 4R_{sun}$ (spectral type: B).

To date no extrasolar Neptune-sized planets have been discovered. This is because the radial velocity method cannot detect planets with $M \leq M_{Neptune}$. However, from Figure 2 it is apparent that the transit method is capable of de-

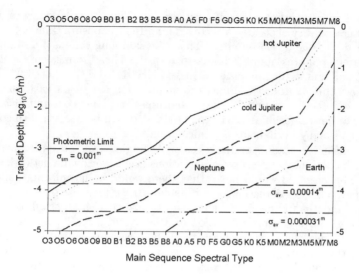

Figure 2.   Transit Depth vs. Main Sequence Spectral Type. Three photometric limits (horizontal lines) are shown - precision for a single RCT measurement ($\sigma_{sm} = 0.001^m$), an average over all measurements made in a single transit window ($\sigma_{av} = 0.00014^m$) and an average over all transit windows during an entire observing season ($\sigma_{av} = 0.000031^m$).

tecting Neptune-sized planets revolving around stars with $R_s \leq 1.1 R_{sun}$ (spectral types: G, K and M). Thus, since Neptune is the smallest of the gas giants of the solar system, the photometric approach can detect all types of the solar system gas giants.

Most importantly Figure 2 makes clear that earth-sized planets can be detected via the photometric method. In fact, apart from the microlensing approach the transit method is the only ground-based method capable of detecting earth-sized planets. However, for single measurements this is only true for $R_s \leq\sim 0.3 R_{sun}$ (late type M stars). And there are two caveats: Firstly, there may be relatively few late type M stars found down to the limiting magnitude of our search (19 mag). Secondly, these stars exhibit intrinsic photometric variability and we must learn to differentiate between their stellar variability and that caused by a planetary transit.

5.1.   Time Averaging and Planet Detection

The types of planets that can be detected with the photometric method depends upon the precision achieved. Time averaging allows one to obtain significantly higher precisions. Consequently, smaller planets or planets at earlier spectral types can be detected. This higher precision is achieved however at the loss of time resolution. We will consider two extreme cases.

First we consider averaging all of the points of a single transit curve. This means no time resolution at all. That is, the actual transit curve cannot be determined, only whether a transit exists or not. Thus, the planetary radius derived will not be very accurate. We consider an RCT exposure time of 2 minutes with a readout time of 1.5 minutes. Thus, the sampling time is 3.5 minutes. A "hot Jupiter" requires about 3 hours to transit. Thus, 51 measurements take place in the 3 hours. The precision achieved is $.001/(51)1/2 = .00014^m$. Figure 2 also shows this photometric limit, 140 mmags. One sees that earth-size planets may be detected around K stars; Neptune-size planets around B and hot Jupiters can be seen even at later type O stars. It will be necessary to compare the results of many periods for a confirmation.

We consider a second extreme case. Averaging over an entire observing season with continuous telescopic observations (via a worldwide network of telescopes). This scenario would mean that there would be no information on the transit curve only on detection of the planet. If we take the observing period to be three months (90 days), a 51 Pegasi like planet with a period of 4.2 days, and again as above a transit duration of 3 hour and a RCT sampling time of 3.5 minutes, then there are 21 transits in the observing period or 1071 observations in the transit widows. Averaging gives a precision of: $.001/(1071)1/2 = .00003^m$. Figure 2 also contains this photometric limit, 30.6 mmags. We see that with this precision Jupiters (both hot and cold) can be detected around all main sequence spectral types. Neptune-size planets up to early B and earth-size planets can be seen even at A type stars. The necessary confirmation must be achieved by comparing the results over a number of years.

The photometric prerequisite of attaining these extremely high precisions is the elimination of systemic errors, which we have showed above, is possible.

6. Probability of Planet Detection

The number of stars, n, with an observable planetary transit at any given instant is (Giampapa 1995):

$$n = \frac{f_p N}{\pi} \left(\frac{R_s}{a}\right)^2 \tag{2}$$

where R_s is the stellar radius, f_p the fraction of stars with at least one planet, N the number of stars in the sample, and a is the orbital radius of a planet. Equation 2 assumes circular planetary orbits. We will use it to estimate the probability of planet detection.

Approximately 3% of solar type stars have giant planets within .1 AU (Cumming et al. 1999). Perhaps 60% of the stars in a field are solar type (late F down to M). Thus about 18 in every 1000 stars will be solar type with a planet.

Clearly, a successful photometric detection is only possible if the planet is in the line of sight (Borucki & Summers 1984). Given that a star possesses a planet, then the probability that the planet produces an observable transit is proportional to the radius of the star and inversely proportional to the radius of the planet's orbit (Koch & Borucki 1996). For a randomly oriented sample of target systems, the probability of a transit is only 0.47% for a Jupiter-size planet in a Jupiter-like orbit. But "hot Jupiters" have a relatively high probability ($\sim$5%, $R_s/a = R_{sun}/.1 = 0.05$) of producing a transit because they are very close to their parent stars. Thus, about 1 in a 1000 stars will have an observable transit.

Now given that a star possesses an observable transit, what is the probability at any given instant of actually seeing the planet in transit? This probability is simply the ratio of transit duration to the orbital period. For a circular orbit this is: $R_s/\pi a$. For a sun-like star this probability is: 0.016. It follows that we have to observe about 100,000 stars at any given instant just to find 1-2 stars with planets that are transiting.

We conclude that in order to find an appreciable number of planets we need to observe a large number of stars *continuously*.

7. The STARBASE Network

The above section makes it clear that we need to observe star fields continuously in order to maximize the number of planetary detections. Obviously, a single telescope cannot observe continuously because stars cannot be seen in the daytime. What is required is a network of longitudinally-spaced, robotically-controlled, imaging telescopes. WKU is in the process of creating such a network. It is called the STARBASE (Students Training for Achievement in Research Based on Analytical Space-Science Experiences) network because students will observe and analyze the data under professional supervision. Initially, the network will consist of three telescopes, which we describe below.

WKU possesses the largest (.6 m), research-grade optical telescope in Kentucky, and one of the largest local facilities of any state or private institution of higher education in the southeast. We already use this facility for student training and research experiences at WKU, and it will serve as the starting point for expanded access by the STARBASE network.

The telescope is located (longitude: 86°36'42.2" W, latitude: 36°55'10.9" N) at a dark-sky site on a rural hilltop about 12 miles southwest of WKU and Bowling Green, a small city with over 45,000 inhabitants. It is a Cassegrain telescope with an $f/11$ focal ratio. Group 128 in Waltham, MA, built it in 1975. The telescope is equatorially mounted and housed in a permanent structure, with an Ash Dome. The primary scientific instrument is an Axiom K-2 thermoelectrically cooled CCD camera, with a field of view of 5' by 7' and a set of BVRI filters. The telescope was refurbished and automated as the initial node of the STARBASE Network by Astronomical Consulting and Equipment, Inc. (ACE).

The most important telescope in the network is the 1.3 m telescope located at the KPNO. In an open competition held last year, NSF and NOAO requested bids for ownership and use of the 1.3 m telescope on KPNO. WKU as the lead institution along with its partners - PSI, Boston University, South Carolina State University, and Lawrence Livermore National Laboratory submitted a proposal. In December 1999 WKU was informed of the success of our peer-reviewed proposal to assume responsibility for the operation of this telescope.

NASA and NSF commissioned this telescope in the 1960's. It was to be the first remote-controlled telescope (and therefore called the RCT) using phone lines for control and data transfer from Tucson. Initially the purpose was to develop techniques for controlling orbiting space telescopes and later it was employed as an attempt to enhance the productivity of moderate aperture telescopes (Maran 1967). We have contracted with EOS Technologies to refurbish and fully robotize this instrument. We still call it the RCT, whereby the letters now stand for "robotically controlled telescope".

We have already purchased a SIT2 2048 X 2048 CCD with pixel well depths of 363,000 electrons, a read noise of 5 electrons, and a quantum efficiency of 80% between 400 and 700 nm. We will be able to image a 20' X 20' field of view with 0.6 arcsec/pixel to provide well-sampled PSFs. The telescope is slated for engineering and camera tests in early 2001, initial science observations and "at telescope" preliminary data reduction in fall 2001, with full operations occurring in early 2002.

Finally, WKU plans to place a robotic telescope (at least .6 m aperture) at the Wise Observatory in Israel. The site (34° 45'48" E, 30° 35'45" N, 875 m altitude, UT+2:00) is about 200 km south of Tel Aviv, located on a high plateau in the central part of the Negev desert, 5 km west of Mitzpe Ramon, a town of 6000.

References

Borucki, W.J. & Summers, A.L. 1984, Icarus 58, 121

Charbonneau, D., Brown, T., Latham, D., & Mayor, M., 2000, ApJ, 529, L45

Charbonneau, D., Noyes, R.W., Korzennik, S.G., Nisenson, P., Jha, S., & Vogt, S.S., & Kibrick, R.I. 1999, ApJ, 522, L145

Cumming, A., Marcy, G., & Butler, R.P. 1999, ApJ, 526, 890

Everett, M.E. & Howell, S.B. 2001, PASP, submitted

Giampapa, M.S., Craine, E.R. & Hott, D.A. 1995, Icarus 118, 199

Guillot, T., Burrows, A., Hubbard, W. B., Lunine, J. I., & Saumon, D. 1996, ApJ, 459, L35

Henry, G., Marcy, G.W., Butler, R.P. & Vogt, S.S. 2000, ApJ, 529, L41

Howell, S. B., Mitchell, K. J., & Warnock, A., III, 1988, AJ, 95, 247

Huang, S.S. 1963, The problem of life in the universe and mode of star formation. In *Interstellar Communication*, Cameron Ed. Benjamin, New York

Koch, D. & Borucki, W., 1996, A Search For Earth-Sized Planets In Habitable Zones Using Photometry, First International Conf on Circumstellar Habitable Zones, Travis House Pub., 229

Lardner, D. 1858, *Lardner's Handbooks of Natural Philosophy and Astronomy*, Vol. 3 Meteorology and Astronomy

Maran, S. P., 1967, Science, 158, 867

Mayor M., & Queloz D., 1995, Nature 378, 355

McGruder, III, C.H., Everett, M., Howell, S.B., & Barnaby, D., 2000, The Detection of Extrasolar Planets via a Newwork of Robotic Imaging Telescopes. In *"Instruments, Methods and Missions for Astrobiology III"*, Richard B. Hoover, Editor, Proceedings of SPIE, 4137

Rosenblatt, F. 1971, Icarus, 14, 71

Sartoretti, P., & Schneider, J. 1999, A&AS, 134, 553

Sturve, O. 1952. Observatory 72, 199

Small-Telescope Astronomy on Global Scales
ASP Conference Series, Vol. 246, 2001
W.P. Chen, C. Lemme, B. Paczyński

STARBASE: A Network of Fully Autonomous Telescopes for Hands-on Science Education

Richard Gelderman, David Barnaby, Michael Carini, Karen Hackney, Richard Hackney, Charles McGruder, and Roger Scott

Western Kentucky University, Dept. of Physics and Astronomy, Bowling Green, Kentucky, USA 42101

Abstract. Students Training for Achievement in Research Based on Analytical Space-science Experiences (**STARBASE**) is being established to provide exciting hands-on research opportunities for students. STARBASE is a network of networks, consisting of dedicated hardware, universities, professional astronomers, teachers, and students all working together in scientific investigations. Funded through the NASA Office of Space Science, the STARBASE network is working to bring major science research projects to motivated students all over the globe.

Western Kentucky University is the lead institution for **STARBASE** – **S**tudents **T**raining for **A**chievement in **R**esearch **B**ased on **A**nalytical **S**pace-science **E**xperiences. In 1999, WKU received funding from NASA's Office of Space Science, allowing us to begin to put into place the networks of hardware, universities, teachers, and students which comprise STARBASE.

The Network of Hardware: Fully autonomous observatories have recently been made possible through the revolutionary advances in computing technology. Advances in computing power and networking allow even remote institutions with limited budgets to operate a small telescope at a world-class observing site. The STARBASE network of observatories – with the first of three fully autonomous meter-class telescopes coming on-line – is another example of the recent efforts to establish global networks of small telescopes.

The 0.6-meter Western Kentucky University telescope is located near Bowling Green, Kentucky (latitude = +36°55, longitude = +86°36.7, elevation 225m). The telescope system has been refurbished and automated by Astronomical Consultants and Equipment, Inc. of Tucson, Arizona. Commissioning tests were being performed at the time of this conference, with the intention of achieving robotic operation in the spring of 2001.

The 1.3-meter Robotically Controlled Telescope (RCT) is located at the Kitt Peak National Observatory (latitude = +31°57, longitude = +111°35.7, elevation 2064m). The refurbishment and automation of the former KPNO 1.3-meter is being undertaken by EOS Technologies, Inc. of Tucson, Arizona, and recommissioning of the RCT is scheduled for early 2002.

Negotiations to locate a new meter-class fully autonomous telescope at the Wise Observatory in Israel (latitude = +30°36, longitude = −31°45.8, elevation 900m) are progressing well, though no start date has been established.

An additional observatory at a site in eastern Asia is desired, but funding had not yet been secured as of this conference.

Other hardware requirements include the need for a powerful computer network to archive and distribute the data. Again, the remarkable advances in computing technology make STARBASE possible – allowing storage of terrabytes of data and daily transfers to the end users of gigabytes of data (both of which were only recently unthinkable volumes). We are relying on a combination of Unix, Linux, and WinNT machines to host the hard disks, tapes, and CDRs which comprise our data retrieval system.

The Network of Universities: The research programs which are the foundation of STARBASE include: photometric detection of extrasolar planets transiting their parent star (Everett & Howell 2001); photometric monitoring of quasars and blazars (Clements & Carini 2001); and emission line imaging of AGN, starburst galaxies, and Galactic nebulosities (Buckalew et al. 2000). Shared research interests and a joint commitment to basic science education bind our partners in a common purpose. The professional astronomers in the STARBASE network of universities firmly believe that the education of both secondary and college students is best served by directly involving them in scientific investigation.

The Network of Teachers and Students: Our educational plans address the need for hands-on, inquiry-based learning in order for students to develop the ability to think and work as scientists. When all the networks are operating together, the standardized hardware and software will enable the teachers and students in the STARBASE network to develop meaningful research projects and to collect, reduce, and analyze their own data in order to carry out their projects. There will be no 'canned' STARBASE investigations, each group of students will have a unique and original experience as they plan, execute, and present the results of their own scientific projects.

The STARBASE network of teachers and students is initially being recruited from rural school districts in the southeast United States: Kentucky, South Carolina, Tennessee, and West Virginia. Training workshops are planned for the Kentucky and Arizona telescopes to introduce teachers to observational astronomy and the procedures required to establish a student-originated investigation or to participate in existing STARBASE research programs. In addition, we are developing web-based 'smart software' to assists students in formulating successful observing plans, as well as to provide instruction in observational astronomy as each student realizes the need for that information.

STARBASE students are already assuming responsible roles in the development of facilities, and will eventually be involved in all phases of the project. They are working with modern mechanical, electronic, and computer systems in support of the hardware development, and presenting the results of their work at local, regional, and national conferences.

References

Buckalew, B., Dufour, R.J., Shopbell, P., and Walter, D.K. 2000, AJ, 120, 2402

Clements, S.D. and Carini, M.T. 2001, AJ, 121, 90

Everett, M.E. and Howell, S.B. 2001, PASP, submitted

Small–Telescope Astronomy on Global Scales
ASP Conference Series, Vol. 246, 2001
W.P. Chen, C. Lemme, B. Paczyński

The NORT: Network of Oriental Robotic Telescopes

Roger Hajjar

Department of Sciences, Notre-Dame University, Po-Box 72 Zouk Mikael, Lebanon (e-mail: rhajjar@ndu.edu.lb)

François R. Querci & Monique Querci

Observatoire Midi-Pyrénées, 14 Av. Edouard Belin, 31400 Toulouse, France (e-mail: querci@ast.obs-mip.fr)

Abstract. In this paper, we present the NORT, a network of oriental robotic telescopes originally proposed in 1993 by Querci & Querci to study stellar variability. The NORT project covers all of northern Africa and Asia near the tropic of Cancer. The interest generated among a number of astronomers in countries from the Arab World and Asia offers a solid basis for the rebirth of astronomy and astrophysics and space sciences in developing countries. The number of national projects for telescopes in different countries makes the NORT a framework of choice to coordinate and stimulate these separate efforts by the creation of a NORT steering committee.

1. Introduction

The NORT was initially proposed in 1993 by Querci *et al.* as a complement to the GNAT. The aim was to have a dedicated network of robotic telescopes completely devoted to the study of long-term variability of stellar objects. They planned initially to use stations of three 1-m telescopes with photometers. The development of large CCDs lead to the idea of a network of small (0.6-1m) to medium (1m-2m) class telescopes.

The idea of establishing stations at or near the tropic of Cancer comes from the interesting astronomical possibilities, that will be discussed later, presented by this region of the world. Thus locations from North Africa to Asia where sought for the establishment of the Network as a complement to GNAT, and a French-Arab project was initiated. It is interesting to note that this longitude range covers mostly developing countries (DCs). In this paper we seek to develop interest in a network and the NORT in particular which will lead to the development of astronomy in these countries. The current status of the project is shown as well as its future prospects.

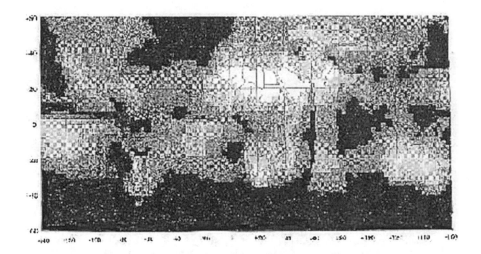

Figure 1.   Mean annual nebulosity obtained from meteorological satellite data. White represents nebulosity > 30%, Black < 70%. Map produced with a 250 km square mesh.

2.   The NORT

The NORT proposes to link a number of stations placed in excellent astronomical sites to monitor photometrically, spectroscopically and polarimetrically variable objects. The NORT would like to achieve 24-hour, year round monitoring of selected targets. Robotic telescopes are an excellent choice for that task. By robotic, we mean a fully automated observatory where all functions are operator independent, a system that might basically need a kick once every few weeks to run.

2.1.   Locations

To determine the best possible sites, meteorological data have been used to build a map of mean nebulosity for the whole world as shown in Querci & Querci (1998). The map is reproduced from their article in Figure 1. White area represents very low nebulosity. The mesh used to average data is 250 × 250 km. The data then should be taken as indicative of possible excellent astronomical sites. It is noticeable that both North Africa and the Middle East show compelling evidence for exceptional conditions, that is dry, semi-arid and with high peaks (the Atlas mountains in North Africa, the peaks south of Jordan, Lebanon and the Anti-Lebanon mountain ranges, Mount Sinai, the South-Arabia mountain chain, etc...).

The presence of DCs in this range makes it interesting because the current state of development makes preservation of good astronomical conditions at these sites quite possible. Based on Figure 1 and on geographical information one can determine possible locations for observatories, as shown in Figure 2

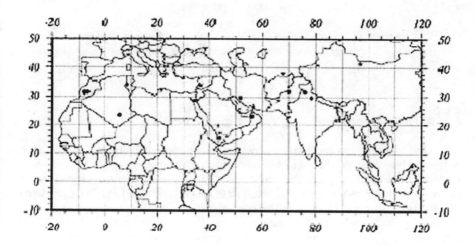

Figure 2.   Sites chosen based on the meteorological map shown in Fig.1 and geographical information (presence of peaks). Requires *in situ* site testing to confirm.

(Querci & Querci, 1998). Filled circles represent exceptional sites, stars represent good astronomical locations. Future steps will involve studying satellite data with a much smaller mesh (3-5 km) and in three wavebands, as well as the implementation of preservation measures for selected sites (Querci & Querci, 1999). Some countries have already started or are planning such studies. Tunisia has already completed the site selection process (Jaidane, 2000), as well as Syria (Al-Mousli, 2000). Libya is planning such a study.

2.2. Objectives

Although all countries in this longitude range are considered to be DCs, almost all have had a long tradition in (ancient) astronomy. Since it provides grounds for a rebirth and/or development of astronomy, NORT would like then to contribute to astronomy education as well as research development in DCs.
Its aims are:
• Building capacity in basic space sciences: education in universities, technical staff training, and building up research potential.
• To help establish national observatories within the framework of NORT.
• A permanent network for the study of long-term stellar variability by spectroscopy, photometry, and polarimetry, and exo-planet research as chief objectives to increase collaboration among partner countries.
• Providing ground for collaboration with and observations at larger facilities (NGST, VLTI, Keck ...) based on NORT observations.
• To produce a large database of variable-star observations for studies like astroseismology and circumstellar environments. All data collected by the NORT

will be shared in real time by all network partners.

The interest in robotic telescopes and networks by education and science institutions is now largely shared by different groups and people (*e.g.*, Hessman, this volume; Querci & Querci, 2000).

3. A DC Perspective

DCs can benefit immensely from a network as opposed to individual, isolated national observatories. For selected sites (Fig. 2), one could easily expect more than 300 clear nights a year. Efficient use of that time will require a large number of projects as well as a substantial number of astronomers and graduate students in any given country. In view of the current ratio of astronomers to the general population in these countries such an aim is currently unachievable. Three astronomers/astrophysicists are currently working at Lebanese universities with a population of little less than 4 million (2000 estimate). Three of our students are currently pursuing graduate studies in France.

Furthermore, science budgets in DCs are very low compared to GDP. Allocation of funds is more competitive. Publication rates and their impact will thus play a major role. Arab League budgets in science amount to no more than USD 550 million for 23 countries. The Lebanese National Council for Scientific Research is granted an annual budget of USD 3-4 million, with an estimated GDP of USD 16 billion. An isolated observatory with a very small group would have difficulties using all the available clear nights and justifying expenses.

Isolation is also one of the main problems facing researchers in DCs. Lack of funds makes it difficult to travel extensively. Access to literature and important scientific information is also reduced because of its cost.

4. NORT Benefits

A network helps alleviate all or part of the above mentioned problems. Networked robotic telescopes with defined tasks will efficiently use all available clear nights. Furthermore, the focused research topic approach for a permanent network provides a reasonable framework for the development of astronomy at the national level, while benefitting from all the resources available due to other collaborators. The NORT would thus help at both the national level and the multinational level.

4.1. National Observatories

Small robotic telescopes (0.6-1m) have many advantages:
• They are now widely available in standard configurations.
• Their low cost makes them accessible to national research budgets.
• Their manageable size does not require a large technical workforce which lowers their maintenance cost.
• They can be made available to advanced amateurs thus increasing their user base.

NORT–organized summer schools will use local facilities to reduce training expenses.

4.2. Networking

Multinational collaboration will help to sustain national projects. Collaboration within a focused research approach will bring about an increased productivity through the sharing of resources and data. Publication rates of national groups will improve.

Observatories evolve and grow. Technological challenges are to be expected that benefit other fields of astronomy. Development of instrumentation and automatic systems, as well as data dissemination, distribution and storage are some of the main problems facing networks of robotic telescopes due to the large volume of data produced per night, per telescope. Estimates have yet to be done for the NORT.

5. Current Status of NORT

NORT has already generated interest in different groups and with individual researchers in different countries in this longitude range. Collaborators are from France, India, Jordan, Lebanon, Libya, Malaysia, Morocco, Pakistan, Syria, Tunisia, and Yemen. Iranians have also shown some interest in the project. More details can be found in Querci et al. (2000).

Since 1993, the initial proposal for NORT, a number of observatory projects have been in various stages of development. A non-exhaustive list includes (from Querci et al., 2000):
- 1.93 m Kottamya Observatory, Egypt, refurbished in 1996,
- Iraq has a German made 3.5-m mirror that was going to be installed in a nearly complete building before the war,
- Jordan has a 40-cm operational observatory and a project for a 1.5-m one,
- Kuwait is planning for a 1-m to 2-m class observatory to serve the Gulf area,
- Lebanese astronomers and scientists are proposing a 1-m class telescope,
- Libya is discussing the implementation of a 2-m class telescope,
- A 50-cm robotic telescope is being commissioned in Malaysia,
- Morocco has recently received a 55-cm Newton telescope from France that is being used for education,
- Pakistan has completed site selection studies for a 1.5-m project,
- Syria is planning for two university observatories or a 1-m class project,
- Tunisia has also completed site selection for 1-m class telescope,
- Yemen is considering a 1-m class national observatory.

NORT was approved by the sixth UN/ESA Workshop (Bonn, Germany, Sept. 1996), and subsequently endorsed by the United Nations General Assembly in December 1996, so it forms a follow-up project of the UN/ESA Workshops. Again, this decision was encouraged by the UN General Assembly on October 2000, following the recommendations issued by the ninth UN/ESA Workshop (Toulouse, France, June 2000). It is also considered by the Arab Union for Astronomy and Space Sciences (1998) as a key project for the development of Astronomy and Astrophysics in Arab countries.

6. Future Prospects

We are currently in the process of establishing a steering and coordinating committee to synchronize efforts toward the actual establishment of some or all of the projects mentioned above. The committee will also implement plans to network already existing telescopes and to foster scientific collaboration between national groups. One of the main goals is to study the means for rapid and automated dissemination of data.

7. Conclusions

The NORT plans a permanent network to study stellar variability with stations located at/or near the tropic of Cancer. It will help to foster astronomy in its partner countries while contributing to the regional and international development of this science. In view of the number of observatory projects that have arisen, the NORT is now ready to move toward the creation of a steering committee to further national projects. One of the main challenges lying ahead will be to make sure that the observatories, while contributing to NORT, can still be used for training and educational purposes locally. This requires a hybrid mode of operation combining both manual and robotic techniques.

References

Al-Mousli, M. 2000, in Fourth Arab Conference for Astronomy and Space Sciences, Mafraq, Jordan, August 28-31, 2000

Hessman, F.V. 2001, this volume

Jaidane, N. 2000, in Ninth UN/ESA Workshop on *Basic Space Science: Satellites and Networks of Telescopes*, Toulouse, France, 27-30 June, 2000

Querci, F.R., Querci, M., Hajjar, R., Al-Naimy, H., Konsul, K. 2000, in Ninth UN/ESA Workshop on *Basic Space Science: Satellites and Networks of Telescopes*, Toulouse, France, 27-30 June, 2000

Querci, F.R., Querci, M., Kadiri, S., de Rancourt, L. 1993, in IAU Colloquium 136 on *Stellar Photometry*, eds. E. Elliott and C.J. Butler (Dublin Institute for Advanced Studies), 122

Querci, F.R., Querci, M. 2000, Astrophys. & Space Sc., 273, 257

Querci, F.R., Querci, M. 1999, in IAU Symp. 196 on *Preserving the Astronomical Sky: an IAU/COSPAR/UN Special Environment Symposium*, eds. R.J. Cohen and W.T. Sullivan, 12-16 July 1999, Vienna, Austria, in press

Querci, F.R., Querci, M. 1998, in A.S.P. Conf. Series on *Preserving the Astronomical Windows*, eds. S. Isobe and T. Hirayama, 139, 135

Small-Telescope Astronomy on Global Scales
ASP Conference Series, Vol. 246, 2001
W.P. Chen, C. Lemme, B. Paczyński

Taiwan Oscillation Network and Small-Telescope Research at Tsing Hua University

Dean-Yi Chou

Physics Department, Tsing Hua University, Hsinchu, 30043, Taiwan, R.O.C.

Abstract. Two projects, the Taiwan Oscillation Network (TON) project and the earthshine project, at Tsing Hua University will be discussed. The TON is a ground-based network to measure solar intensity oscillations to study the solar interior. Four telescopes have been installed in Tenerife (Spain), Big Bear (USA), Huairou (PRC), and Tashkent (Uzbekistan). The recent scientific results from the TON data will be briefly discussed. The earthshine project is to measure the brightness of the dark portion of the lunar disk to obtain the Earth's global albedo. The dark portion of the Moon is lit by the sunlight reflected from the Earth. The global albedo is linked to the global temperature of the Earth. The long-term measurement of earthshine will provide information on the long-term variation of the global temperature. An automated earthshine telescope is being developed at Tsing Hua University. It will be installed at Lulin Mountain in central Taiwan. The ultimate goal is to build a ground-based global network to measure the long-term variation of earthshine to learn about the long-term variation of the global temperature.

1. The Taiwan Oscillation Network Project

The Taiwan Oscillation Network (TON) is a ground-based network measuring solar K-line intensity oscillation for the study of the internal structure of the Sun. The TON project has been funded by the National Research Council of ROC since July of 1991. The plan of the TON project is to install several telescopes at appropriate longitudes around the world to continuously measure the solar oscillations. So far four telescopes have been installed. The first telescope was installed at the Teide Observatory, Canary Islands, Spain in August of 1993. The second and third telescopes were installed at the Huairou Solar Observing Station near Beijing and the Big Bear Solar Observatory, California, USA in 1994. The fourth telescope was installed in Tashkent, Uzbekistan in July of 1996. The locations and pictures of four TON telescopes are shown in Figure 1. We are planning to install an automated TON telescope at Lulin Mountain.

The TON is designed to obtain informations on high-degree solar p-mode oscillations, along with intermediate-degree modes. A discussion of the TON project and its instrumentation has been given by Chou et al. (1995). Here we give a brief description. The TON telescope system uses a 3.5-inch Maksutov-type telescope mounted on a German-type equatorial mount. The annual average diameter of the Sun is set 1000 pixels. A K-line filter, centered at $3934\mathring{A}$, of FWHM $= 10\mathring{A}$ and a prefilter of FWHM $= 100\mathring{A}$ are placed near the focal plane. The measured amplitude of intensity oscillation is about 2.5%. A 16-bit water-cooled CCD is used to take images. The image size is 1080 by 1080 pixels. The exposure time is set to 800-1500 ms, depending on the solar brightness. The photon noise, about 0.2%, is greater than the thermal noise of the CCD and circuit. The TON telescope is a semi-automated system. At the beginning of day, observers have to open the cover, point the telescope to the Sun, and enter control commands on two computers. The telescope will guide itself and take images at a rate of one per minute. The image data are recorded by two 8-mm Exabyte tape drives.

The TON full-disk images have a spatial sampling window of 1.8 arcseconds per pixel, and they can provide information of modes up to $l \approx 1000$. The TON high-resolution data is specially suitable to study local properties of the solar interior.

Recently, we have developed a new method, acoustic imaging, to study local inhomogeneities in the solar interior (Chang, Chou, & LaBonte 1997; Chen et al. 1998; Chou et al. 1999; Chou 2000). A resonant p-mode is trapped and multiply reflected in a cavity between the surface and a layer in the solar interior. The acoustic signal emanating from a point at the surface propagates downward to the bottom of the cavity and back to the surface at a different horizontal distance from the original point. Different p-modes have different paths and arrive at the surface with different travel times and different distances from the original point. The modes with the same angular phase velocity ω/l have approximately the same ray path and form a wave packet, where ω is the mode angular frequency and l is the spherical harmonic degree. The relation between the travel time and the travel distance of the wave packet can be measured in the time-distance analysis (Duvall et al. 1993). Acoustic imaging uses the time-distance relation to construct the acoustic signals at the target point. For a target point on the surface, one can use the measured time-distance relation. For a target point located below the surface, one has to use the time-distance relation computed with a standard solar model and the ray approximation (D'Silva and Duvall, 1995; Chang et al. 1997).

The constructed signals in acoustic imaging contain information on intensity and phase. The first acoustic intensity maps of the solar interior were constructed with the high-resolution TON data (Chang et al. 1997). They show that the acoustic intensity is lower in magnetic regions. The phase-shift maps, first derived by Chen et al. (1998), show that the sound speed is smaller in magnetic regions. To correctly interpret the phase-shift maps, the inversion of measured phase shifts is necessary (Chou, Sun, & The TON Team 2001). We have developed a kernel for phase shifts which links the measured phase shifts with the phase-speed perturbation distribution (Chou & Sun 2001). An inversion method with regularized least-square-fit has been developed by Sun & Chou (2001). The preliminary inversion results show that the depth of the phase-speed perturbation distribution for a typical active region is about 20,000-30,000 km (Chou et al. 2001).

2. The Earthshine Project

Earthshine is sunlight reflected by the Earth which is visible as a dim image of dark portion of the lunar disk. The intensity of earthshine relates to the average of the Earth albedo, so it relates to the global temperature of the earth. A fraction (albedo, about 30%) of sunlight incident on the Earth is reflected by the earth atmosphere and surface back into space. The rest of solar energy is absorbed by the Earth (atmosphere and surface) and converted into heat. This energy is re-emitted into space in the infrared range as an approximate black-body radiation of 225°K (The spectrum of this re-emitted energy is well separated from the incident sunlight whose effect temperature is about 5780°K). Thus the albedo relates to the global temperature of the earth. A change in albedo relates to a change in the global temperature T.

More than seventy years ago, Danjon first tried to determine the average earth albedo from earthshine measurements (Danjon 1928, 1936, 1954). He used the ratio of the intensity of the dark portion to the intensity of the bright portion to reduce the effects of the atmosphere and solar intensity. However, the error bar of his measurements, about 5%, was too large to determine the variation of global temperature. Recently the earthshine measurements at the Big Bear Solar Observatory show that the error bar of A can be as small as 1%, which corresponds to about 0.25% change in global temperature (Goode et al. 2001). Thus earthshine measurements with modern technology becomes a promising method to measure the variation of the global temperature.

Since earthshine measurements at one site can yield the albedo averaged over only a part of the Earth. Our ultimate goal is to to build a ground-based global network to measure the long-term variation of earthshine to learn the long-term variation of the global temperature. To make long-term observations feasible with a low cost, we need to use automated telescopes. The goal of our design is to make the telescopes operate automatically without human care for a reasonable period, for example, a couple of weeks. The first step of the project is to build a prototype automated telescope at Lulin Mountain. Then we will build a few more telescopes to install at suitable sites around the world. The design of the prototype telescope is briefly described as follows.

2.1. Optical and Imaging System

A 3.5-inch ruggedized Questar telescope (Maksutov type) and a German-type equatorial mount are used. A 10^{-5} neutral density filter is used to reduce the intensity of the bright portion of the lunar disk so that the dark portion and the bright portion can be measured simultaneously. A heat-block filter will also be used to remove the near-infrared energy. A 16-bit 1024 × 1024 air-cooled CCD will be used to take images.

2.2. Tracking and Control System

Tracking the Moon is more difficult than tracking stars or the Sun because the shape of the dominant bright portion of the lunar disk changes with time. We will use both the passive tracking (using the pre-determined coordinates) and the active tracking (using the observed lunar images to adjust pointing).

Pointing of the telescope will be measured with an optical decoder accurate to sub-arcseconds.

To make the telescope robust, we minimize electronic hardware by using a digital signal processor (DSP) to control the telescope. This allow us to improve or update the control system by modifying the software of the DSP. The DSP will be controlled by a Linux-based computer. A GPS will be used to provide accurate time.

Acknowledgments. We collaborate with Ming-Tsung Sun of Chung Gung University on both projects. DYC and the TON project are supported by the NSC of ROC under grants NSC-89-2112-M-007-038. MTS is supported by the NSC under grant NSC-89-2112-M-182-001. The earthshine project is supported by the MOE of ROC under grant 89-N-FA01-1-4-5. We thank all members of the TON Team for their efforts to keep the TON project working. The TON Team includes: Ming-Tsung Sun (Taiwan), Antonio Jimenez (Spain), Guoxiang Ai and Honqi Zhang (PRC), Philip Goode and William Marquette (USA), Shuhrat Ehgamberdiev and Oleg Ladenkov (Uzbekistan).

References

Chang, H.-K., Chou, D.-Y., LaBonte, B., and the TON Team 1997, Nature 389, 825

Chen, H.-R., Chou, D.-Y., Chang, H.-K., Sun, M.-T., Yeh, S.-J, LaBonte, B., and the TON Team 1998, Astrophys. J. 501, L139

Chou, D.-Y., Sun, M.-T., Huang, T.-Y. et al. 1995, Solar Phys. 160, 237

Chou, D.-Y., Chang, H.-K., Sun, M.-T., LaBonte, B., Chen, H.-R., Yeh, S.-J., and the TON Team 1999, Astrophys. J. 514, 979

Chou, D.-Y. 2000 Solar Phys. 192, 241

Chou, D.-Y., Sun, M.-T., & the TON Team 2001, in Proc. SOHO10/GONG2000: Helio- and Astero-seismology at the Dawan of the Millennium, ed. P.L. Palle, ESA SP-464 (Noordwijk: ESA), 157

Chou, D.-Y. and Sun, M.-T. 2001, Astrophys. J., submitted

Danjon, A. 1928, Ann. Obs. Strasbourg, 2, 165

Danjon, A. 1936, Ann. Obs. Strasbourg, 3, 139

Danjon, A. 1954, in The Earth as a Planet, ed. G. P. Kuiper (Chicago: Univ. of Chicago Press), 726

D'Silva, S. and Duvall, T. L. Jr. 1995, Astrophys. J. 438, 454

Duvall, T. L. Jr., Jefferies, S. M., Harvey, J. W., and Pomerantz, M. A. 1993, Nature 362 430

Goode, P. R., Qiu, J., Yurchyshyn, V., Hickey, J., Chu, M.-C., Kolbe, E., Brown, C. T., and Koonin, S. E. 2001, GRL, in press

Sun, M.-T. and Chou, D.-Y. 2001, in Proc. SOHO10/GONG2000: Helio- and Astero-seismology at the Dawan of the Millennium, ed. P.L. Palle, ESA SP-464 (Noordwijk: ESA), 251

台灣全球日震觀測網

Taiwan Oscillation Network

Tenerife   Beijing

Big Bear   Tashkent

Figure 1.   Locations and pictures of four TON telescopes.

Small–Telescope Astronomy on Global Scales
ASP Conference Series, Vol. 246, 2001
W.P. Chen, C. Lemme, B. Paczyński

Monitoring Variability of the Sky

Bohdan Paczyński

Princeton University, Princeton, NJ 08544-1001, USA

Abstract. Variability in the sky has been known for centuries, even millennia, but our knowledge of it is very incomplete even at the bright end. Current technology makes it possible to built small, robotic optical instruments, to record images and to process data in real time, and to archive them on-line, all at a low cost. In addition to obtaining complete catalogs of all kinds of variable objects, spectacular discoveries can be made, like the optical flash associated with GRB 990123 and a planetary transit in front of HD 209458. While prototypes of parts of such robotic instruments have been in operation for several years, it is not possible to purchase a complete system at this time. I expect (hope) that complete systems will become available 'off the shelf' in the near future, as monitoring bright sky for variability has a great scientific, educational and public outreach potential.

1. Introduction

During the last decade several billion dollars have been spent worldwide to build 6.5-10 meter class telescopes, and there are about 15 of those giants in operation or under construction. With ever larger apertures and ever more sophisticated detectors it is possible to study the universe not only in ever greater detail but also to make entirely new and very important discoveries. However, in this very expensive race to reach the faintest objects, with the highest angular and spectral resolution and over the widest spectral range, a broad area of research has been largely neglected: the monitoring of optical sky for variability.

The all sky monitors were known in X-ray and gamma-ray domains for many decades. The two examples are: Compton GRO (exists no more)

http://cossc.gsfc.nasa.gov/cossc/

and Rossi XTE (still in operation)

http://heasarc.gsfc.nasa.gov/docs/xte/xte_1st.html

capable of monitoring gamma-ray and X-ray variability on time scales from milliseconds to years over the whole sky. And there is BACODINE system with the GCN electronic circulars

http://gcn.gsfc.nasa.gov/gcn/gcn_main.html

which provides worldwide distribution of 'what is new' in the X-ray and gamma-ray sky. But there is no such rapid discovery and distribution system in the optical domain. True, BACODINE/GCN is used by optical and radio observers to report their follow-up work triggered by X-ray and gamma-ray events. Also, gravitational microlensing, supernovae and asteroid searches provide optical alerts for very limited areas in the sky. Yet, there is no optical system capable of recognizing that something new and unexpected is happening anywhere in the sky, and to instantly verify the discovery. The only all sky optical monitoring is done by amateurs using the naked eye, and therefore it is limited to 4 mag, or so, with the verification and follow-up possible on a time scale of hours or days, but not seconds, as it is the case with the BACODINE/GCN.

Professional astronomers do not appreciate how under-explored is the sky variability, even at the bright domain. The ASAS (Pojmański 2000) and ROTSE (Akerlof et al. 2000) projects demonstrated that by using a 10 cm aperture it is possible to increase the number of known variables brighter than 13 mag by a factor of 10. ROTSE (Akerlof et al. 1999) and STARE (Charbonneau et al. 2000) demonstrated that it is possible to make very important discoveries with such small apertures: the optical flash from a redshift $z = 1.6$, and the planetary transit in front of a star. But note: these two spectacular discoveries were made in a follow-up mode, with the target area in the sky, and the target star, selected with expensive space instruments (BATSE, BeppoSAX) or a large optical telescope. The existing hardware: small, inexpensive robotic instruments, can detect optical flashes and planetary transits, but the software required to make independent discoveries does not exist, and there appears to be little will to develop it, though check McGruder (2001) and STARE:

http://www.hao.ucar.edu/public/research/stare/stare.html

In this presentation I am making a case for small instruments. There is a lot of science to be done, but to be efficient, and therefore effective, the small instruments must be fully robotic. It is a great pleasure to develop and operate such instruments: no need to struggle with TACs (time allocation committees), no need to write observing proposals. A large team with all the managerial and funding problems is not necessary, as a full system can be developed with very modest funds by a competent individual or a small group, as demonstrated by ASAS (Pojmański 2001):

http://archive.princeton.edu/~ asas/

Still, it is not trivial to develop fully robotic hardware, and robust software is the main bottleneck.

2. Today's Systems

There are several projects which use small robotic instruments to image almost all sky every clear night, or every few nights. Almost all of them are focused on specific targets, usually searching for optical flashes associated with gamma-ray bursts, and archiving data with no serious attempt to analyze it. The volume is huge, in some cases several terabytes, so data handling is not easy, and data analysis seems beyond the capability or interest of large teams involved, like ROTSE, LOTIS, TAROT, STARE. Links to their Web sites may be found at

http://www.astro.princeton.edu/faculty/bp.html

http://alpha.uni-sw.gwdg.de/~ hessman/MONET/

There are many projects taking data, but as far as I know only one of them monitors everything that varies within its field of view: the All Sky Automated Survey. Interestingly, ASAS is a single person undertaking (Pojmański 2000, 2001). Unfortunately, even ASAS is still processing data off-line. Another small

group, OGLE (Udalski et al. 1997):
http://bulge.princeton.edu/~ ogle/
is processing all data almost in real time, but its Early Warning System (EWS) alerts on microlensing events only, on a time scale of a day or so, and it monitors less than 0.001 of the sky. I expect that within a year or so ASAS's software will be like OGLE's, perhaps even faster, with the alert time scale of minutes, and near real time verification of anomalous photometric and/or astrometric variation of any type. Perhaps some other team will reach this capability ahead of ASAS. It would be great if this became a common mode of operation.

Unfortunately, the present small robotic instruments are little more than prototypes. It is not possible to order and purchase a complete system, or just a complete hardware, for a known price. The only exception is CONCAM, a very compact camera (cf. Nemiroff et al. 2000, Pereira et al. 2000, Perez-Ramirez et al. 2000):
http://concam.net/
capable of imaging all sky every few minutes using a CCD detector with a relatively small number of pixels, controlled by a lap-top computer. Unfortunately, at this time no photometric/astrometric pipeline software exists for CONCAM. In general, no complete and portable software package is available for any small instrument. This means that most of the data are just archived, and never fully processed. I suppose this is not unusual for a field which is in its early stages of development, with relatively few people involved, and even fewer people convinced that it is scientifically useful to have open-minded rather than narrowly focused observing projects.

3. Scientific and Educational Goals

The list of known types of variable objects is long. It includes eclipsing, pulsating, and exploding stars, active galactic nuclei (AGNs), asteroids, comets, and a large diversity of optical flares or flashes. Scientific goals are very diverse. Complete catalogs of variable stars are needed for studies of galactic structure and stellar evolution. Calibration of various distance indicators can be done with the nearest, and therefore apparently the brightest objects. AGNs are still poorly understood, and they vary on all time scales longer than an hour or so. Comets and asteroids are important for studies of the solar system, while 'killer asteroids' have a great potential for entertainment, as most of them are not deadly at all, just spectacular. Finally, with so many big telescopes in operation and under construction it is useful to have a variety of targets of opportunity detected in real time (cf. Paczyński 1997, 2000, Nemiroff & Rafert 1999, and references therein).

What makes small instruments scientifically interesting is the very high data rate which can be generated and processed at low cost, provided suitable software is available. With the gradual decrease of detector prices it is possible to have a large number of pixels. Computer power is increasing and its cost is falling all the time. The operating expenses of OGLE hardware translate to over 100,000 photometric measurements per $1. By the time these proceedings are published the cost of 1 terabyte of IDE disk will be about $2,000, making it possible to have huge data sets on-line. So, it makes sense to use a 'vacuum cleaner' approach and to process all CCD frames and convert 'pixel data' into 'catalog data', which can be analyzed by a much broader range of astronomers, even amateurs. The diversity of data types is small, as all variables are point sources, but the diversity of phenomena is large, making the database interesting for a variety of scientific as well as educational projects.

With an avalanche of data a small team like OGLE or ASAS cannot possibly analyze it all. A question comes up: should the data be kept in a closet for future analysis, or should it be made public domain so other astronomers can do science with it now? My view, as well as the view of the OGLE and ASAS teams, is that the latter solution is preferable; with the data rate increasing exponentially there will never be time to analyze it all internally. An example of this policy is a recent publication by the OGLE team of almost 1,000,000 photometric measurements for almost 4,000 Cepheids in the LMC and SMC (Udalski et at. 1999a,b). The data posted on the web was analyzed by Dr. D. S. Graff of the Ohio State University, who noticed a small but clear systematic error reaching several hundreds of a magnitude near the edges of OGLE images. The error was verified and the electronic archive was revised on April 1, 2000. Subsequently, the data was successfully used to study the geometry of LMC and SMC (Groenewegen 2000). While this work was being done elsewhere the OGLE team had time to work on other projects, and Dr. A. Udalski had time to work on a new, large CCD camera for OGLE.

Some focused projects may have very diverse applications. Let me give two examples.

The very successful Katzman Automatic Imaging Telescope (Filippenko 2001) discovers dozens of relatively nearby supernovae every year. It is not an all sky system, but it monitors several thousand galaxies. It would be great if the Katzman system could be copied, and the detection rate of supernovae increased to $\sim 1,000$ per year. Who needs so many events? One of the most outstanding unsolved problems in modern astrophysics is a relation between supernovae and gamma-ray bursts (GRBs). It is likely that GRBs are strongly beamed (cf. Frail et al. 2001, and references therein). If the true GRB rate is $\sim 1,000$ times higher than the observed rate, then up to 1% of all supernovae may generate a gamma-ray burst which in most cases is not beamed at us. However, its afterglow may become detectable as a bipolar radio-supernova remnant several years after the explosion. Some supernovae are followed by a strong radio signal which can be detected with the VLA. The few sources which are strong enough could be followed-up with the VLBA with a sub-milli-arcsecond resolution. The explosions related to GRBs which are not beamed at us could be recognized by their bipolar structure and relativistic expansion. We need as many nearby supernovae as possible to have a chance to discover those hypothetical bipolar radio remnants (Paczyński 2001)

A search for near Earth asteroids (NEAs) is aimed at the discovery of all (or at least most) 'killer asteroids'. These are objects which could, upon impact, cause global catastrophe. The minimum diameter is estimated to be ~ 1 km. Hundreds of such asteroids were already discovered by many projects, like

LINEAR, http://www.ll.mit.edu/LINEAR/

LONEOS, http://asteroid.lowell.edu/asteroid/loneos/loneos_disc.html

NEAT, http://neat.jpl.nasa.gov/

SPACEWATCH, http://www.lpl.Arizona.edu/spacewatch/
An up to date information may be found at
MPC, http://cfa-www.harvard.edu/cfa/ps/mpc.html
and a recent review was written by Ceplecha et al. (1998). The searches discovered also a large number of smaller objects, down to several meter diameter. It turns out that about once a month an asteroid 12 meters across, with a mass of $\sim 1,000$ tons collides with Earth and releases in the upper atmosphere ~ 10 kilotons of TNT equivalent. Ten times more energetic events happen once a year. These are spectacular fireballs with strong acoustic effects, but no harm is done at the ground level. About once a century a Tunguska-like event releases up to 10 megaton in an explosion which is locally devastating.

The searches for near Earth asteroids are done with 1 meter class telescopes, which implies that a major part of the sky is covered once every week or so. This is frequent enough to discover a broad range of asteroid sizes, and to make statistical estimates of the probability of impacts, but not frequent enough to recognize those few which are about to collide with Earth. Naturally, the huge data archive contains information about many types of objects with variable brightness, but they are ignored unless they also change their position. This is a huge untapped treasure with the information of general variability of thousands, perhaps millions of stars and AGNs. In the next section I shall discuss modest extensions of the current asteroid searches which could provide alerts about impending impacts as well as alerts for any unusual variability in the sky.

Educational opportunities offered by small telescopes are very well described elsewhere in these proceedings (Hessman 2001). Unfortunately, neither OGLE nor ASAS has an active educational program so far.

4. Prospects for the Future

With all the current, very diverse activity one goal has not been achieved so far: we do not know what is happening in the sky in real time, even at the bright end. This is a huge gap in astronomical research, which can be filled only with small, wide angle instruments. The goal is to monitor all sky at the shortest possible time intervals down to whatever magnitude limit is technically and financially feasible, to process the data as soon as it is acquired, to send alerts, and to archive the results in public domain, so that broad scientific analysis can be accessible to many users, who could be called virtual observers.

There are several obvious steps to be made for this idea to become viable. First, a complete system with fully automated hardware and software pipeline and real time alert system should become operational - none exists at this time. Next, hardware should be made easy to duplicate, to allow for a relatively simple expansion to various sites and various groups. The availability and the cost of all the components should be known, and it should be low. I expect software to be public domain and free. Once the systems spread it will be necessary to find a way to coordinate the data flow, the diversity of alerts and the ever growing on-line archive. I think there will be many problems which are impossible to predict and we should be open-minded and flexible. It is very likely that several distinctly different systems will be developed, with a broad range of costs, data rates, depths and the scope of surveys.

Notice that optical variability may have a time scale as short as a microsecond, as long as the age of the universe, and anything in between. It is obviously impossible to cover the whole sky every second to 24 mag. But it is relatively easy to cover it down to 10 mag every minute, and this may be a good start, or a good followup on the CONCAM. The search space is multi dimensional: how large area in the sky is monitored, how often, how deep, in which filters, with what accuracy? All past and current searches operate in some area of this parameter space, and there is no way to know where the most spectacular and unexpected discoveries are to be made. Consider an example: for over a century enigmatic super-flares were observed on normal main sequence, single, slowly rotating stars of F8 - G8 type (Schaefer 1989, Schaefer et al. 2000). However, an instant follow-up was never done and their nature, or even their reality is not known. Perhaps they are not actually on the stars, but on companion planets (Rubenstein 2001)?

A serious discussion is under way to define science to be done with the LSST (Large Synoptic Survey Telescope = Dark Matter Telescope, Tyson et al. 2000). This is a project to build a telescope with a fast 8.4 meter mirror, a field of view 3 degrees across, and 1.4 Giga pixels. If built, it will be able to image all sky in just a few nights, reaching 24 mag, and saturating at 15 mag. As powerful as it will be, LSST will not be likely to discover an optical flash like the one associated with GRB 990123, as at any given time LSST will image only $\sim$ 0.0001 of the sky. Of course, LSST will detect a large number of very interesting faint transients. But note: it is much easier to follow-up an optical flash which peaks at 10 mag than one that peaks at 24 mag, yet the bright sky variability on a time scale of seconds or minutes is not explored at all.

A detection of small asteroids about to collide with Earth should be possible several hours or even days prior to their impact. The alerts would be useful not only scientifically, but they also would be great for public outreach and entertainment if the time and location of the next explosive fireball in the upper atmosphere could be predicted. Such alerts are not possible now, but several near flybys were reported. A few years ago a graduate student in Tucson, Timothy Spahr, discovered a 300 meter diameter asteroid 1996 JA1, passing within 450,000 kilometers of Earth (Spahr 1996). It reached 11 mag at the closest approach:

http://cfa-www.harvard.edu/cfa/ps/mpec/J96/J96K06.html

A typical relative velocity of an approaching asteroid is $\sim$ 14 km/s, which implies $\sim$ 8 hour time to reach Earth from the Moon distance. A rock with a 30 meter diameter appears as a $\sim$ 15 mag object at 400,000 km. A 12 meter rock would be $\sim$ 14 mag two hours prior to its impact. Such objects collide with Earth once per year and once per month, respectively. Obviously, detecting them is not easy, but it is not outside the range of current technology. If recognized as heading our way the follow-up observations and the determination of their trajectory would have to be done very quickly, and presumably automatically, in order to make a prediction of the time and the location of their impact. The publicity would be justified, even for near misses, i.e. near Earth flybys. A major asteroid or comet on a Tunguska scale, with a $\sim$ 100 meter diameter, might be detected several days prior to its impact, providing enough time to evacuate the 'ground zero'.

There is no obvious limit to the expansion of all sky monitoring. Gradual reduction of detector and computer costs will make it possible to cover all sky every night, every hour, every minute, to ever lower flux limits. Modest scientific returns are to be expected even for a project reaching 14 mag every night (like ROTSE) or 10 mag every minute (a bit better than CONCAM) provided the data analysis is automated and real time alerts of any unusual variability are implemented.

Acknowledgments. It is a pleasure to acknowledge the support by NSF grants AST-9819787 and AST-9820314.

References

Akerlof, C. et al. (ROTSE collaboration) 1999, Nature, 398, 400

Akerlof, C. et al. (ROTSE collaboration) 2000, AJ, 119, 1901

Ceplecha, Z. et al. 1999, Space Sci. Rev., 84, 327

Charbonneau, D. et al. 2000, ApJ, 529, L45

Filippenko, A. 2001, these proceedings

Frail, D. A. et al. 2001, astro-ph/0102282

Groenewegen, M. A. T. 2000, astro-ph/0010298

Hessman, F. V. 2001, these proceedings

McGruder, C. 2001, these proceedings

Nemiroff, R. J. & Rafert, J. B. 1999, PASP, 111, 886

Nemiroff, R. J. et al. 2000, AAS Meeting 197, 120.04

Paczyński B. (1997) Proc. 12th IAP Colloquium: 'Variable Stars and the Astrophysical Returns of Microlensing Searches', Paris (Ed. R. Ferlet), p. 357

Paczyński, B. 2000, PASP, 112, 1281

Paczyński, B. 2001, AcA, 51, 1

Perez-Ramirez, D. et al. 2000, AAS Meeting 197, 115.08

Pereira, W. E. et al. 2000, AAS Meeting 197, 115.10

Pojmański, G. (ASAS) (2000) AcA, 50, 177

Pojmański, G. (ASAS) (2001) these proceedings

Rubenstein, E. P. 2001, astro-ph/0101573

Schaefer, B. E. 1989, ApJ, 337, 927

Schaefer, B. E., King, J. R. & Deliyannis, C. P. 2000, ApJ, 529, 1026

Spahr, T. 1996, IAUC No. 6402

Tyson, J. A., Wittman, D. & Angel, J. R. P. 2000, astro-ph/0005381

Udalski, A., Kubiak, M. & Szymański, M. 1997, AcA, 47, 319

Udalski, A. et al. 1999a, AcA, 49, 223

Udalski, A. et al. 1999b, AcA, 49, 437

Small–Telescope Astronomy on Global Scales
ASP Conference Series, Vol. 246, 2001
W.P. Chen, C. Lemme, B. Paczyński

The All Sky Automated Survey (ASAS-3) System – Its Operation and Preliminary Data

G. Pojmański

Warsaw University Astronomical Observatory, Al Ujazdowskie 4, 00-478 Warszawa, Poland

Abstract. ASAS-3 is the next step in the All Sky Automated Survey. This paper describes the new hardware, consisting of two wide field (9×9 deg) telephoto instruments and one $F = 750$ mm $D = 250$ mm telescope with 2×2 deg FOV, each equipped with a 2K×2K CCD camera. Wide field instruments are now observing the whole southern sky (almost 30,000 sq. deg) at the average rate of 0.5 measurements per day. A narrow field instrument is connected to the GCN network and is ready to respond to the GRB alerts in tens of seconds. Preliminary photometric data are presented here and briefly discussed.

1. Introduction

The All Sky Automated Survey (Pojmański 1997, 1998, 2000) is an observing project which ultimate goal is the photometric monitoring of various objects all over the sky (Paczyński 1997, 2001). In the years from 1997-2000 a small, automated prototype instrument, equipped with a 768×512 MEADE Pictor 416 CCD camera, a 135 mm f/1.8 telephoto lens and an I-band (Schott RG-9, 3mm) filter has been monitored 0.7% of the sky, to the limiting magnitude of $I \sim 13$. During 3 years of operation it has collected over 50×10^6 measurements of over 150,000 stars, detecting almost 4000 variable stars.

This prototype instrument, its data acquisition and reduction pipeline and the ASAS Catalog have been described by Pojmański (1997). Some first results of the search for short time scale periodic variables were presented by Pojmański (1998) while results of the search for long term variables in the ASAS Selected Fields, using data obtained during the first two years of the prototype instrument's operation, were summarized by Pojmański (2000).

After completing the prototype phase of the project in early 2000 we installed an upgraded ASAS-3 system in the dome of the 10" astrograph in the Las Campanas Observatory (operated by the Carnegie Institution of Washington). The new hardware consists of three independent instruments (Fig. 1), each equipped with an automated parallactic mount, imaging optics with a standard filter, a 2K×2K CCD camera (AP-10 from Apogee) and a dedicated computer (Dual Pentium-3 650 MHz Linux box, 128 RAM, 35 GB hard disk). There is one DAT-3 tape storage for all three systems, and a simple surveillance camera connected to the Internet.

Figure 1.   ASAS-3 in the 10" astrograph dome. From the left: 10" astrograph, $F = 750\ mm$ telescope, 2 wide-field instruments with 200/2.8 telephoto lenses.

2.   Instrumentation

2.1.   Mount

The equatorial mount has a compact (500 mm side) horseshoe design driven in both axes by stepper motors, intermediary sprocket gears and friction rollers. It is controlled by a dedicated electronic box containing power supplies, microprocessor controller and translator drives. Both stepper motors are controlled over the same serial link.

The RA drive transmission rate was designed for 1.5 arcsec steps, so it requires about 10 steps per second to track the sky rotation. The sprocket gears and friction roller form a non-backlash drive with a very good positioning repeatability. The declination drive has a much smaller gear rate, so that each step corresponds to about 12.5 arcsec on the sky.

The maximum slewing rate is about 5 deg/sec in hour angle and 40 deg/sec in declination. The average pointing time during a regular observing run is about 10 seconds.

The instrument bay is 280 mm in diameter and the total weight of the imaging equipment is less then 15 kg.

2.2.   CCD Camera

Three 2K×2K (14 um) commercial cameras were purchased for the project. High read-out noise ($> 10e^-$) and 14 bit ADC were accepted as a trade-off for a 5 sec read-out time. Each camera is thermoelectrically cooled to -20 C during night operation. Unfortunately two of the three systems do not yet meet factory

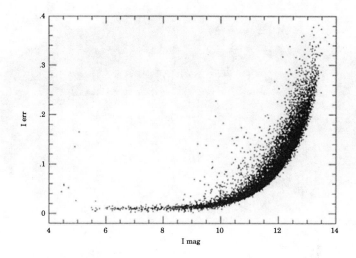

Figure 2. Standard deviation σ_I of the stellar magnitudes *vs* I-band magnitudes for LMC stars observed with the wide-field I camera and 1-minute exposures.

specifications and show twice as much read-out noise as desired, hence reducing system sensitivity. Each camera system is linked to a dedicated Dual Pentium-3 650 MHz Linux box, which handles both telescope and camera control, data acquisition and data reduction.

2.3. Imaging Optics

The two wide-field systems are equipped with the Minolta 200/2.8 APO-G telephoto lenses. These provide superb sharpness (in fact better than needed: the $FWHM$ of stellar images is only 1.3-1.8 pixels) but unfortunately also strong vignetting (40-50% in the corners). With the 2K×2K CCD, each camera captures 8.8×8.8 deg of the sky. One system is equipped with a standard I filter and the other with V.

The narrow field instrument is a $D = 250$ mm, $F = 750$ mm Cassegrain system with a three element, Wyne-type field corrector. It gives sharp images ($FWHM < 2.2$ pixels) within a 2 deg diameter field. With the 2K×2K CCD, the field of view is 2.2×2.2 deg. This system has an I filter in its optical path.

3. Performance

For 1 minute exposures the limiting magnitudes for the wide-field V, wide-field I and Cassegrain I systems are 13.5, 13.2 and 14.8, respectively (Fig. 2). The last two numbers should improve by 0.5-1 magnitudes once problems with the high read-out noise are fixed.

The mechanical quality of the mount allows for exposures as long as 5-10 minutes for wide-field systems and about 3 minutes for the Cassegrain telescope,

Figure 3.   Coverage of the sky after three months of ASAS-3 operation. The white color corresponds to areas observed at least 70 times.

without any substantial degradation in the PSF of the stellar images. The
relatively large Cassegrain system is sensitive to moderate winds (> 7 m/s), in
which case slightly elongated images are produced.

The pointing accuracy of the mounts (without a pointing model being applied) is a few arc minutes in both axes.

4. Observations

The whole sky has been divided into 709 8×8 deg fields, of which 500 (70%)
can be observed from Las Campanas in the I-band and about 420 (60%) in the
V-band (the difference is due to dome obscuration). During a single summer
night one can observe up to 300 different fields.

Weather indications to start/stop observations are obtained automatically
from the Polish OGLE telescope project (Udalski et al. 1997). If that telescope is
not operating, we use the nearby La Silla (ESO) WWW weather service instead.

A routine observing schedule for each wide-field camera consists of cycling
through the list of fields, and picking the one that is the most suitable for
observation at the moment.

The default exposure time for the I systems is 2 minutes, while for the
wide-field V system it is 3 minutes. Taking into account the overhead read-out
and pointing time we can take between 160 and 200 frames per night with the V
system and between 230 and 300 frames per night with the I systems (depending
on the season). This means that we can cover the whole available sky with both
filters in one or two days.

After the initial whole-sky observations, we will change the above schedule
slightly to search for short-period variability in each field.

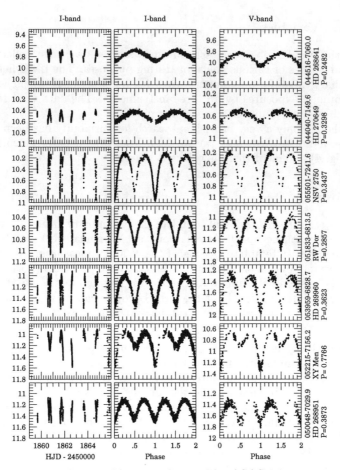

Figure 4.   Periodic variable stars observed by ASAS-3 instruments in the LMC and SMC fields. The ASAS name, cross-identification and periods are printed on the right side of the panel.

The narrow-field instrument is connected directly to the GRB Coordinates Network (Barthelmy et al. 1998). Follow-up observations will start about ten seconds after receiving a GRB alert. Between alerts a set of preselected fields of interest is observed.

The ASAS-3 system currently takes about 600 frames per night, which means a raw data stream of 5 GB per night. The loss-less compression reduces this stream to 2 GB per night (more than one DAT-3 tape per week). We are therefore considering a compression tolerating loss in future.

Up to February 2001, after the first months of unattended operation, over 400 fields were observed. Some 15000 frames from each of the wide-field *I* and Cassegrain *I* systems and 11000 frames from the wide-field *V* system were collected, covering 95% of the observable sky, on an average of nearly 40 times (Fig. 3).

5. First Variables

Before the regular observing schedule has started, we have collected several thousand images of various fields for testing purposes. The LMC field was used as a primary target for photometric tests. For a few nights this field was observed with 1 minute exposures by all three instruments. The data were reduced and analyzed by the default data processing pipe-line.

Although only a very provisional search for variability has been performed, a few periodic and irregular variable stars were immediately identified in the data (Fig. 4) - some of them new bright binaries.

6. Prospects

In a few weeks the ASAS-3 will complete the preliminary data collection process for the entire southern sky. At this point some manual work will be necessary to create the best template catalog of the V and I measurements. We plan to stop observing for a short period, fix the hardware problems (noise) and finally restart the system with full on-line access to the photometric database.

Current information about the ASAS project can be found on the WWW home page at *http://www.astrouw.edu.pl/~gp/asas/asas.html* or *http://archive.princeton.edu/~asas/*

Acknowledgments. This project was made possible by a generous gift from Mr. William Golden to Dr. Bohdan Paczyński, and funds from Princeton University. It is a great pleasure to thank Dr. B. Paczyński for his initiative, interest, valuable discussions, and the funding of this project.

I am indebted to the OGLE collaboration for the support and maintenance of the ASAS instrumentation, and to the Observatories of the Carnegie Institution of Washington for providing the excellent site for the observations.

This work was partly supported by the KBN 2P03D01416 grant.

References

Barthelmy, S.D., et al. 1998, in Gamma-Ray Bursts: 4th Huntsville Symposium; ed. Charles A. Meegan, Robert D. Preece, and Thomas M. Koshut, 99

Paczyński, B. 1997, in Proceedings of 12th IAP Colloquium, Variable Stars and the Astrophysical Returns of Microlensing Searches, ed. R. Ferlet, 357

Paczyński, B. 2001, these proceedings

Pojmański, G. 1997, Acta Astr., 47, 467

Pojmański, G. 1998, Acta Astr., 48, 35

Pojmański, G. 2000, Acta Astr., 50, 177

Udalski, A. Kubiak, M., and Szymański, M. 1997, Acta Astr.,47, 319

Small–Telescope Astronomy on Global Scales
ASP Conference Series, Vol. 246, 2001
W.P. Chen, C. Lemme, B. Paczyński

RTLinux Driven Hungarian Automated Telescope for All Sky Monitoring

Gáspár Á. Bakos

Konkoly Observatory, Budapest, H-1525, P.O. Box 67

Abstract. Massive variability searches initiated the design of HAT (Hungarian Automated Telescope), an autonomous observatory[1]. HAT consists of a horseshoe mount similar to the one used in the Automated All Sky Survey (ASAS, Pojmański 1997), a clamshell dome with various utilities, a telephoto lens and a CCD. Expensive hardware elements have been substituted by software running under Realtime Linux: a multitask operating system which can handle processes in real time. A virtual observer – modeled as a finite state machine – is responsible for managing the observatory. The modular structure allows of running virtually any kind of observing program for small telescopes. We describe technical aspects, as well as present test results carried out by HAT1 (Budapest).

1. Introduction

An effective way of survey mode observations is to use a completely autonomous observatory placed on a perfect site, thus minimizing manpower, usually a bottleneck in recent projects. This observing philosophy reduces the size of teams by substituting observers with software controlled operation and data reduction. Quality software also plays a central role because it can supersede unnecessary hardware elements and makes communication between telescopes well-controlled.

Perhaps 90% of bright variable stars are unknown, or their light curves have not been followed (Paczyński, 2000). The ASAS project has found thousands of new variables in the southern hemisphere with a small telephoto lens an amateur CCD camera attached to a horseshoe mount (Pojmański 2000).

HAT1 is intended to carry out the northern counterpart of Pojmański's study, but it is also suitable for any project which needs automatization of a small telescope. The system was tested in the fall of 2000 at Konkoly Observatory, Budapest, and has recently been relocated to Steward Observatory, Kitt Peak, Arizona.

---

[1]Members of the HAT construction team are József Lázár (software), István Papp (electric engineering), Pál Sári (structural design) and the author. Continuously updated information on this project can be found at the following website: http://www.konkoly.hu/staff/bakos/HAT.

2. Hardware Overview

We use a horseshoe mount which is friction driven by five phase stepper motors (both for the right ascension and declination). The stepper motors are controlled by a simple electronic converter card, which receives signals from the PC's parallel port. We would like to emphasize that the card is simply a converter, and that the motordrive/clockdrive is the PC's central processing unit (CPU). Proximity sensors on both axes detect home and end positions in order to ensure fail-safe operation, so that the telescope can never loose orientation, even though we use an open-loop control system (no costly encoder employed). The RA and Dec resolutions are $1''$/step and $5''$/step, respectively. The system is backlash-free due to the use of sprocket gears and friction drive.

The telescope and electronics are enclosed in a weather-proof box, which has a special, double-axis opening structure minimizing resistance against wind in any position. Attached devices include lens-heating, rain-detector, domeflat light and power on/off, all controlled by the second parallel port of the PC, including dome opening/closing and status information.

The test CCD was an amateur-class Meade Pictor 416xt camera with 512×768, $9\mu m$ pixels. This camera yielded noisy images with bad interference pattern and showed erratic functioning, but still eligible for test-mode operation. The Kitt Peak setup will try to use an Apogee AP10 $2K \times 2K$, $14\mu m$ camera, which also shows some operational anomalies. A Bessel I filter is selected due to its high transmission at the CCDs sensitivity maximum and because the sky background in I band is low even at full moon.

We chose a Nikon 180mm f/2.8 ED lens, which yields a sharp and relatively non-varying point spread function (PSF) all over the 9 degree field of view with the AP10 camera, but has rather strong vignetting (60% intensity on the edge compared to the center). Although under normal conditions the PC controls the dome, it is possible to switch to manual mode for servicing.

3. Software System

All the devices are controlled by one single PC running RTLinux, a multitask operating system (OS) which allows using realtime applications. Generally speaking, running tasks (eg. receiving an email) on a multitask OS trigger interrupts and might disturb the scheduling of other important tasks (such as emitting clock signals to drive a mount with sidereal rate). The RTLinux kernel (core of the OS) treats these realtime processes as independent threads with signal handlers with response times better than $15\mu s$. The Linux environment, the user's interface, runs as the lowest priority thread, but on relatively fast PCs and assuming only a few real-time processes, there is no noticeable difference to the user. The software environment consists of four main parts: **drivers** (low level programs), **mountserver**, **central database (DB)** and **virtual observer**.

While the mount is operated as a realtime application (via kernel driver), other devices are run by low-level C programs, behaving similarly to drivers. Communication with the devices is done via the mountserver, which receives complex commands from a TCP/IP port (ie. anywhere on the "internet"), parses them, and distributes them to the appropriate resort (dome, mount, ccd, etc.).

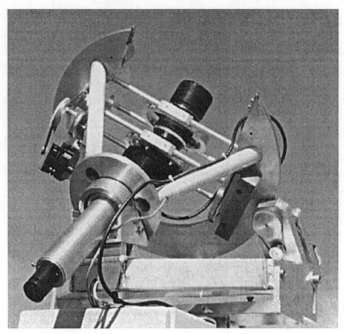

Figure 1.   The HAT horseshoe at Budapest during the tests with attached devices: Nikon lens, Meade Pictor CCD and polar telescope.

The mountserver also has the ability to receive packages from the GRB Coordinates Network (GCN).

We use a fast and reliable relational SQL database to store associated parameters of the stations, such as information on the mount, telescope, CCD, scheduled tasks, archived images, and various other properties, all of them arranged into tables. One central DB can maintain several sites set up in any preferred topology, which greatly simplifies management. Calibration information of any station can be updated straight away by modifying the database, which involves automatic re-computation of all the derived parameters.

The virtual observer is responsible for managing the observatory by communicating with the mountserver, the central database and the running (observer) programs. The observer programs (such as taking bias, dark, flatfield frames, all-sky monitoring) are run as independent threads, so the observer is never stuck waiting for a time-consuming operation to finish. Tasks have starting times (relative to sunset) and priorities – both taken from the DB, which properties are used by a task-manager (part of the observer) to launch or stop them and to suspend or interrupt other tasks. Similar to LaTeX text processing language, there is a template, which makes writing new tasks very simple for the user without requiring a profound understanding of the underlying structure. In a simplified picture, the observer is modeled as a finite state machine with transitions between the 5 states: run, wthsleep, daysleep, closeexit and service. The observer resides in "run" state if there is any task to be executed and if the weather conditions are adequate. Otherwise, if bad weather hinders observa-

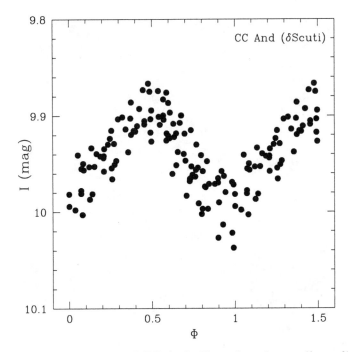

Figure 2.   Light curve of CC And. Note that the small amplitude, and the width of the curve are partly due to the multiperiodic behavior of the variable. The measurements were carried out at Budapest with a 6cm lens and a Meade Pictor Camera.

tions, the program enters "wthsleep" state, basically waiting for clearing, with devices turned off. If no tasks are specified within a given time interval, the observer completely shuts down the observatory (eg. daytime). The "closeexit" and "service" states are self-explanatory. Other stations can monitor the status of the observatory, and with supervisor/service entitlement there is also the possibility of interacting with the observer.

The observatory automatically starts operation after booting the PC, and it is properly shut down in case of rebooting (eg. due to a power outage). The status of the programs is regularly checked (Linux cron daemon), so as to detect and to try to recover from anomalies.

4.   Test Results from Budapest

During three months of test period 14 "clear" nights were used to monitor selected variables under the light-polluted sky of Budapest with the Meade CCD camera attached to HAT. One example is featured in this paper: CC And, a δ Scu variable with $< I > \approx 10^m.0$, $A \approx 0.2^m$, $P \approx 0.12^d$ (Fitch, 1960). Simple aperture photometry was used without removing any bad data point to yield the light curve shown in Fig. 2. The phase was derived from the the main frequency $f_0 = 8.00620$ cyc/day of the star after fitting a double sine.

As the figure indicates, the width of the light-curve (residual $\approx 0.02^m$) is not purely due photometric scattering. Fourier analysis (cf. Fig. 3) revealed the following frequencies (values in braces are from Fitch, 1960): $f_0 = 8.00620$ (8.00591), $f_1 = 7.81920$ (7.81480) and $f_2 = 6.59466[c/d]$ (new, with a lower significance). The final residual after whitening with the three Fourier components was $\sigma = 0.^m015$.

Previous observations were performed using 0.5-1m class telescopes with photometers. This is to demonstrate that a 0.06m instrument is capable of valuable results as well.

Acknowledgments. This project is done in collaboration with Prof. Bohdan Paczyński, to whom we are indebted for his enthusiasm, help and funds, partly from a generous gift of Mr. William Golden. The mount and part of the software were developed using Dr. Grzegorz Pojmański's plans, which he kindly shared with our group. We are grateful to Dr. Lajos Balázs (Konkoly Observatory) for providing facilities to our tests carried out at Budapest. The author would like to thank Dr. Géza Kovács and Dr. Johanna Jurcsik for enlightening discussions and for partial support of the project (OTKA-T24022 and T30954 grants). Participation in the conference was also sponsored by the Graduate School of Eötvös Lóránd University (Institute of Physics), the International Astronomical Union and the Local Organizing Committee.

References

Fitch, W. S. 1960, ApJ, 132, 701
Paczyński, B. 2000, PASP, 112, 1281
Pojmański, G. 1997, Acta Astronomica, 47, 467
Pojmański, G. 2000, Acta Astronomica, 50, 177

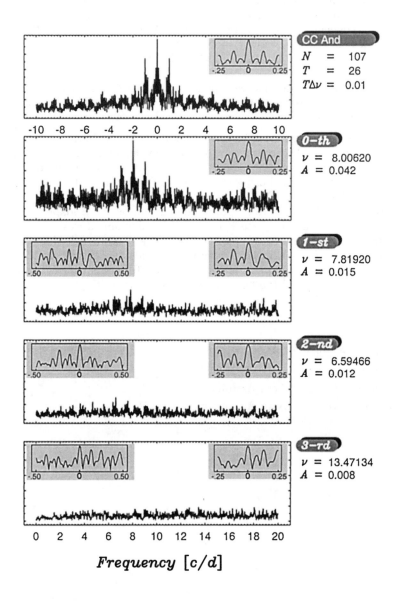

Figure 3.   Amplitude spectrum of CC And. The upper panel shows the spectral window, the second panel the raw spectrum, consecutive panels the spectra prewhitened with all the previous main frequencies. The 40000 spectrum points are binned, and only the maxima of each bin are shown (this type of compression yields no loss of information at the resolution of the figure). Blow-ups of the maxima are shown as inserts in the corresponding panels.

Small–Telescope Astronomy on Global Scales
ASP Conference Series, Vol. 246, 2001
W.P. Chen, C. Lemme, B. Paczyński

The Grid Giant Star Survey for the Space Interferometry Mission

Richard J. Patterson, Steven R. Majewski, Catherine L. Slesnick, Jaehyon Rhee, Jeffrey D. Crane, Allyson A. Polak

Astronomy Department, University of Virginia, Charlottesville, VA 22903, USA

Arunav Kundu

Astronomy Department, Yale University, New Haven, CT 06520, USA

William E. Kunkel

Las Campanas Observatory, Casilla 601, La Serena, CHILE

Douglas Geisler, Ricardo Muñoz, Jose Arenas, Juan Seguel, Wolfgang Gieren

Departamento de Fisica, Universidad de Concepción, Concepción, CHILE

Verne V. Smith

Department of Physics, University of Texas, El Paso, TX 79968 USA

G. Fritz Benedict

McDonald Observatory, University of Texas, Austin, TX 78712 USA

Kathryn V. Johnston

Astronomy Department, Wesleyan University, Middleton, CT 06459 USA

Abstract. NASA's Space Interferometry Mission (SIM), scheduled for launch in 2009, will determine the positions of thousands of stars as faint as $V = 20$ to a precision better than 4 microarcseconds (μas). A key part of the mission is the Astrometric Grid, which is a reference frame of several thousand stars with $V \leq 13$ against which all relative measurements will be calibrated. To serve as a reliable inertial reference frame, the Grid must be astrometrically stable against photocenter jitter (from planets, binary companions, flaring or spotting) at the $\sim 4\mu$as level. Sub–solar metallicity giant stars, by virtue of their intrinsic luminosity, can probe the Galaxy to greater distances than almost any other stellar type at the same apparent magnitude. Thus, distant (> 3 kpc) giants with $V < 13$ will have proportionately smaller astrometric jitter compared to other potential Astrometric Grid star candidates. The Grid Giant Star Survey is a patchwork all-sky survey to find sub–solar metallicity K giants for the Grid, and to provide a unique database for studies of Galactic stellar populations. We describe here the survey characteristics and give examples of results to date.

1. Introduction

The Space Interferometry Mission (SIM) is a micro-arcsecond precision interferometer to be launched by NASA into earth-trailing orbit in 2009. SIM performs astrometry by measuring the additional path length taken by starlight to the slightly more distant of a pair of collecting apertures (located at the ends of a 10-m baseline). Absolute positions of stars over many degrees ("wide–angle mode") are obtained via comparison of angles from target stars to a well-defined grid of several thousand astrometric standard stars. Picometer accuracy in delay line measures translates to a single measure positional precision (for wide angle astrometry) of 10 μas. Observations between stars are repeated over the whole sky during the 5 year mission, resulting in mission–end precisions better than 4 μas. The complete list of reference star positions (including parallactic and proper motions) will constitute the mission Astrometric Grid. All SIM science programs requiring absolute astrometry will make use of this grid.

The high degree of astrometric precision required from the Grid places severe limits on the needed stability of individual Grid stars. A further requirement, that the majority of Grid stars have $V < 13$ (with the exception of a relatively small sample of quasars for an extragalactic frame tie), arises from a desire to minimize the time spent on the Grid to less than 20% of the total mission time; the goal is to achieve the minimum mission fraction as set only by the time to acquire stars, with little contribution by integration time, and this is the case for stars with $V < 13$. Grid candidates of all stellar types suffer to various degrees from the following sources of astrometric instability: 1) binary companions, 2) brown dwarfs, 3) planets, 4) photocenter wander from starspots and flares (SIM resolves the photospheres of $V < 13$ stars). In principle, SIM itself can be used to detect astrometrically unstable stars in the Grid, but the entire mission will be jeopardized if too many stars drop out of the Grid. Thus, before launch it is critical to prepare an Astrometric Grid with stars for which the stability problems are minimized. This requires: a) a wise choice of stellar types for the Astrometric Grid, and b) an ambitious pre-launch observational campaign to weed out astrometrically unstable, problem stars.

We have been sponsored by NASA to conduct the Grid Giant Star Survey (GGSS) to find bright ($V < 13$), sub–solar metallicity giant stars for the SIM Astrometric Grid. Giant stars have been selected for the Grid because (1) they are common in all stellar populations and along all lines of sight, and (2) compared to almost any other class of star, giants at the same apparent magnitude will be more distant and therefore contribute a compensatory reduction in any astrometric jitter. To further decrease astrometric jitter, we search for sub-solar abundance K giants from the Galactic thick disk and halo with apparent magnitudes near the Astrometric Grid limit of $V \sim 13$. The net effects of extreme age and low metallicity increase the intrinsic brightness of giants, so that greater distances are attainable at the same apparent magnitude limit. We outline these and other advantages of sub–solar metallicity giants in §2. The details of the photometric technique and the logistics of the GGSS are detailed in §3.

Apart from their use in the SIM Astrometric Grid, stars from the GGSS will be useful for a host of Galactic structure studies, even before a subsample of the stars (those selected for the SIM Grid, and perhaps others submitted as SIM strategic targets) will have extremely precise, SIM-measured parallaxes and proper motions. We discuss the properties and some uses of the GGSS stars in §4.

2. Population II Giant Stars as Grid Stars

2.1. Distance and Age

Intermediate to Extreme Population II giant stars are excellent candidates for the All–Sky Astrometric Grid. A primary advantage is that at $0 \gtrsim M_V \gtrsim -3$, giants are at $\sim 1 - 10$ kpc for $V = 9 - 12$. Searching to $V \sim 13$ ensures that we can find numerous giant stars from the Galactic Intermediate Population II and halo. It is now established that the Intermediate Population II is likely to be as old as the halo (Nissen & Schuster 1991), and both populations contain members that are extremely metal-poor (Carney et al. 1996). With a judicious selection of targets from a thorough search, and pushing the mean magnitude of the sample to the Astrometric Grid limit of $V \sim 13$, a sample of giants with median distance of ~ 3 kpc, and with at least 25% reaching to > 6 kpc, is being obtained.

Moreover, older stellar systems are less likely to have well established planetary systems, a possible source of astrometric jitter. Of course, inflated giant stars will not have an inner zone of terrestrial planets, but old, Population II giants are also less likely to have Jovian type planets, according to the standard notion that even giant gas planets begin life by accreting around rocky cores of ten Earth mass size (Stevenson 1985; Cameron 1988). The lack of observed planetary systems in the [Fe/H]$= -0.7$ globular cluster 47 Tuc (Gilliland et al. 2000) appears to strengthen this argument. Indeed, stars with known planetary systems are observed to be anomalously metal-rich when compared to stars of similar age (Gonzalez 1999).

2.2. Binary Fraction

As with most types of stars, little is known with certainty regarding the binary fraction in Population II giants. The binary fraction of Population II giant star progenitors is expected to be $\sim 30\%$, roughly equivalent to that for Population I dwarfs (Duquennoy & Mayor 1991, Ryan 1992). For example, low density globular clusters (where most primordial binaries would *not* have been tidally disrupted) appear to have a binary frequency close to that of Population I stars (Hut et al. 1992). For K giant stars in particular, the binary fraction for short period binaries is $\sim 15 - 20\%$ for both Population I and II (Harris & McClure 1983). Close binaries, those most likely to create significant photocenter movement around the barycenter in the SIM lifetime, will have likely been accreted during the red giant swelling phase of the primary.

2.3. Starspotting and Flaring

Even at 10 kpc the photosphere of a K3 III giant subtends an angle of 25 μarcsecs, so starspotting and chromospheric activity (flaring) could affect SIM photocenters. However, rotationally induced chromospheric activity (as measured by CII and CIV emission line fluxes) is strongly correlated with $B - V$ color, and falls rapidly from a peak in line flux for early G giants to very low levels for colors redder than $B - V \sim 0.9$ corresponding to late G or early K giants (Simon & Drake 1989). Similar trends are seen in the CaII H and K lines. Gray (1989) finds that as giants evolve across the H-R diagram, surface rotation declines abruptly for types later than G3 III. This is thought to be caused by a dynamo-generated magnetic braking mechanism. In addition, few slowly rotating, later–type giants show signs of variability (Hall 1991). We therefore expect lower amounts of *rotationally induced* chromospheric activity and spotting in later-type giants.

Less evolved, solitary G and K subgiants that *do* exhibit activity are thought to be located near the base of the red giant branch as they enter a phase of stellar evolution where their convective zones encroach on their more rapidly rotating radiative cores, leading to dynamo driven stellar activity. These less evolved, less desirable grid candidate stars should be identifiable by their bluer colors.

In addition to any emission line flux that can be attributed to a magnetic dynamo driven by rotation, there also appears to be a "basal" component, present in all late–type stars (Rutten et al. 1991). This basal flux level is independent of the stellar metallicity (Cuntz, Rammacher & Ulmschneider 1994), and it appears to indicate the onset of a hot outflow from stars near the tip of the red giant branch, at $M_V < -1.7$ (Dupree & Smith 1995). This wind from the non–pulsating, inactive late–type giants is thought to be caused by acoustic shock heating of the chromosphere. Because this is a symmetric outflow from a slowly rotating, non–pulsating, inactive star, it should not affect the utility of halo K giants as astrometric reference stars.

In summary, astrometric jitter from binarity, flaring or starspotting should be no worse a problem for Population II giants than other types of stars, and giant stars, particularly late G and K type giants, may conceivably be better in this regard. In any case, several groups are monitoring variability in giant star brightness and radial velocity specifically to understand these issues as they pertain to the stability of SIM reference stars. The results of this work will provide an important check on the above claims.

3. Observing Strategy and Status

We (Majewski et al. 2000a) have developed, tested and begun using an efficient technique for finding giant stars photometrically. It is a variant of the combined Washington/DDO51 filter system by Geisler (1984) – simplified to a three filter system for the specific purpose of identifying distant giant stars – and relies on the gravity sensitivity of the Mg I triplet + MgH feature near 5150 Å in F-K stars (Figure 1). Imaging in the Washington M and T_2 filters (Canterna 1976) provides a reliable measure of stellar T_{eff}, while the continuum subtracted flux in the MgH+Mgb region is given by $M - DDO51$. In Majewski et al. (2000a) we demonstrate the effectiveness of the $(M - T_2, M - DDO51)$ diagram for separating dwarf and giant stars (Figure 2). With this filter system, we have nearly finished the photometric search phase of our southern GGSS (SGGSS) with the Swope 1-m telescope (§3.1).

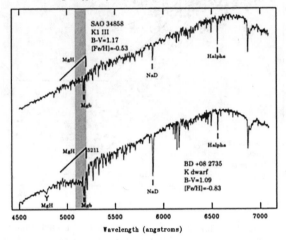

Figure 1. Comparison of spectra for K giant and dwarf stars of similar color and abundance, illustrating the dependence of the MgH + Mgb triplet on luminosity class. The location of the *DDO*51 filter bandpass is indicated by the shaded region.

Targeted GGSS fields (or *bricks*) are shown in Figure 3, with the filled circles indicating the bricks that have been photometrically observed to date. Along each line of declination the bricks are separated by 7.°06 in RA. Declination bands are separated by 3.°53 in declination, and the brick/gap spacing along each declination band is staggered (brick centers shifted by 3.°53 in RA) with respect to those on the adjacent band. This results in a total of 1306 bricks on the celestial sphere, with a typical mean spacing between the bricks of $\sim$ 6°. With this density, there are about 6–7 bricks to be found in each SIM 15° diameter field of regard. Our goal is to find at least one good giant candidate in each brick, and thereby supply about 6–7 Grid stars per field of regard. To reach this goal, we are starting with a several times larger initial sample, from which we will pick the best candidate per brick after further study (§3.2-3.3).

3.1. Photometric Imaging

South of $\delta = +20°$, we have used the Swope 1-m telescope at Las Campanas for the initial imaging, as well as the follow–up, low resolution spectroscopy. A SITe 2K×2K chip provides a 23.'7 × 23.'7 field of view at a plate scale of 0.''696/pixel. To ensure that we obtain several bright, sub–solar metallicity giant candidates per brick, the southern GGSS bricks are constructed of sets of three adjacent and overlapping pointings of the Swope telescope, for a total of 1° × 0.33° per brick.

   In the north, we are using the 0.8-m telescope and Prime Focus Camera (PFC) at McDonald Observatory, with a field of view of 46.'5 × 46.'5, and a plate scale of 1.''355/pixel. The large PFC field of view has allowed us to construct

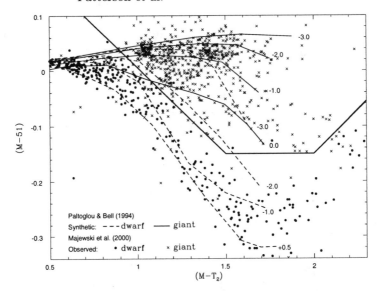

Figure 2.   Color-color diagram of all giant and dwarf data in Majewski et al. (2000a), showing how our filter system separates giants from most dwarf stars for late G – early M spectral types. Data from 42 open and globular clusters and a sample of field dwarf and giant stars are shown. ×'s represent giant stars, while •'s represent dwarf stars. Giant stars are, for a large range of $(M - T_2)_o$ colors, well separated from the dwarfs, for all metallicities.

the northern bricks from a single telescope pointing. The coarse pixel scale does not affect our ability to identify the bright giants for the GGSS, and crowding does not appear to be a problem at $V < 13$ even close to the Galactic plane.

We have completed imaging in 830 of the 1306 GGSS bricks (64%). Imaging for the southern GGSS ($\delta < +20°$) is 80% complete (Figure 3).

3.2.   Low Resolution Spectroscopy

Once $V < 13$, sub–solar metallicity giant candidates are identified photometrically, we obtain follow-up, low-resolution (1Å/pixel) spectroscopy to verify them. A spectral range that we have found to be useful for spectroscopic dwarf/giant separation is 4800 – 6800 Å, which, of course, contains the gravity sensitive MgH band and Mgb triplet, but the gravity sensitive Na D doublet as well. We are experimenting with various schemes to mask the spectra into three sets of lines with principal sensitivities to $\log g$, [Fe/H], or T_{eff}, in order to give more precision in these quantities for our stars. With about 1 Å/pixel scales, and 3 – 4 Å resolution we also get 10 – 15 km/s accuracy radial velocities, which provide useful information on the population membership of the stars. In the south, we have been using the Modular Spectrograph on the Swope 1-m for this work, and have obtained spectra of over 1000 giant candidates to date. For low resolution spectroscopy of the northern GGSS, we are building a fiber-fed bench spectrograph for the University of Virginia's Fan Mtn. 1-m telescope.

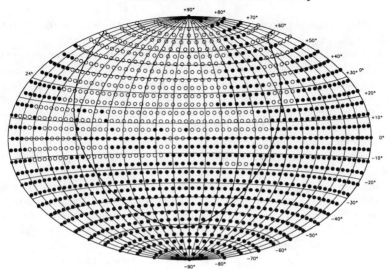

Figure 3. Aitoff projection in equatorial coordinates showing the distribution of GGSS fields to be observed. Filled circles indicate fields for which photometric data has been obtained as of 1 Feb 2001. The Galactic Equator is shown as the solid curve.

3.3. High Resolution Spectroscopy

The final phase of the project (being conducted in collaboration with several groups, including V. Smith using the McDonald 82-inch and D. Geisler using the FEROS on the ESO 1.52-m) involves (1) high resolution ($\sim$ 20 m/s) spectroscopic radial velocity monitoring and (2) photometric monitoring (being conducted by Geisler with the Las Campanas 1-m and CTIO 0.9-m telescopes, and Smith with the McDonald 0.8-m). Both of these monitoring projects will help in the elimination of binaries and stars with large planets, as well as chromospherically active stars, from the SIM Grid. Frink et al. (2001) have found that the intrinsic radial velocity variations for single K giants peaks around 20 m/s; reflex velocity sensitivity at this level will allow K giants with stellar, brown dwarf (which are in any case expected to be rare due to the observed "Brown Dwarf Desert"; Marcy & Butler 2000) and Jupiter-mass planetary companions with a semimajor axis $a \lesssim 2$ AU to be identified, and left out of the SIM Grid. Lower mass planetary companions to distant sub–solar metallicity K giants will not affect the astrometric stability at the 4 μas level. Therefore, most companions that will affect the Grid can be ruled out with a moderate number ($\sim$ 5) of radial velocity visits, and nearly all of the remaining contaminants can be ruled out with SIM itself. We expect a contamination level of undetectable astrometrically unstable stars in the Grid of less than 1%.

4. Characteristics of and Uses for Stars in the Survey

We have already demonstrated (Majewski et al. 1999,2000b) how giants can be used to find and potentially map halo substructure in the form of tidal debris of disrupted satellite galaxies and clusters. Even if no distance-RV substructure should be found in our giant sample and a more homogeneously structured halo and thick disk are implied, we may (with consideration of the many stars fainter than the $V = 13$ SIM candidates that will be in our CCD frames) still explore uniquely and systematically the shape and size of the halo to larger R_{GC} than many (e.g., starcount) analyses. While the 1306 GGSS fields only sample about 4% of the entire sky, their isotropic placement allows an exploration of Galactic stellar populations free of the usual problems faced by studies focused on just a small number of selected lines of sight.

Galactic structure surveys the size of the GGSS are typically limited in scope to broadband color and magnitude data. Interpretation of such surveys must rely on various approaches to the classical starcount analysis via von Seeliger's (1898) Fundamental Equation of Stellar Statistics. The problem is not well constrained because of the inability to invert the von Seeliger convolution of luminosity function and density law, which is further complicated by the presence of superposed stellar populations with different luminosity functions and density laws, and suffers additionally from the inability to discriminate stellar luminosity classes using broadband colors alone. Unless the survey keys on a rather special, and readily identifiable, class of star (e.g., RR Lyrae stars, Cepheids), considerable effort must be expended to ascertain the luminosity class of the sample stars (e.g., through spectroscopy or trigonometric parallaxes). In large or faint surveys for which such effort presents a daunting challenge, it is often simpler to assume a luminosity class for the stars under study and recognize (or model) the expected systematic errors on the survey results. For example, universally asserting that all stars are dwarfs introduces a systematic underestimate of distances and transverse velocities, since *some* stars will be giants or subgiants.

An advantage of the GGSS survey is that not only does our photometry yield the usual colors and magnitudes of a large area photometric survey, but the inclusion of the $DDO51$ intermediate band filter provides us with unprecedented leverage on luminosity classes for stars in such a huge dataset: We can rather reliably separate late type dwarfs from evolved stars on the red giant branch, breaking a particularly troublesome color-magnitude degeneracy. This will allow us to obtain photometric parallaxes for *individual stars* directly. Thus, we avoid the usual *modus operandi* of treating this problem in a statistical way across the survey.

Surface gravity alone is insufficient to determine precisely the intrinsic luminosity of a star; both metallicity and surface gravity determine the absolute magnitude of a star of given surface temperature. Fortunately, our filter system provides some additional sensitivity to a wide range of abundances in late G to early M giants (see Figure 2). The combined gravity and abundance data improve the accuracy of our photometric parallaxes, as the giant branch luminosities at a given color vary by as much as 3 magnitudes over the range $-2 <$ [Fe/H]< 0.

A primary goal of SIM is to understand the size and structure of the dark matter halo of the Milky Way. Our fainter (i.e., $V = 13 - 17$), halo giants from the GGSS are primary targets for just such a program because they are both relatively distant, but bright. Stars that can be verified as members of coherent

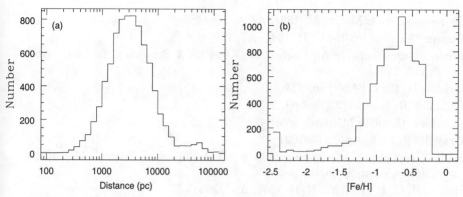

Figure 4. (a) Photometric parallax and (b) metallicity distributions for $\sim 8,000$ stars with $M_o < 15.5$ selected as K giants from about 580 photometrically completed GGSS bricks (Figure 3). Only stars in fields with $E(B-V) <$ 0.2 are shown here, and photometric error cuts of $(\sigma_M, \sigma_T, \sigma_D) < 0.04$ have also been applied. All stars with photometric [Fe/H]< -2.5 are combined in the first [Fe/H] bin.

tidal structures have increased value for this work; Johnston et al. (1999) discuss an algorithm that can recover the circular velocity and flattening of the Milky Way to within 2.5% using SIM measurements of just 100 stars known to be members of a tidal tail associated with any one of the Galactic satellites. In order to understand the mass distribution of the inner Galaxy and its disk, SIM will need to measure parallaxes and proper motions for stars to at least 20 kpc in all directions as a means to undertake definitively both the classical K_z problem – determining the dependence of the Galactic potential on distance from the Galactic plane – as well as for measuring the Galactic rotational velocity to $2R_o$. The GGSS will provide several times more than the number of giant stars in the desired range of distances (to 20 kpc) required for such studies (Figure 4a). Finally, the upcoming availability of the unique GGSS dataset has already spawned a new study into the binary and variable fraction of Pop II stars by Smith, Geisler and collaborators, and is providing lists of bright, metal–poor stars (see Figure 4b) that are valuable targets for high resolution spectroscopic studies of chemical abundances in the early Galaxy.

Acknowledgments. It is a pleasure to thank Steven Unwin and Ann Wehrle of JPL for their assistance in setting up this project. This work has been supported by NASA/JPL grants 1201670 and 1222563, a David and Lucille Packard Foundation Fellowship, and a Cottrell Scholar Award from The Research Corporation.

References

Cameron, A. G. W. 1988, ARA&A, 26, 441

Canterna, R. 1976, AJ, 81, 228

Carney, B. W., Laird, J. B., Latham, D. W. & Aguilar, L. A. 1996, AJ, 112, 668

Cuntz, M., Rammacher, W. & Ulmschneider, P. 1994, ApJ, 432, 690

Duquennoy, A. & Mayor, M. 1991, A&A, 248, 485

Dupree, A. K. & Smith, G. H. 1995, AJ, 110, 405

Frink, S., Quirrenbach, A., Fischer, D., Röser, S. & Schilbach, E. 2001, PASP, 113, 173

Geisler, D. 1984, PASP, 96, 723

Gilliland, R. L. et al. 2000, ApJ, 545, L47

Gonzalez, G. 1999, MNRAS, 308, 447

Gray, D. F. 1989, ApJ, 347, 1021

Hall, D. 1991, in IAU Colloquium 130, eds. I. Tuominen, D. Moss & G. Rüdiger, (New York: Springer), p 353

Harris, H. C. & McClure, R. D. 1983, ApJ, 265, L77

Hut, P. et al. 1992, PASP, 104, 981

Johnston, K. V., Zhao, H., Spergel, D., & Hernquist, L. 1999b, ApJ, 512, L109

Majewski, S. R., Ostheimer, J. C., Kunkel, W. E. Johnston, K. V., Patterson, R. J. & Palma, C. 1999, IAU Symp. 190, 508

Majewski, S. R., Ostheimer, J. C., Kunkel, W. E. & Patterson, R. J. 2000a, AJ, 120, 2550

Majewski, S. R., Ostheimer, J. C., Patterson, R. J., Kunkel, W. E., Johnston, K. V. & Geisler, D. 2000b, AJ, 119, 760

Marcy, G. W. & Butler, R. P. 2000, PASP, 112, 137

Nissen, P. A. & Schuster, W. J. 1991, A&A, 251, 457

Rutten R. G. M., Schrijver, C. J., Lemmens, A. F. P., & Zwaan C. 1991, A&A, 252, 203

Ryan, S. G. 1992, AJ, 104, 1144

Simon, T. & Drake, S. A. 1989, ApJ, 346, 303

Stevenson, D. J. 1985, Icarus, 62, 4

von Seeliger, H. H. 1898, in *Papers of the Mathematical–Physics Class of the Bavarian Academy of Sciences*, 19, 565

Small–Telescope Astronomy on Global Scales
ASP Conference Series, Vol. 246, 2001
W.P. Chen, C. Lemme, B. Paczyński

A Robotic Wide-Angle Hα Survey of the Southern Sky

John E. Gaustad

Dept. of Phys. & Astr., Swarthmore Coll., Swarthmore PA 19081, USA

Wayne Rosing

Las Cumbres Obs., 1500 Miramar Beach, Montecito CA 93108, USA

Peter McCullough

Dept. of Astronomy, Univ. of Illinois, Urbana IL 51801, USA

Dave Van Buren

Extrasolar Research Corporation, Niskayuna NY 12309, USA

Abstract. We are conducting a robotic wide-angle imaging survey of the southern sky at 656.3 nm wavelength, the Hα emission line of hydrogen. Each image of the survey covers an area of the sky 13° square at an angular resolution of approximately one arcminute, and reaches a sensitivity level of 0.5 Rayleigh ($3 \times 10^{-18} erg\ cm^{-2}\ s^{-1}\ arcsec^{-2}$), corresponding to an emission measure of 1 $cm^{-6}\ pc$, and to a brightness temperature for microwave free-free emission of 3 μK at 30 GHz.

1. Scientific Purpose

This survey will produce detailed information on the structure of the diffuse, warm, ionized component of the interstellar medium, information necessary for understanding the dynamics and evolutionary history of the interstellar gas.

The intensity of the microwave free-free emission from Galactic hydrogen is directly proportional to the brightness at Hα. Thus these images will also allow measurement of anisotropies in the Galactic free-free emission at microwave wavelengths (or proof that these are negligible), emission which must be subtracted to obtain the true cosmic background anisotropies.

2. Instrumentation and Observations

Our CCD camera, located at Cerro Tololo Inter-American Observatory, contains a 1024 × 1024 TI chip with 12-μm pixels. The sky is imaged onto the chip with a 52-mm f.l. lens operated at f/1.6, yielding a field of view of 13° × 13° and a scale of 48″/pixel. A filter wheel mounted in front of the lens contains an Hα filter of 3-nm bandwidth and a dual-band notch filter, which excludes Hα but transmits two 6-nm bands of continuum radiation on either side of Hα. The

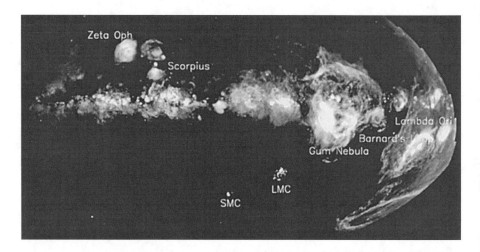

Figure 1.   A mosaic of the southern sky in Hα. See also front cover of the volume.

unguided tracking accuracy of our Byers Series 2 German mount is better than 2 arcseconds per hour. Thus no external guider is needed.

We can inquire of the status of the robot via email, and make changes in the program or observing schedule via ftp. But we do not operate the camera in real time – it is a true robot. The robot has its own weather station, but as an extra precaution it asks permission (via email) of the 4-meter telescope operator before opening the dome.

All observations are taken after twilight and with the moon below the horizon. A normal set of science observations consists of five 20-minute exposures through the Hα filter interspersed among six 5-minute exposures through the dual-band continuum filter.

3.   Survey Status

The survey consists of 283 fields covering the sky from −90° to +15° declination, with the same centers as those in the IRAS Sky Survey Atlas. In order to reduce the noise and to confirm objects discovered at the faintest brightness levels, we repeat all fields with 5° offsets. The data-gathering phase of the project was completed in October 2000, and the survey is expected to be published by summer 2001. A mosaic, in galactic coordinates, is shown in Figure 1.

4.   Acknowledgements

The project has been supported by grants from Las Cumbres Observatory, the National Science Foundation, NASA, Dudley Observatory, the Fund for Astrophysical Research, Research Corporation, the University of Illinois, and Swarthmore College.

Small–Telescope Astronomy on Global Scales
ASP Conference Series, Vol. 246, 2001
W.P. Chen, C. Lemme, B. Paczyński

MOA Extra-Solar Planet Research via Cluster Supercomputing

Nicholas Rattenbury, Ian Bond, Phil Yock

Department of Physics, University of Auckland, Private Bag 92019, Auckland, New Zealand

Jovan Skuljan

Department of Physics and Astronomy, University of Canterbury, New Zealand

Abstract. Developments in the search for extra-solar planets via gravitational microlensing by the Japan/New Zealand group MOA are discussed. The use of the Kaláka cluster computer is introduced and preliminary results presented.

1. Introduction

The Microlensing Observations in Astrophysics (MOA) collaboration is a joint Japanese/New Zealand experiment established in 1995. The project is designed to perform large scale photometry for the detection and analysis of microlensing events. The research goals of the MOA project include the investigation of dark matter, extra-solar planets, difference imaging, GRBs and variable stars. The telescope used by the MOA collaboration is a 61cm Boller & Chivens Cassegrain with optics modified to f/6.25. The telescope is one of three at the Mount John University Observatory, Lake Tekapo, New Zealand (170.465 E, 44 S). The MOA camera uses three 2k × 4k SITe CCDs with $15\mu m \times 15\mu m$ pixels, each covering $0.81''$ on the sky. The overall FOV is $1.38° \times 0.92°$. The target fields are in the Galactic Bulge, and the Magellanic Clouds. Searches for Gamma Ray Burst optical transients are made within a Target of Opportunity observation mode.

2. Extra-Solar Planet Search

The detection of extra-solar planets via gravitational microlensing was first suggested by Mao & Paczynski (1991). A planet in orbit around the lens star may introduce a perturbation to the standard single lens light curve. In events of high magnification, planets, including those with masses less than that of Jupiter, can be detected with high efficiency (Griest & Safizadeh, 1998).

2.1. Modelling of Planetary Systems

Light from a background source star is deflected to an observer by the gravitational effect of an intervening lens star system. The position in the source plane of a ray originating in the lens plane is given by:

$$\mathbf{y} = \mathbf{x} - \frac{m_1\mathbf{x}}{|\mathbf{x}|^2} - \sum_{i=2}^{n} \left\{ \frac{m_i(\mathbf{x} - \mathbf{x_i})}{|\mathbf{x} - \mathbf{x_i}|^2} \right\}$$

where $\mathbf{x}$ and $\mathbf{y}$ are positions in the lens and source plane respectively, in units of the Einstein ring radius. m_1 is the mass of the lens star, and m_i are the masses of planets orbiting the lens star. The positions of the planets (projected onto the lens plane) are given by $\mathbf{x_i}$. To obtain a theoretical lightcurve for a microlensing event we need to invert the above equation. This is not possible for multi-planet lens systems. We used the inverse ray shooting technique of Wambsganss (1997) to model planetary systems of arbitrary complexity. Rays are shot from the observer, through the lens plane and their position in the source plane computed via the above equation. Rays that fall within the radius of a source star are counted and are used to generate a model lightcurve.

Using the data from each high magnification microlensing event, we compute the χ^2 values for a wide range of model light curves, generated using the inverse ray shooting method. For each model we assume the following about the lens and source stars: Lens star mass, $M_L = 0.3 M_\odot$; observer-lens distance, $D_{OL} = 6$ kpc; lens-source distance $D_{LS} = 2$ kpc and source star radius, $R_S = R_\odot$.

The Kaláka computer cluster   The inverse ray shooting technique allows us to model lens systems with any number of planets. It also allows us to vary the source star radius. It is a numerical technique which is relatively computationally expensive. In order to analyse a large number of systems, we have made use of a computer cluster at the University of Auckland. The Kaláka cluster was created by P. Dobcsányi and comprises around 200 Pentium II 350MHz computers. These machines are used in the undergraduate computer science laboratories. When these laboratories are not in use, the computers are linked together to form the cluster. Using the Kaláka cluster, we are able to search the parameter spaces for lens systems of arbitrary complexity. Details of the cluster and the ray shooting technique are in Rattenbury et al (2001).

3. Results

Three high magnification events, MACHO 98-BLG-35, MACHO 99-LMC-2 and OGLE 00-BUL-12 are currently under analysis. Images obtained by the MOA and OGLE groups for these events were analysed using the image subtraction method of Alard (1999) as modified by Bond (2001). We present the results from one of these events, MACHO 98-BLG-35.

Data for this event originated from the microlensing groups MOA, MPS and PLANET. The Kaláka cluster was used to compute light curves for several million lens configurations. A search for a single planet lens system was performed initially. The light curves were generated at observation times in the interval $[-t_E, t_E]$ where t_E is the Einstein ring radius crossing time. The planet position co-ordinates (x_p, y_p) were allowed to vary across the interval $[-2R_E, 2R_E]$ where R_E is the Einstein ring radius. A total of 129 steps for each of x_p and y_p were used. Given the above assumptions about the system geometry, $R_E = 1.9$ AU.

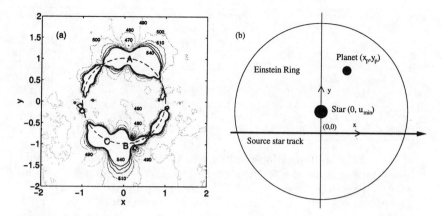

Figure 1.   (a) Contour χ^2 map for MACHO event 98-BLG-35. The mass ratio used was $\epsilon = 10^{-5}$. The projected planet positions are in units of R_E. Three pairs of minima are indicated. Deeper χ^2 minima can be found at the different positions by varying ϵ. (b) Co-ordinate system used for the inverse ray shooting method. In the lens plane, the projected planet positions are at (x_p, y_p). The lens star is placed at position $(0, u_{min})$. The source star's position of minimum projected distance to the lens star is thus at co-ordinates $(0, 0)$.

The planet-lens mass ratio, ϵ, was allowed to vary in 33 logarithmically spaced steps across the interval $[10^{-7}, 10^{-3}]$. A χ^2 map was generated for each of the mass ratios used. Minima in these maps indicate the approximate mass and position (projected onto the lens plane) of possible planets around the lens star at the time of the microlensing event. Figure 1 shows the χ^2 map for mass ratio $\epsilon = 10^{-5}$. This mass ratio corresponds to an Earth-mass planet. There are three pairs of minima shown in Figure 1. This pairing of minima is due to a degeneracy inherent in the microlensing method. Deviations in a lightcurve due to a planet at projected planet-lens distance a are indistinguishable from a planet at distance a^{-1} (Griest and Safizadeh 1998). We create multi-planet models by placing planets at two (or all three) of the χ^2 minima. We use these planet position and mass parameters as starting values for a further χ^2 minimisation using the Simplex method. Table 1 shows the results from this minimisation. The parameters u_{min}, t_0 and t_E where also allowed to vary in this minimisation.

Model "A" appears to be the most likely single planet model. The improvement in χ^2 of this model over a single star (no planet) lens model corresponds to about 9σ. Model "B+C" appears to be the best two planet model. The lens system of model "B" appears to correspond with that previously found by Rhie et al (2000). Model "B+C" of this work seems to be a generalisation of the Rhie model, to which a second, low mass planet has been added. The χ^2 value for model "A" improves only slightly for models "B+C" and "A+B+C". A statistical argument would favour the simplest model, model "A".

The 2σ limits on planet mass ratio for model "A" are $(0.4 - 1.5) \times 10^{-5}$ which corresponds to a planet of mass $\sim (0.4 - 1.5) \times M_\oplus$. The orbital radius

Model	Planet A			Planet B			Planet C			χ^2/dof
	x_p	y_p	ϵ	x_p	y_p	ϵ	x_p	y_p	ϵ	
S	-	-	-	-	-	-	-	-	-	487.9/296
A	0.11	1.22	1.3	-	-	-	-	-	-	402.0/293
B	-	-	-	0.30	-1.11	2.8	-	-	-	432.9/293
C	-	-	-	-	-	-	-0.37	-0.86	0.17	461.7/293
A+B	0.16	1.25	0.79	0.35	-1.14	2.8	-	-	-	419.4/290
B+C	-	-	-	0.30	-1.12	2.6	-0.34	-0.86	0.19	400.8/290
A+C	0.15	1.21	0.99	-	-	-	-0.33	-0.84	0.18	412.7/290
A+B+C	0.19	1.28	0.30	0.34	-1.15	2.9	-0.35	-0.87	0.17	398.2/287

Table 1. χ^2 minimisation results for MACHO 98-BLG-35. The combined data set from the MOA, MPS and PLANET collaborations from $-t_E$ to $+t_E$ were used. Model "S" denotes a single star lens model. The planet positions are in units of R_E. The mass fraction ϵ is in units of 10^{-5}. A typical χ^2 map and the co-ordinate system used is shown in Figure 1.

of the planet in this model is either $0.82R_E$ or $1.22R_E$. This corresponds to either ~ 1.5 or ~ 2.3 AU. Figures 2 and 3 show the light curves for models "A" and "B+C" respectively. The planetary perturbation is negative for model A because the source star track does not pass between the lens star and the planet for this model (see Fig. 1).

4. Further Work

In the 2000 Galactic Bulge season, the MOA project monitored several interesting transient events. Table 2 lists some of these events. The majority of these events are probably due to microlensing, but some are nova-like events. Additional events with $A \geq 10$ are candidates for planetary searches.

5. Discussion and Conclusion

The detection of extra-solar planets in high magnification gravitational microlensing events has been demonstrated. Earth-mass planets with orbit radii ~ 2 AU can be detected by terrestrial 1-m class telescopes. The critical requirement for this research is the dense sampling of the peak of a high magnification microlensing event. This can be done by several co-ordinated telescopes across the world. Using the image subtraction process can give higher accuracy in the analysis of crowded field images, such as those encountered in microlensing surveys. The inverse ray shooting method has been successfully implemented on a cluster computer to sample densely the parameter spaces for multi-planet lens systems.

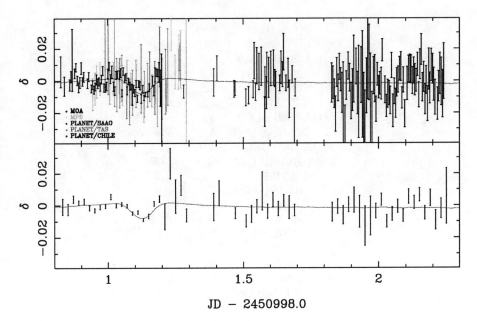

Figure 2.   MACHO 98-BLG-35 observed data and fitted lightcurve
for the single planet lens model "A". The bottom plot shows the same
data weight averaged and binned into 0.02 day intervals. The y-axis of
each plot shows δ, the fractional deviation from a single lens system.

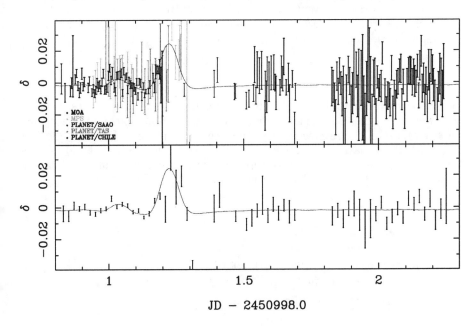

Figure 3.   As Figure 2 but showing the two planet model "B+C".

MOA ID	A_{max}	
	$2\sigma_l$	$2\sigma_u$
MOA-2000-BLG7	14	-
MOA-2000-BLG11	7.4	8.8
1-3-2540	8.2	46
1-3-2548	5.1	-
MOA-2000-BLG3	9.6	-
2-2-1648	4.0	7.4
MOA-2000-BLG9	3.0	-
MOA-2000-BLG13	14.8	20.5
4-2-2197	-	1.7
4-3-159	1.2	5.9
5-1-1616	-	-
5-1-1629	38.6	(101.7)
5-1-1668	2.0	14
5-1-1672	-	3
5-1-1673	22	-
MOA-2000-BLG12	6.2	11.6
MOA-2000-BLG8	1.6	2.2
9-3-841	-	4.0
MOA-2000-BLG10	9.0	11.2
12-2-1052	7.7	21.3

Table 2. A sample of Galactic bulge events observed by MOA in the year 2000 season. Events with a MOA-2000-BLGn designation were issued as transient alerts. Upper and lower 2σ limits are given for the amplification of each event.

Acknowledgements

We are very grateful to the OGLE collaboration for making their images available to us and to Andrzej Udalski for comments and discussion. Joachim Wambsganss for his advice on the inverse ray shooting method and Peter Dobcsányi for his help with the Kaláka computer cluster. We thank the Marsden Fund of New Zealand, and the Ministry of Education, Science, Sports and Culture of Japan. NJR thanks the Graduate Research Fund of Auckland University.

References

Alard, C., 1999, A&A, 342, 10

Bond, I.A. et al, 2001, submitted to MNRAS

Griest, K. & Safizadeh N., 1998, ApJ, 500, 37

Mao, S. & Paczynski B., 1991, ApJ, 374, L37

Rattenbury, N.J., et al, 2001, in preparation

Rhie, S.H., et al, 2000, ApJ, 533, 378

Wambsganss, J., 1997, MNRAS, 284, 172

Small–Telescope Astronomy on Global Scales
ASP Conference Series, Vol. 246, 2001
W.P. Chen, C. Lemme, B. Paczyński

YSTAR :
Yonsei Survey Telescopes for Astronomical Research

Yong-Ik Byun[1], Won-Yong Han[2], Yong-Woo Kang[1], Moo-Young Chun[2],
Sung-Yeol Yu[1], Sun-Youp Park[1], Sang-Chul Kim[1], Hong-Kyu Moon[2],
Jong-Seop Shin[2], Kyu-Baek Lee[1]

[1] Department of Astronomy, Yonsei University, Seoul 120-749, Korea
[2] Korea Astronomy Observatory, Taejon, 305-348, Korea

Abstract. The YSTAR program is a general sky survey looking for
variability. The main equipments are three 0.5-m telescopes. These
telescopes have fast F/2 optics covering nearly 3.5 square degree field
onto a 2K CCD. They also have very fast slew capability, which exceeds
10 degrees per second. These two factors make them most suitable for
rapid target acquisition and wide-field surveys of various kinds. Our primary objective is to identify and monitor variable stars down to 18th
R-magnitude, and our observing mode allows the same data set to be
also useful in identifying asteroids. Our first telescope has just begun
regular automated operation, and the second telescope will be installed
in South Africa within this year to provide coverage of the southern sky.

1. Introduction

We have been preparing an observational program to catalog and monitor variable events in both northern and southern hemispheres. The program is starting
with three 0.5-m wide field robotic telescopes and, contingent on funding, we intend to increase the number of our survey telescopes and place them around the
globe. The observational and computational facilities involved in this project
are named as YSTAR (Yonsei Survey Telescopes for Astronomical Research).
This will be the major observational campaign of Yonsei University Observatory (YUO) in the coming years.

Sky surveys have been made in various wavelengths in the past, both from
the ground and from the orbit. Variability of different kinds have also been investigated from such survey data. However, as expressed by several contributions
in this conference, there is still a very large area left for further explorations.
In fact, the vast majority of variable phenomena has never been properly investigated. Instead of going into the details here, we would like to encourage
the readers to read Paczynski (2000) and references therein, which are in our
opinion the best starting places for the subject.

Variability monitoring requires time-series observations, which give the opportunity to look for both photometric variability and positional variability. The
former involves identification and light curve study of variable stars and galaxies,
as well as transient events such as Gamma Ray Bursts (GRB), planet transits,

Figure 1.   YSTAR No.1 telescope installed in the Chun-an site

and microlensing events. The latter usually involves the detection and tracking of asteroids and comets.

The advanced digital detector technology, computer controlled robotic operation, and high speed computing capability for photometric analysis are now making it possible to carry out an all sky survey with unprecedented photometric accuracy. Such attempts have recently been made with small aperture instruments, no larger than 10-cm in diameter, nevertheless still producing thousands of new variable star discoveries (see Pojmanski 2000, Akerlof et al. 2000a) and the first optical detection of a GRB during the actual burst (Akerlof et al. 2000b).

YSTAR project shares the same goal as these studies, i.e. monitoring of all sky for variability, but aims to reach somewhat fainter photometric depth with larger aperture. Our telescopes will fill the gap between existing small aperture surveys and future big aperture surveys, which will cover the much fainter and more distant universe.

2.   Telescopes

Our 0.5-m telescopes, manufactured by Torus Technology in Iowa, are identical to those designed by the TAOS project team (Taiwan American Occultation Survey, see King 2001) except in minor details. The optical configuration of this hybrid Cassegrain uses an F/1.5 parabola primary, a large spherical secondary, and a field corrector consisting of 5 lenses. The FOV onto a 2K CCD of 14-micron pixel is nearly 3.5 square degrees. The CCD is attached to a 10 position filter wheel. The filter wheel contains B, V, R, I broad band filters and three interference filters of H-alpha, [SII], and continuum. We intend to carry out most of our routine observations in R-band, occasionally using other broad bandpass for standard star calibrations.

The telescope drive system, based on friction drive with DC servo motor, has a capability of very fast slew speed over 10 degrees per second. This feature makes the telescope ideal for observations which require immediate pointing. GRB followup observations are good examples. It also contributes to reduce the down-time for our scheduled survey observations; the telescope can usually point and stabilize at the next target position during the few seconds it takes to read out the CCD.

The telescope control system is connected to a GPS, a weather station, and an enclosure control. We are presently developing a cloud monitoring device as an added safety measure. The control computer is connected via internal Ethernet network to a group of on-site computers for near real-time data processing and mass storage.

3. Observing Strategy

With the present FOV of YSTAR telescopes, the entire sky can be covered by 14916 Target Fields (TF) with slight overlap between them. This is still too large a number for efficient observation management. Therefore we grouped them into 452 Target Clusters (TC) according to their positions, each TC containing 33 TFs. The scheduling and survey progress monitoring are automated based on the TC list and their mean coordinates.

Each observation of a TC involves three consecutive coverages of all TFs in the given TC. With a typical exposure time of 30 seconds for each frame, the total time required to finish a TC is somewhat more than 1 hour, and this can be adjusted by changing the exposure time or by grouping two TCs into one as needed. Three exposures with nearly equal time intervals become useful data to search for moving objects, while providing opportunity to detect brightness variabilities of relatively small time scales.

The data processing pipeline is divided into an on-site operation and an off-site operation. On-site operation includes basic image preprocessing such as bias, dark, and flats. The astrometric plate solution and estimate of image quality are made almost immediately after each exposure. The on-site operation includes data pipelines for photometric and astrometric analysis. This consists of two independent phases; the first being based on Sextractor (Bertin & Arnouts 1996) for near real time data reduction and the second procedure based on DAOPHOT (Stetson 1987). The Sextractor provides quick and robust source detection and photometric data, which are very useful to identify candidates for moving objects and also for objects with significant brightness changes. Alerts are generated based on the Sextractor outputs, and transmitted to the YSTAR home institute via Internet together with small postage images extracted from the data frames for visual verification. The DAOPHOT routines run during daytime only, as it requires more extensive computing time, in order to derive more accurate photometric information especially for crowded fields. The resulting photometry files as well as the original images are archived at the telescope site. We use a tape library using SONY AIT tapes as a local data storage device. The tapes

Figure 2.   Opening of Yonsei University Survey Observatory on November 24, 2000.

are regularly forwarded to the home institute for the main data archive of RAID arrays, and then sent back to the site to be used again.

The off-site data pipeline includes the construction of light curves for individual sources, image difference photometry for very crowded fields in low galactic latitudes, and general database management tools. It is our plan to release the data to the general astronomy community once proper calibration has been done.

4. Project Status

The first YSTAR telescope was temporarily installed near the YUO site in Chunan, Korea, on March 2000 and used as a testbed for various experiments. These included telescope setup procedures, automated observations, and CCD integration and enclosure operations. The site was however without any infra structure except for a single power line, making the experiments harder to perform.

In November 2000, the YUO Survey Observatory was formally opened. Modest living quarter was also built together with the observatory building which houses three piers. Internet connection based on a dedicated 256 Kbps optical fiber line was also established, enabling us to monitor and control the facility remotely from the campus in Seoul. The first telescope was moved to the new observatory and is presently being used to generate commissioning data. Data pipelines and automatic survey routines are being tested and constantly improved using the facility.

Not everything went as planned. The fast nature of our optics made it very difficult to collimate the telescope. The unreliable pointing and tracking of the telescope also caused much delay. The list of problems grew very long as we strove to make progress with the program. However, in order to resolve these issues, we have been working closely with the telescope manufacturers and also with astronomers experienced in similar fast optics and automated operation. Most problems have already been solved at the time of writing.

One of the major progresses in the year 2000 is that YSTAR is no longer just a university project. Korea Astronomy Observatory (KAO) and YUO have decided to work together for the common goal of efficient sky surveying. While KAO's main interest lies in detecting near-earth objects and comets, the observational strategy and software development are almost identical to ours. By sharing experiences and existing facilities of each party, we believe that we can maximize the chance of success in general sky monitoring.

5. 2001 and Beyond

Good sky condition is crucial for the efficiency of a survey project. Since ours is not a target oriented short term observation but a general long term survey of entire sky, the number of clear nights directly dictates the sky coverage and coverage frequency of our observational campaign. Unfortunately the climate of Korea is not suitable for our purpose and, from the start of this project, we have been planning to locate our survey telescopes elsewhere.

With the kind invitation of South African Astronomical Observatory (SAAO) we are now preparing to set up YSTAR No. 2 telescope in Sutherland within this year. Large number of clear nights as well as its dark sky condition make our facility much more competitive in all sky monitoring. The access to the southern sky is also a very important factor, as nearly all photometric and astrometric monitoring programs are being run in the northern hemisphere. After successful commissioning at SAAO, we will expand the facility to other places around the globe.

KAO and YUO are now working together to secure more survey telescopes of somewhat larger aperture. Five or more telescopes will be constructed and placed around the world making a network of survey telescopes working 24 hours a day. International cooperation is essential in this effort, and we welcome anyone who is interested in working with us.

References

Akerlof, C., Amrose, S. Balsano, R., Bloch, J. et al. 2000a, AJ, 119, 1901

Akerlof, C., Balsano, R., Barthelmy, S., Bloch, J. et al. 2000b, ApJ, 532, 25

Bertin, E. and Arnouts, S. 1996, A&Ap, 117, 393

King, S. K. et al. 2001, in this proceeding

Paczynski, B. 2000, PASP, 112, 1281

Pojmanski, G. 2000, Acta Astronomica, 50, 177

Stetson, P. B. 1987, PASP, 99, 191

(From top left clockwise)
McGruder, Byun, A. Herbst and W. Herbst, Mikołajewska

Small–Telescope Astronomy on Global Scales
ASP Conference Series, Vol. 246, 2001
W.P. Chen, C. Lemme, B. Paczyński

The 1.3-meter Robotically Controlled Telescope: Developing a Fully Autonomous Observatory

Richard Gelderman

*Western Kentucky University, Dept. of Physics and Astronomy,
Bowling Green, Kentucky, USA 42101*

Abstract. The former KPNO 1.3-meter telescope is being refurbished and automated by a consortium of U.S. institutions, headed by Western Kentucky University, with the goal of a 2002 recommissioning as the Robotically Controlled Telescope (RCT). The 1.3-meter RCT will operate in fully autonomous mode to obtain guided images for a variety of research and education programs. Distinctions between a fully autonomous versus robotic observatory are presented, along with a discussion of why fully autonomous operation is necessary for increased productivity of small telescopes.

1. The RCT – Honoring the Past

In 1964, with NASA funding, a new 1.3-meter telescope was commissioned on Kitt Peak, in Arizona, and named the **"Remotely Controlled Telescope."** As the name implies, the **RCT** was designed to be operated remotely with no people actually at the telescope. The project was originally headed by the Kitt Peak National Observatory (KPNO) Space Sciences Division, intended to develop techniques for operating space-based telescopes. This effort included state-of-the-art (for the mid 1960's) motion encoders, a 'powerful' computing system (so compact that it fit inside a single large room), and a communication network based on phone line connections from the mountain to the KPNO headquarters roughly 75 km away in Tucson. However the poor results from engineering tests performed at the downtown headquarters before the telescope was installed on the mountain prompted KPNO to reevaluate its goals.

In response to the Whitford Committee (1964) report, the purpose of the 1.3-meter's automation was shifted to an attempt to enhance the productivity of small telescopes through remote operation. In 1969, control of the telescope was shifted to KPNO's Stellar Division and the attempts at remote operation were directed toward demonstrating that a ground-based observatory could be operated from a distance of many miles. Eventually, this experiment was also not deemed to be successful and the KPNO 1.3-meter telescope was refurbished for classical, on-site operations. Shortly thereafter, the troublesome aluminum mirror was replaced with a honeycombed lightweight Cer-Vit primary. For the next 25 years the 1.3-meter was a productive small telescope, first with a photoelectric photometer and later as a testbed for pace-setting infrared instrumentation. Finally, in 1995, due to pressure from a decreasing budget, the 1.3-meter was closed by KPNO.

In 1999, the National Optical Astronomical Observatory announced an opportunity to "assume responsibility for operation of the Kitt Peak 1.3-meter telescope." A collaboration of US astronomers, headed by Western Kentucky University, successfully proposed to refurbish the 1.3-meter and automate the observatory for operation as the **"Robotically Controlled Telescope."** The RCT Consortium was established to manage the refurbishment and establish scientifically and educationally productive operations. Funding has been provided by contributions from the RCT Consortium partner institutions and a generous award from the NASA Office of Space Science. The contract to refurbish, automate, and provide maintenance for the new **RCT** was awarded in the spring of 2000 to EOS Technologies, Inc. of Tucson, Arizona.

2.   Defining a Fully Autonomous (*vs.* Robotic) Observatory

In general, a robot is a mechanism guided by automatic controls but is not necessarily a mechanism which operates fully autonomously. While the ultimate robot might be fully autonomous (or even self-aware, as in Karel Capek's 1920 science fiction play *Rossum's Universal Robots*, which first introduced the term "robot") the term is commonly used for any device which performs a repetitive task without constant human intervention. In astronomy, there are many robotic observatories but very few fully autonomous ones.

With a hope that our discussions will benefit from a common vocabulary we offer the following definitions for stages of robotic observing.

Classical On-site observing involves having all observers and operators at the site, with the observatory completely under local control.

Remote observing allows the observer to have some level of instrument control from a distant location but still requires a local telescope operator at the site. Many facilities routinely offer remote observing (*e.g.,* the ARC 3.5-meter at Apache Point, New Mexico and most of the large telescopes on Mauna Kea).

Unmanned observing requires enough computer control that remote observers/operators can control all aspects of the observatory functions. Verification and tasks such as acquiring guide stars are performed manually. Many ground based observatories and some early orbiting observatories (*e.g.,* the International Ultraviolet Explorer) operated in unmanned mode, without being robotic.

Robotic observing requires automatic control of the instruments but involves pre-scheduled observations submitted as a queue and the need for pre-selected guide stars. Modern space-based observatories (*e.g.,* the Hubble Space Telescope) and a number of small ground based telescopes (*e.g.,* the Fairborn Observatory or the Katzmann Automated Imaging Telescope) are productive robotic telescopes.

Fully Autonomous observations involve the ability to constantly rearrange observing schedules to optimize performance, much as a well-trained human astronomer would adjust to changing sky conditions.

The refurbishment of the 1.3-meter will increase its capabilities to allow for fully autonomous operation, surpassing both "remote" and "robotic" modes of operation. We chose, however, to avoid the name "Fully Autonomous Telescope" in favor of the more attention grabbing term "robotic" and the chance to honor the past by reclaiming the **RCT** acronym. Despite our misuse of the terms, we argue that robotic and fully autonomous are not synonymous and that that fully autonomous is a significantly more stringent definition than is robotic.

The key to fully autonomous observing is software which acts as a scheduler, allowing for flexibility in response to changing conditions and complex science requirements. A robotic telescope, with each night's observing scripted in advance, will be productive with a well defined and repetitive program. An optimized, fully autonomous scheduler, however, will allow a diverse collection of targets to be observed without having to specify their exact place in the program's sequence. It is not hard to imagine situations where a robotic observatory will lose productivity whenever the observing program is no longer well defined and repetitive or for some other reason an efficient schedule cannot be predetermined. For instance, robotic telescopes are rarely flexible enough to overcome the mechanical and/or weather problems which challenge any observing program; and generally can not deal with the desirable complexities of observations coordinated among multiple telescopes.

3. Global Networks of Fully Autonomous Telescopes

One interpretation of *Small Telescope Astronomy on Global Scales* is the establishment of global networks of small aperture telescopes. Coordinating such a network imposes all the demands of operating a single observatory, while adding concerns related to distance, non-standard hardware, and the needs of a diverse collection of users. To be successful, a global network must allow for realistically complex observing programs while keeping the costs low enough for long term and large scale projects to be feasible. New observers should not have to immerse themselves in specialized procedures in order to make use of the network. For instance, an observer should be able to schedule a single observation to run each night for weeks, or years, without devoting weeks, or years, of her/his life to monitoring the progress of the observing.

A global network of small telescopes will be most efficient and effective if the telescopes can operate in fully autonomous mode. A major hurdle to establishing such a network is the personnel costs of coordinating a full time network with on-site operators. Examples of successful global networks tend to be either well-funded efforts (at least compared to the typical small telescope initiative; *e.g.,* the Global Oscillation Network Group or the Very Long Baseline Interferometry (VLBI) radio network) or low- to non-funded collaborations coordinated on a part-time basis (*e.g.,* the Whole Earth Telescope). However, many astronomers have a full-time load of responsibilities beyond research and are unable to commit to an additional load of observing multiple times each week. The availability of eager and capable graduate student researchers can solve the personnel dilemma for some observatories. But as a general rule, hiring and retaining a group of top notch telescope operators can be prohibitively expensive. Thus a global network should at minimum be based upon robotic observatories and only a fully autonomous system can hope to come close to replacing the adaptability of human observers.

4. Refurbishing and Automating the RCT

The 1.3-meter RCT is being refurbished and automated as one piece of the STARBASE global network of telescopes. The RCT is a Cassegrain telescope built by Boller & Chivens according to some interesting specifications, many of which presage the design of modern observatories such as WIYN, Keck, or Gemini. The mount and dome for the RCT are relatively undersized. Its german equatorial mount was designed for a 0.9-meter telescope with an open truss support structure designed for its lightweight (originally spun aluminum) 1.3-meter primary mirror. The dome also is typical of a 0.9-meter observatory, since the intended remote operations could be accomplished with less extra space than human observers might require.

After an engineering assessment, the RCT Consortium chose to keep the existing $f/13.5$ optics, the original mount, and the 36 year old worm gear drives. We also chose to keep the telescope in its original dome at the existing site, despite its proximity to the KPNO administration building and asphalt parking lots. Instead, we have focussed our efforts on refurbishing the encoders and computer interfaces to provide input to custom software for fully autonomous observing.

Observatory Control Computer: The central nervous system of the RCT is the Observatory Control computer, which accepts input, makes decisions, and outputs commands and messages as it coordinates and controls the activities within the entire observatory. For instance, based on data provided by the weather station and sky monitors, the Observatory Control computer makes the determination to open or close the dome, and then issues commands to the dome slit controller and updates the status in the event log.

The Observatory Control computer incorporates the Scheduler program, the event log, and searchable star catalogs; as well as interfacing through the local area network (LAN) to all other observatory functions. Self-contained instruments (*e.g.,* the meteorological station, 15 MHz clock, and uninterruptible power supply) interface directly to the Observatory Control computer. Other instruments, such as the CCD camera, the autoguider, or the controller for the telescope systems, interface with the Observatory Control through their own separate computers.

Scheduler: Fully autonomous operation is derived from the function of the Scheduler. The Scheduler accepts observation requests from users and, during operation, determines the next observation based on various constraints.

Requests consist of 1) details about the user's identity and the target, 2) scheduling parameters (*e.g.,* range of dates to complete observations, worst acceptable seeing conditions, maximum airmass, maximum moon phase, & minimum separation from the moon), and 3) observation requirements (*e.g.,* integration time, filter, & calibration needs). A Scheduler unit consists of a single observation of a target. The ability to request an observing sequence of multiple images of the same target (*e.g.,* with varying integration times, through different filters, or as identical observations) is easily accomplished with the request editor; however, once submitted, such a request is treated as multiple observations by the Scheduler.

The operation of the Scheduler is tuned by adjustment of weighting parameters which relate to various observing decisions such as quality, quantity, timeliness, and fairness. The **quality** weighting factor compares the scheduling parameters submitted by the user to the current conditions. Potential observations may be excluded if the conditions do not meet the requirements or may be given a lower weight depending on how far the current conditions are from the requested optimum conditions. The **quantity** weighting factor controls the emphasis that the Scheduler places on maximizing the total number of observations completed within a given time period. Potential observations can be given a higher weight if they have shorter integration times or require a minimal slew time from the previous position. The **timeliness** weighting factor allows the Scheduler to prefer observations which will not be possible to obtain at a later date. The **fairness** weighting factor ensures that over a given accounting period the allocation of time within the Consortium is distributed according to either a fixed number of hours or predetermined percentages. The observing request which is returned to the telescope for execution is the one with the maximum total for all of the weighting factors. If the telescope is unable to complete the observation for any reason then the request is returned to the Scheduler for reconsideration.

Telescope Control Computer: The Telescope Control computer manages the operation of the telescope drive motion and limits, focus, temperature sensors, filter wheels, calibration lights, observatory security lights, mirror covers, dome shutter, and dome azimuth motion. The various components of the telescope system interface to the Telescope Control computer through new digital I/O gate arrays. The existing DC servo drive motors and worm gears are monitored and controlled with new PMAC motion control cards, switch mode drive amplifiers, and on-axis incremental encoders. Axis position and drive motor velocity are used as feedback to establish a stable and accurate servo loop. Multiple levels of limit protection are provided by the software and hardware. These range from a warning to human operators who might issue a command to move beyond the predefined range of operation, through a software commanded reduction in speed as a limit is approached, to an interruption of power to prevent catastrophic failures. The Telescope Control computer also controls the possibility of german flips and establish zenith and stow positions for the scope. The focus can be monitored and either adjusted remotely or as a part of fully autonomous operations. The secondary mirror position is controlled automatically to compensate for expansion of the telescope's open truss structure using input from the various temperature sensors. There are 16 available slots for broadband and narrowband filters, distributed between two filter wheels. The brightness level of flat field calibration lamps and security lights within the observatory are adjustable through a computer interface. The dome shutter and mirror covers can open and close the observatory automatically in threatening conditions; such as high wind, threat of precipitation, or imminent power failure. The dome can be moved independently of the telescope, as required during flat-field calibration. However, for normal operation, as the telescope slews or tracks across the sky, the dome azimuth is monitored and controlled to keep the RCT's relatively undersized dome slit aligned with the telescope. A virtual control panel allows remote users to easily interface with the Telescope Control computer, either taking control of the observatory systems or simply monitoring its robotic operation.

Autoguider: The RCT's off-axis autoguider employs a pick-off mirror, driven by an X-Y stage to the position selected within the available field of view, to feed the image to the guide camera. Suitable guide stars are selected by the Observatory Control computer from its searchable star catalogs. The autoguider computer moves the autoguider's X-Y stage to the expected position for the selected guide star and executes a search pattern to center the guide probe on the selected star. It is possible to focus the telescope from the image of the guide star, with the seeing (measured from the size of the guide star's point spread function) returned to the Observatory Control computer for use by the Scheduler. When enabled, the autoguider computer controls and monitors the position of the probe while the output from the autoguider program provides the guider offsets to the Telescope Control computer.

Camera: The RCT is a single instrument telescope. We have an imager with a thinned, back illuminated SITe 2048×2048 pixel CCD. It is operated through its own computer and linked to the RCT Observatory Control computer.

5.   Research and Educational Goals for the RCT

The RCT Consortium is organized around common educational goals and common research projects. Our overall goal is to provide high quality astrophysical images for use in these education and research programs. The four main research areas are: broadband photometry of moderately dense star fields to identify extrasolar planets (Everett & Howell 2001); broadband photometry of quasars and blazars (Clements & Carini 2001); narrowband imaging of Galactic and extragalactic nebulae (Buckalew et al. 2000); and rapid response to transient events such as gamma ray bursts.

Our educational program addresses the need for hands-on, inquiry-based learning. All of the projects will require students to work through the entire scientific process – from learning to frame productive questions, to performing good observations or experiments, to presenting to an audience the results of the project. Teacher-student teams will either participate in one of our main research programs or will be able to establish their own investigation with help from a mentor.

References

Buckalew, B., Dufour, R.J., Shopbell, P., and Walter, D.K. 2000, AJ, 120, 2402

Clements, S.D. and Carini, M.T. 2001, AJ, 121, 90

Everett, M.E. and Howell, S.B. 2001, PASP, submitted

Whitford et al. 1964, Ground-Based Astronomy: a Ten-Year Program, (Washington: National Academy of Sciences), Pub. No. 1234

Small–Telescope Astronomy on Global Scales
ASP Conference Series, Vol. 246, 2001
W.P. Chen, C. Lemme, B. Paczyński

The Development of Advanced-Technology Automated/Robotic Telescope Systems and the Future of Small-Telescope Astronomy

Richard J. Williams & James Mulherin

Torus Technologies, 3007 Sierra Court, Iowa City, IA 52244, USA

Abstract. During the 1990s groups at universities around the world developed small working automated/robotic telescopes that proved the feasibility of using such systems for education and research projects. A few of the more successful projects such as the Bradford Robotic Observatory in the United Kingdom and the University of Iowa's Automated Telescope Facility (AFT) and Iowa Robotic Observatory (IRO) programs proved how useful and powerful these systems can be in practice. This paper describes how one company, Torus Technologies, developed hardware and software technologies to create the most advanced integrated small automated/robotic telescope systems in the world. These systems were designed from the "bottom up" to be automated/robotic telescopes capable of operating an entire observatory including domes, CCD cameras, and other peripheral equipment.

Automated/robotic telescopes can play a major role in enabling small colleges and universities, especially in developing countries, to actively participate in serious "hands on" research and education projects that otherwise would not be practical. A commercially available affordable, high-precision, and proven turnkey automated/robotic small telescope system capable of operating remotely via the Internet is crucial for bringing this technology into widespread use. Today Torus Technologies telescopes are installed at locations worldwide as primary instruments for research programs, discovery and monitoring programs, and education programs. This paper describes some of the current applications for using these telescopes and how these telescope systems will be used in the future in standalone installations and in global networks.

1. Introduction

This paper describes the history of the development of robotic telescopes by Torus Technologies and what new technologies and products are planned for the near future. Before describing the development of these technologies, we need to explain our definition of what a robotic telescope actually is.

1.1.   What are Automated and Robotic Telescopes?

We define an automated telescope as a telescope capable of automatically following a scripted observing run pointing to objects and collecting data that are stored for later analysis. The hardware and software requirements to operate an automated telescope are modest and one or more people normally oversee the operation of the observatory. A robotic telescope extends the capabilities of an automated telescope and has the following characteristics:

- full automated control of an observatory and instruments without the need for human intervention

- system acts and reacts to feedback from environmental monitoring sensors to fulfill observing requests and protect the telescope and other equipment that are part of the observatory

- software is flexible and able to adapt to changes in the observing schedule as new events or input require new action

- remote access and control of the observatory and acquired data

A robotic telescope is a complex system that incorporates high-precision hardware, advanced electronics, and sophisticated software all of which must interoperate effectively in a system that is flexible and somewhat autonomous.

2.   The Development and Evolution of Torus Technologies Telescopes

James Mulherin began fabricating high quality Newtonian and Cassegrain optics in the early 1990's. James and his brothers Toney Mulherin and John Mulherin designed and built their first high-precision telescopes in 1994. These first telescopes used gears and other hardware designs that were standard at the time for telescopes in the half-meter size. The Mulherin brothers used a commercially available telescope control system called PC/TCS made by COMSOFT that drove stepper motors and computer software that ran on the MS-DOS operating system. These telescopes worked well with an observer in attendance in the observatory.

2.1.   Torus Technologies' First Automated Telescope System

In 1995, after spending more than a year researching what was available in the industry and talking with many people knowledgeable about telescope technology, Rich Williams decided to contract James Mulherin and Torus Precision Optics (later to be renamed Torus Technologies) to design and build a 16-inch telescope for a personal observatory on his property in Western Washington. Rich planned to use his telescope and observatory to search for supernova and other monitoring projects that would require using a CCD camera for imaging and a control system that enabled automating sequences of operation. During the design phase of the project James and Rich discussed many possible configurations for the telescope design and decided on the following features:

- Classical Cassegrain F/10 optics

- Friction rim drive system driven by stepper motors

- Equatorial fork mount

- Open-truss optical tube assembly

- A focusing mechanism that moved the secondary mirror with repeatable positioning for accurate focusing

- Components manufactured to the highest precision possible

- A control system capable of high precision tracking and pointing

Torus manufactured the telescope to the designed specifications and installed the telescope at Rich's observatory in Buckley Washington. The telescope had excellent optics and was very well made.

The original software used to control the telescope was PC/TCS, which was one of the best systems available commercially at the time. The software ran on the MS-DOS and the user interface was all text based. Rich wanted a more friendly graphical user interface with planetarium-style view to point the telescope in addition to entering object names in a form field. The owner of the software was not immediately receptive to the idea but eventually he added that capability to point the telescope using a commercially available program that ran on Microsoft Windows. However this required an additional computer to run Windows. In addition PC/TCS could not control a CCD camera and thus a third control computer was required. Although PC/TCS proved capable of running scheduled runs automatically, there were many shortcomings that soon became apparent with the overall hardware/software control system. Some of the issues that needed to be addressed included:

- *Focusing and maintaining focus of the CCD camera*
  Focusing a CCD camera on a telescope manually can be a time consuming and tedious task. PC/TCS had no capability to automatically focus a camera. Also the focus of the camera would change over time as the ambient temperature changed. During an evening scheduled run a perfect focus would gradually (sometimes rapidly) degrade until the images became poor or useless.

- *Separate computers to control the telescope, CCD camera, and filter wheel*
  Having to use separate computers with different operating systems increased the complexity and the opportunity for a failure of the system.

- *Tracking and pointing of the telescope not as accurate as desired*
  The original version of the telescope operated in an open control loop with the position being determined by counting the steps of the stepper motors. There were no encoders, hardware limit switches, or a discreet home switch. The pointing and tracking accuracy were calculated from a pointing model generated from individual stars that were centered manually during the initial setup. Overall the tracking and pointing were adequate, but not as precise as was hoped for.

2.2.   OCAAS and the Evolution of the Automated/Robotic Telescope

In February 1997 Rich became a partner in Torus Technologies and began working for the company full time. In early 1997 Torus decided to use the observatory control software called Observatory Control and Astronomical Analysis System (OCAAS) that the University of Iowa developed for their Automated Telescope Facility (ATF) as the telescope control system for Torus telescopes. OCAAS was a suite of software programs that controlled the ATF telescope, a CCD camera, filter wheel, and dome while checking the environment using a weather monitoring system. OCAAS operated on a single computer running the Linux operating system. In addition to its observatory control functions, OCAAS also had many image processing and data analysis capabilities and the user interface was graphical running several programs simultaneously in separate windows. The ATF was a great success and hundreds of students at the University of Iowa used the ATF as a laboratory for astronomy courses. By 1997 the University of Iowa was expanding on the AFT and proceeded with plans for the Iowa Robotic Observatory (IRO), which was a 0.5-meter automated/robotic telescope on an alt-az mount running OCAAS for observatory control. Unlike the ATF, which was located on campus in Iowa, the IRO would be located at the Winer Mobile Observatory in Sonita Arizona and operated remotely from the University of Iowa campus. Torus worked with the development team of the IRO project fabricating the optics and consulting on some of the engineering aspects of the IRO telescope.

Rich named his observatory the Torus Observatory and the sixteen-inch Cassegrain telescope became the prototype for an evolving telescope design. OCAAS required additional hardware and electronics that the current telescopes did not have. Thus Torus retrofitted the telescope with optical encoders, hardware limit switches, and home switches for the RA, Dec, and secondary-mirror focus positions. After months of testing and debugging the telescope and CCD camera system, OCAAS proved to be a very powerful and effective control system. The Torus Observatory used an Apogee AP7 CCD camera with a 512 x 512 SITe back-illuminated CCD chip with 24-micron pixels, which provided a 10.5 arc-minute by 10.5 arc-minute field of view. Although the weather in Western Washington is notoriously cloudy, Rich managed to get impressive results from the system and began doing asteroid astrometry work for the Minor Planet Center, working on other projects, and demonstrating the telescope to prospective customers from around.

2.3.   Birth of the Classical Cassegrain Series Telescopes

The Torus Observatory telescope proved to be so successful that Torus decided to use OCAAS and the features incorporated in the telescope as the basis for the development of all the company's telescopes and the standard was set for future research and development. Torus decided to market a line of telescopes based on OCAAS and the prototype system built for the Torus Observatory. This line of telescopes was called the TORUS Classical Cassegrain Series and the telescopes have common features: OCAAS, F/10 Classical Cassegrain optical systems, equatorial fork mounts fabricated from aluminum, friction-rim drives, and so on. Standardizing on major parts of the telescope systems was an important business decision for Torus. Rather than producing each telescope as a "one off" development project with a large investment in engineering time for research and development, the main components would be designed and the production blueprints would be used over and over again. This approach cut the development and production time for manufacturing telescopes considerably and reduced the cost making the company more profitable for investors.

Beginning in 1997 members of the Torus staff started attending and exhibiting at major astronomy meetings and workshops around the world. With the success of the Torus Observatory telescope and OCAAS, the astronomy community started to take notice of what Torus was doing. Torus sold the first TORUS Classical Cassegrain Series telescope in early 1998, and then other innovative customers that believed in our ideas and telescope designs, showed interest and ordered telescopes. To handle the increased production, Torus moved to a larger facility. Eventually Torus standardized on three different mount designs for the telescopes in the Classical Cassegrain Series ranging from 0.4-meters to 1.0-meters in diameter. The general specifications of the hardware and software are common to all telescopes in the series.

2.4. Development of the Wide-Field Hybrid Cassegrain Telescope System

Rich Williams and John Mulherin attended the UN/ESA Workshop on Basic Space Science in Tegucigalpa Honduras in March 1997. At the workshop they met Dr. Syuzo Isobe from Japan who talked with them about his plans for a project that would require a 0.5-meter and a 1.0-meter wide field telescope to search for and monitor near-earth objects (NEOs). Ultimately Torus won the bid for the telescopes and designed a new type of catadioptric Cassegrain telescope. The optical configurations of the telescopes were a Cassegrain design with several refractive elements (lenses) to achieve focal ratios of F/1.9 for the 0.5-meter telescope and F/3 for the 1.0-meter telescope. The telescopes were designed to achieve a very wide filed of view to cover as much of the sky as possible in the shortest amount of time using new CCD cameras with a mosaic of CCD chips with 2k x 4k pixels. The 0.5-meter telescope was designed for a 2-degree field of view and the 1.0-meter telescope was designed for a 3-degree field of view. Because we needed a name for this new design, we decided to call them wide-field hybrid Cassegrain telescopes. Shortly after starting Dr. Isobe's project for the Bisei Spaceguard Center telescopes, Torus discovered that there was a great deal of interest in the wide-field hybrid Cassegrain telescope design. These new telescopes were ideal for doing survey and monitoring projects automatically. In 1999 Torus won contracts to build several more 0.5-meter wide-field hybrid Cassegrain telescopes for the Taiwan-America Occultation Survey (TAOS) project and the Yonsei Survey Telescopes for Astronomical Research (YSTAR) project. Because we had several orders for these telescopes virtually at the same time, we treated them as a "production run" using as much of the engineering as possible from the Classical Cassegrain Series designs. Because the optical tube assembly part of these telescopes required new research and development, producing several of these telescopes at more or less the same time caused problems that we did not foresee. The optical design of the wide-field hybrid Cassegrain telescopes is very challenging to fabricate, align, and test requiring the development of new assembly and test equipment. The actual implementation of the design proved to be much more difficult than we anticipated. However, we eventually resolved the problems. One of the greatest challenges with this optical design is collimating the optics. Because the optical system is very fast and incorporates refractive elements, we had to devise new collimation procedures for these systems.

2.5. Project Rigel

Dr. Robert Mutel is the principal astronomer responsible for the idea and implementation of the very successful ATF and IRO projects at the University of Iowa. Dr. Mutel is one of the leading experts on using robotic telescope systems for education and research. Hundreds of students have used the ATF and now the IRO for astronomy laboratory work. Based on the success of these projects, Dr. Mutel applied for and received an NSF grant to contract with Torus Technologies to produce an affordable turnkey robotic observatory system based on the work and lessons learned from the ATF and the IRO projects. The purpose of the project is to produce the prototype for a commercial product that Torus can sell and make widely available to institutions and people around the world. Project Rigel is a robotic astronomy laboratory. The following is a description of the Rigel Project published on the University of Iowa's web site:

"The Rigel Project is a robotic observatory designed for use in undergraduate astronomy education and research. It is a complete turnkey system consisting of a 14.5-inch f/14 Cassegrain telescope, large-format CCD camera, filter wheel, spectrometer, dome, and weather station. The observatory control software is Web-based and highly user-friendly. The observatory can be operated in 'real-time' or by queue-based scheduling. The prototype Rigel system is under construction now, and will be tested during the summer and fall of 2001. Torus expects to begin delivery of the first production Rigel systems in the spring of 2002.

A companion curriculum is being developed which will allow instructors to integrate Rigel observations into astronomical laboratories. Although the primary focus will be undergraduate laboratories, parts of the curriculum will be suitable for high school courses, as well as more advanced undergraduate research projects."

As part of the project Rigel development, OCAAS will be enhanced and made more powerful and new web based software tools will make operating and using Torus robotic telescopes even easier.

2.6. A Worldwide Network of Robotic Telescopes and Other Future Developments

Although the Rigel project is in development and will not be an actual product until early 2002, Torus is looking beyond Rigel to the next phase of the evolution of our telescope technology. We plan to expand on the web-based and OCAAS software from the Rigel project to develop a fully integrated worldwide telescope network based on OCAAS and the web browser tools developed for the Rigel project. Starting with the new Torus Observatory in California and our current customers, we will test and build a flexible and robust global telescope network. We will not limit this network solely to Torus Technologies telescopes. Part of our plan to make the global network grow and thrive is to make OCAAS readily available to retrofit existing telescopes and make them compatible to operate among the telescopes in the global network. Ultimately at the highest level of the worldwide network we want to enable any telescope to "plug into" the network independent of what telescope control system it uses. This will most likely be accomplished using a platform independent protocol such as the proposed Robotic Telescope Markup Language (RTML), which is a superset of XML. As new technologies and opportunities arise, Torus Technologies will develop new technologies and products and make them commercially available to the astronomy community.

References

Torus Technologies: www.torusoptics.com

University of Iowa AFT and IRO projects: denali.physics.uiowa.edu

Bisei Spaceguard Center: www.spaceguard.or.jp/bsgc/pamphlet/index.htm

TAOS project: taos.asiaa.sinica.edu.tw

Project Rigel: denali.physics.uiowa.edu/rigel/

(Top from left) You, Rattenbury, Williams, Hajjar, Hojaev; (bottom from left)
Mack, Gelderman (front), Williams, Alcock (back)

Small–Telescope Astronomy on Global Scales
ASP Conference Series, Vol. 246, 2001
W.P. Chen, C. Lemme, B. Paczyński

Progress Report for the KAO 1.0 meter Robotic Telescope

Peter Mack[1], Wonyong Han[2], Matthew Bradstreet[1], Anthony Borstad[1], Jang-Hyun Park[2], Ho Jin[2], Woo-Baik Lee[2] and Chung-Uk Lee[2]

[1]*Astronomical Consultants & Equipment, Inc., P.O. Box 91946, Tucson AZ 85741 USA*

[2]*Korea Astronomy Observatory, Taejon, 305-348, Korea*

Abstract. Korea Astronomy Observatory (KAO) is working to rebuild a 1.0-m robotic telescope in collaboration with a company (Astronomical Consultants & Equipment, Inc. or ACE). The telescope is being totally refurbished to make a fully automatic telescope which can operate in both interactive an fully autonomous robotic modes. This paper describes the design concepts and the work completed. The telescope is an f/7.5 Ritchey-Chretien system mounted on an equatorial fork with friction drives capable of high slewing ($5°/s^2$) and high resolution tracking. The control software manages the entire telescope, instruments and observatory. In interactive local and remote modes the observer can manually enter coordinates or retrieve them from a database. In robotic mode the telescope controller downloads requests from users and creates a schedule. The telescope will be equipped with a CCD camera and will be available over the internet.

1. Introduction

In order for small telescopes to compete with the very largest telescopes they have to be optimized for maximum efficiency, both in terms of optical throughput and the manner in which they are used. A fully autonomous robotic telescope (one that requires no human intervention) can be used to perform repetitive tasks and take advantage of only a few hours of partially clear nights which would not be utilized by conventional observing techniques. The rapid pre-scheduled observing sequence permits many more observations per night than could possibly be hoped for using the telescope manually with a well experienced observer. We present the design concepts for a meter-class telescope and observatory control system that permits interactive, remote and fully autonomous observing. The telescope is equipped with a finder-guider system and a CCD imaging detector with dual filter wheels. A pre-existing telescope was gutted to salvage the optics and servo motors for use in a new system.

2. Control System Logistics

A block diagram of the ACE Robotic Control System^{TM} is presented in Figure 1. The system is normally operated by three computers. The web computer

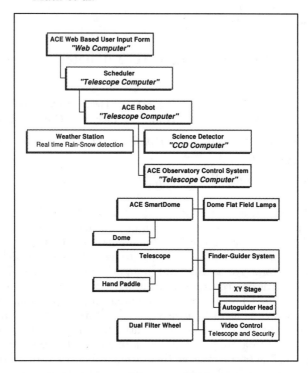

Figure 1.   ACE Robotic Control System^{TM} logistics

is usually physically located at the administrative headquarters. It acts as a server for users to log on and request observations using an internet browser. The telescope computer and instrument computer handle the hardware at the observatory using a fast local area network.

2.1. Web Computer

Users request observations by logging on to a series of web pages. The address of this computer is well published and for security reasons it is not located at the observatory. There are three levels of user (Table 1) designed for "public", "student" and "professor". Public access is restricted to simple requests. Privileged users can request multiple sets of observations (the whole sequence to be repeated at some later time) with mosaics (multiple fields), multiple filters, and multiple exposures per filter. In order to stop any one observer from grabbing the entire resources of the telescope a priority scheme has been devised. The system administrator allocates a certain number of points or credits to each observer. Points are spent according to a simple formula:

$$\text{Points used} = \text{Priority Level} \times \text{observation time.}$$

A given level of observer has a range of Priority Levels. Objects requiring many repeat observations might be given a low priority level and those requiring one observation that is vital would be given the highest possible priority level.

Table 1. User Priorities

User	Priority Level	Repeat Sets of Observations	Multiple Filter	Multiple Exposures per Filter	Mosaic Fields
Public	1	No	Yes	No	No
Student	2–4	Yes	Yes	Yes	Yes
Professor	5–7	Yes	Yes	Yes	Yes
Transient	8	No	Yes	Yes	No
Engineer	9	Yes	Yes	Yes	Yes

There is also a Priority Level for transient events, such as gamma-ray bursts. When a "Level 8" event is received by the system the current observation is halted and the telescope is immediately pointed to the transient event unless an "Engineer" is logged on the system for maintenance work or another Level 8 observation is in progress.

The web pages support solar system observations, such as planets, asteroids and comets. This takes a different route through the pages because the position of the object depends on the time it will be observed. Positions are therefore generated in real time by the telescope computer. For comets and asteroids orbital elements are required.

The web pages also have links to the major astronomical databases. After successful submission of an observing request confirming e-mail is sent to the user. After the observation has been completed the images are achieved on the web computer and further e-mail is sent to the user with retrieval instructions. The address of the telescope and CCD computer are never revealed to users accessing the web pages.

The resolution and format of this paper does not allow us to present sample web pages. Readers are encouraged to inspect the pages by visiting http://www. astronomical.com and following the appropriate links.

Successful completion of the web pages generates an entry in the request database. The requests are also stored in RTML (Robotic Telescope Markup Language) format. The ACE Robotic Control System^{TM} was the first control system to implement and operate a telescope using RTML. The KAO Telescope will be capable of participating in an array of networked telescopes.

2.2. Telescope and Camera Computers

The telescope computer is located in the observatory control room. It receives requests from the web computer. It is also possible for a privileged observer to directly log on to this computer and operate the telescope either from the observatory or remotely using remote access software.

The telescope computer manages the ACE Robot. This generates a schedule, and makes the observations by communicating with the telescope controller and the CCD camera(s) using a series of NT pipes. The ACE software was written using Visual C++ to create native 32-bit applications executing under Microsoft Windows NT. The computer is a dual Pentium motherboard housed in an industrial rack-mount chassis. To maximize the number of interrupts available for instrument control the computer has dual-channel SCSI built into the motherboard. This requires less interrupts than IDE drives, is faster, and easily expandable. The telescope computer is equipped with two motion control cards, each having four axes of control with encoder feedback, a 24 channel digital input-output card, a GPS-based clock card for precision time keeping, and a video capture card for placing real-time video on the computer screen and on the internet. The video capture card uses a PCI slot, so there is just one PCI slot remaining for future expansion. A separate computer manages the scientific detector, usually a charged coupled device (CCD) camera. This is necessary because of computing loads during CCD readouts, and the lack of available slots in the telescope computer.

2.3. ACE Control Crate

The ACE Control Crate is the interface between the computer and the instrumentation. It too is rack-mounted and three cables for the motion control cards and the digital I/O card provide communication between the crate and the computer. The ACE Control Crate contains power supplies and optically isolated relays (input and output) to protect the computer from spurious electrical spikes. All the control cables from the telescope, dome, finder-guider, filter wheel and video control are terminated at this crate.

2.4. ACE Smart Dome^{TM}

The observatory dome is automatically slaved to the telescope. One of the most important aspects of remote and robotic observing is to ensure that the dome can always be closed. An embedded microprocessor, the ACE SmartDome^{TM}, constantly handshakes with the computer. In the event of a computer crash the device automatically sends the dome home and closes the shutter(s). The device has an RS232 communications port to permit full dome status information to be sent to the control computer.

2.5. Video Subsystem

The video subsystem uses a commercial color PAL/NTSC video switcher to permit one of eight cameras to be viewed, or a quad display of the first or last four cameras. The view can be switched in software by sending a simple binary pattern to the switcher using the digital I/O card. The output from the video switcher is also fed into an inexpensive video capture card to permit real-time video to be placed on the internet. The exact arrangement of the cameras is to be determined. Typical uses will be a color cloud camera, a low-light level camera for viewing the telescope at night, a dome camera to view the dome slit, a low light level color camera equipped with a 200 mm lens to view the Sun or Moon during eclipses for public outreach (broadcast live on the internet), and some security cameras. The system is also equipped with relays to toggle low-wattage lights inside the dome for viewing the telescope at night.

2.6. Finder-Guider System

The finder-guider system uses a pick-off mirror to gain off-axis access to three sides of the CCD field. It is held on sliding tables using crossed roller bearings and driven with micro-stepped motors. Internal optics image the focal plane and send parallel light along a compensating arm before being re-imaged by another lens to produce a 2 arc minute field of view at 2.3X magnification of the telescope. An Apogee instruments MegaLisaa CCD camera will be used for autoguiding.

2.7. Filter Wheel

An ACE Dual Filter Wheel is attached to the underside of the finder-guider box. This has two stacked filters wheels, each with ten filter slots to accommodate 3 inch square filters. One slot in each wheel is left open giving a total capacity of 18 filters. It is equipped with an absolute encoder and is driven by stepper motors. One advantage of the dual filter wheel is that neutral density filters can be placed in the upper wheel to permit multicolor photometry of very bright objects using the lower wheel. The filter wheel can take a CCD dewar up to 200mm diameter with a weight of 20 kg.

2.8. GPS Clock

A global positioning satellite (GPS) clock is used to keep the telescope computer clock synchronized to better than 0.05 seconds. It checks once a minute and resets the clock if there has been a drift. The CCD computer is also reset using a Windows-NT time service. The telescope control software calculates sidereal time knowing the universal time from the GPS card.

3. The Telescope

3.1. The Optics Design

The primary mirror was taken from a pre-existing telescope and is a lightweight (70 kg) honeycomb substrate manufactured by HexTek Corporation of Tucson with a f/2.63 focal ratio. The front and back surfaces are approximately 11 mm thick and both are curved such that in cross section they are parallel to each other. The focal ratio of the primary could not be significantly altered. It was decided to open up the size of the central hole and to manufacture a new secondary so as to create a Ritchey-Chretien Cassegrain design with the widest possible field. The effective focal ratio is f/7.5, which gives an image scale of 26.25 arc-seconds/mm. and a useable field diameter of 28 arc minutes

Table 2 gives the optical design parameters. The available field of view of the telescope is limited by two factors: the size of the hole in the primary mirror/baffle system, and, to a lesser extent, the amount of acceptable residual astigmatism. The hole in the primary is only 150 mm. The baffle system was designed to act as both the sky flood baffle and the mirror handling fixture. The wall thickness of the tube has been minimized to allow the largest possible light cone to enter. However, a typical R-C Cassegrain might have a central hole of approximately 225−250 mm diameter. The honeycomb design of the mirror prevents us from opening up the size of the central hole by more than a few millimeters. However, we have gone ahead and done this to help insert the baffles.

Table 2.   Optical design parameters for KAO 1m telescope

System Component	Properties	Value
Primary Mirror	Diameter (physical)	1046
	Diameter (optical)	1046 with 1000 test area
	Radius of curvature	5508.9
	Focal Ratio	2.633
	Conic Constant	−1.1092 (weak hyperbola)
Secondary Mirror	Diameter (physical)	332
	Diameter (optical)	330.4
	Radius of curvature	−2606.7
	Conic Constant	−5.6389
System Characteristics	Effective Focal Length	7845
	Effective focal ratio	f/7.5
	Field of view (diameter)	0.2334 (840 arc seconds)
	Plate Scale	26.25 arcseconds/mm
Spacings	Prime focus intercept	845.7
	Primary-Secondary spacing	1908.7
	Focus ram travel	50
	Back focal distance	500

Figure 2.   The mirror support system is used when the mirror is repolished.

3.2.   Optical Tube Assembly and Mounting

The telescope mount is an equatorial fork, designed for rapid movement during long slew motions across the sky. It uses friction drives rather than the traditional worm drives to minimize backlash. The optical assembly will have an open truss design for optimum seeing performance and minimal mechanical and thermal inertia.

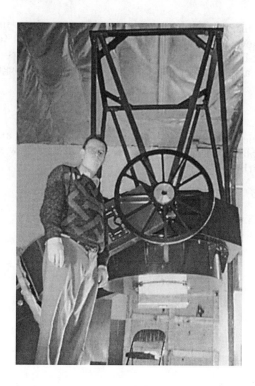

Figure 3.   The telescope during assembly in December, 2000.

The mirror cell supports the mirror on an 18-point elastomeric mounting (Figure 2). It was engineered to provide the optimum support and not to hide any of the mirror chambers which would cause damage when the glass is placed in the aluminizing tank. The support system has been tested by tilting the mirror below the horizon for many days. The support system was also attached to the optical polishing machine so the optical figure was generated and tested as if the mirror were in the telescope. The optical tube assembly is equipped with automatic covers. The top end of the optical assembly is fixed so that only one secondary mirror is used. The secondary is supported using a four-vane spider. The focus ram has a precision ball-screw that drives a crossed-roller slide that is driven by a stepper motor with closed-loop encoder feedback. The control system permits re-initialization of the encoder to the OUT position. Current focus values are retained, even during complete power-down.

The Hour Angle and declination axes are both friction disk drives (Figure 3) employing Compumotor Dynaserve servo motors. These motors have sufficient static torque that they can hold the telescope even with the power off, assuming the telescope is in reasonable balance (within 5Nm). If the telescope is purposefully placed out of balance, such as when changing instrumentation, a set of braces is used to keep the telescope tube pointing at the zenith. The telescope has a mercury tilt switch horizon sensor, which is set to 10 degrees. In additional there are sensors to prevent polar wrap and interference on the southern horizon with the fork arms. A home sensor permits automatic re-initialization of the telescope in the event of a computer malfunction. The disk to roller drive ratio is 53.3:1 for the hour angle axis, resulting in discrete steps on the sky of 0.05 arc seconds and a theoretical maximum slew rate of 13.5 degrees/second.

4. Summary

The KAO robotic telescope has been designed for efficient autonomous observing. It will be placed in active service in 2001 using a new mount and control system and refurbished optics.

Small–Telescope Astronomy on Global Scales
ASP Conference Series, Vol. 246, 2001
W.P. Chen, C. Lemme, B. Paczyński

ACE FlexGrid Telescope Flexure and Pointing Software

Peter Mack[1], John Stein[1,2], Wonyong Han[3]

[1]*Astronomical Consultants & Equipment, Inc., P.O. Box 91946, Tucson AZ 85741, USA*

[2]*Mathematics Department, Geneva College, Beaver Falls, PA 15010, USA*

[3]*Korea Astronomy Observatory, Taejon, 305-348, Korea*

Abstract. We describe ACE FlexGrid, a telescope flexure and pointing model based on an empirical grid of reference points. This software is valid for all types of telescopes and is especially suited to robotic telescopes which repeatedly observe the same object.

1. Introduction

Some pointing models make assumptions about the type of mount and how it will flex at different positions on the sky. We have taken a different approach which assumes nothing about the type of telescope mount. The only assumption is that the telescope flexure is repeatable for a given load and that for significantly different loads a new map will be created.

2. The Flex Grig Database

Using one of the standard catalogs of bright stars with accurate positions and proper motions a set of grid points is created. These points can be randomly placed; they do not need to be in a regular grid pattern. Stars are also observed along the observing horizon to create a set of boundary points (Figure 1).

A database entry is made for each grid or boundary point. The catalog position of the star, the observed Hour Angle and Declination corrections, the temperature and barometric pressure are saved.

3. Flex Grid Interpolation

Once the flex grid has been generated the software makes interpolations between the target star and the reference points (Figure 2). Grid points are selected that "box" the target. Arc, triangular or trapezoidal boxing is created, in increasing order of preference, depending on the geometry. All interpolations are done along great circle arcs. Grid point mapping of the boundaries allows for ease in specifying unusual shapes. The telescope control software can monitor the telescope position and prevent it from crossing exclusion zone boundaries. It

111

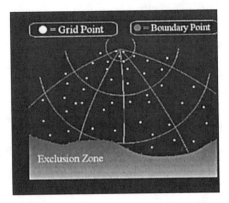

Figure 1.   The Flex Grid

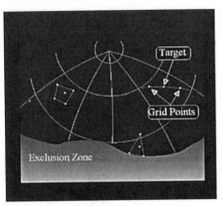

Figure 2.   ACE FlexGrid interpolation

can also report the time before the telescope reaches a boundary, such as the observing horizon or interference with a support pier.

Each time an object is observed it can be added to the database if certain criteria are met (to prevent misidentified stars from being added to the database). Robotic telescopes tend to observe the same object on many nights. By saving the grid points every 10 minutes, an extensive set of grid points can be generated. The system is still to be tested but closure errors of only a few arc seconds are expected. The software is being used with the Korea Astronomy Observatory 1.0-m Robotic Telescope described in these proceedings.

4.   Summary

Pointing model routines suitable for robotic telescopes have been developed. Called the ACE FlexGrid, it is a part of the ACE Observatory Control System. It will also be made available as a library to integrate into other telescope control systems. More information is available at http://www.astronomical.com.

Small–Telescope Astronomy on Global Scales
ASP Conference Series, Vol. 246, 2001
W.P. Chen, C. Lemme, B. Paczyński

Development of the Far-ultraviolet Imaging Spectrograph on KAISTSAT-4

Wonyong Han[1], Kyoung Wook Min[2], Jerry Edelstein[3], Uk-Won Nam[1], Jong-Ho Seon[2], Kwang-Il Seon[1], Jang-Hyun Park[1], In-Soo Yuk[1], Ho Jin[1], Woo-Baik Lee[1], Kyungin Kang[2], Kwangsun Ryu[2], Jin-Geun Rhee[2], Dae-Hee Lee[2], Eric Korpela[3], and W. Van Dyke Dixon[3]

[1]*Korea Astronomy Observatory, 61-1 Whaam, Yusong, Taejon, 305-348, Korea*

[2]*Korea Advanced Institute of Science and Technology, 373-1 Kusong, Yusong, Taejon, 305-701, Korea*

[3]*Space Sciences Laboratory, University of California, Berkeley, California 94720-7450, USA*

Abstract. The Far-ultraviolet IMaging Spectrograph (FIMS) is a small spectrograph optimized for the observations of diffuse hot interstellar medium in far-ultraviolet wavebands (900–1150Å and 1335–1750Å). The instrument is expected to be sensitive to emission line fluxes an order of magnitude fainter than any previous missions. FIMS is currently under development and is scheduled for launch in 2002.

1. Introduction

The preliminary scientific objectives of FIMS are: 1) to trace the energy flow through the hot plasma found on scales ranging from supernova bubbles to galaxies and galactic coronae, 2) to map the distribution of the local and global structures of hot plasma, and 3) to investigate the earth aurora at various FUV wavelengths. FIMS will be the primary payload on the first Korean Science Satellite (KAISTSAT-4), which will provide both an all-sky survey and a targeted pointing program. FIMS system is being developed by a joint research project of KAO, KAIST and U. C. Berkeley. The development of the qualification model of the instrument is in progress.

2. Instrument Development

FIMS is a dual band imaging spectrograph, optimized for faint diffuse radiation. FIMS employs an off-axis parabolic cylinder mirror in front of a slit that guides lights to a diffraction grating (See Figure 1). The reflective grating is an ellipse of rotation, which provides angular resolution. The FIMS design is derived from the two flight-proven EURD instruments (Bowyer, Edelstein, & Lampton 1998). The imaging performance allows for a large field with an imaging resolution (arc minute scales) similar to other important interstellar all-sky surveys.

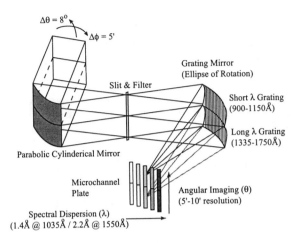

$\Delta\theta = 8^\circ$

$\Delta\phi = 5'$

Grating Mirror
(Ellipse of Rotation)

Slit & Filter

Short λ Grating
(900-1150Å)

Long λ Grating
(1335-1750Å)

Parabolic Cylinderical Mirror

Microchannel
Plate

Angular Imaging (θ)
(5'-10' resolution)

Spectral Dispersion (λ)
(1.4Å @ 1035Å / 2.2Å @ 1550Å)

Figure 1. Schematic Diagram of FIMS Optics

The instrument will yield far-UV emission-line sensitivity that is an order of magnitude fainter than any previous instruments. Monte-Carlo simulation has been performed to verify the detection sensitivity of O_{VI} and C_{IV} emission lines, which are brightest emission lines predicted to occur in galactic plasma cooling models. A simulation of the FIMS observation planning has been performed to ensure spacecraft health and safety and to maximize observation efficiency and instrument performance.

A parabolic cylindrical mirror has been manufactured and its surface profile and micro-roughness is being measured for the final polishing. The design of the ellipsoidal gratings for long and short wavelength bands is finished and the gratings are being manufactured. A drawing of the overall layout of the FIMS system has been completed. The detailed designs of the detector and the shutter modules including the mounting scheme of the detector have been finished.

A Z-stack MCP detector has been adopted because of its high gain. A double delay line (DDL) readout system has been developed and tested for its resolution characteristics. A new cross delay line (XDL) readout system has been designed to overcome resolution limits of the DDL system.

The FIMS electronics system consists of four units, a low voltage power supply, the detector electronics, a digital signal processing unit, and signal monitoring and controlling electronics. Electrical and data interfaces between the spacecraft and FIMS have been designed to meet the specifications . The ETB (Engineering Test Bed) of the FIMS electronics system has been developed and its performance verified. The development of the opto-mechanical and electronics systems of FIMS are in progress for the CDR (Critical Design Review). The mission is scheduled to be launched in 2002 into a low earth polar orbit.

References

Bowyer, S., Edelstein, J., & Lampton, M. 1998, ApJ, 485, 523

Small–Telescope Astronomy on Global Scales
ASP Conference Series, Vol. 246, 2001
W.P. Chen, C. Lemme, B. Paczyński

Design of a Multi-CCD Controller

Yong-Woo Kang, Yong-Ik Byun, Sung-Yeol Yu, Sun-Youp Park &
Yeo-Jin Park

Department of Astronomy, Yonsei University, Seoul, 120-749, Korea

Abstract. We are developing a robust Multi-CCD control system for wide-field surveys of various kinds. Our design concept and possible application modes are introduced.

1. Introduction

The YSTAR(Yonsei Survey Telescopes for Astronomical Research) project aims to monitor the entire night sky for various transient events. As part of the effort, we are developing a new CCD control system to increase the present YSTAR survey capability in the future. The Multiple CCD Imaging System described here is designed to use MPUs(micro processor units) and FPGAs(Field Programmerable gate Arrays). Our basic hardware components are modularized so that each unit can control a CCD and the associated image processing; it is easy to increase the number of modules as needed without modifying the main control unit. Our control system is designed to be very flexible and capable of serving observations of different nature. CCDs of wildly different characteristics can be easily employed as the only needed change involves their clock pattern files.

2. Design Goals

Our design goals are: (1) High speed control of very large image data using DSP (TMS320C31) and MP (i80960), (2) Robust CCD control regardless of the working wavelength and/or the characteristics of the CCDs, (3) Module structure that allows for easy expansion as needed, (4) Real time basic data process by hardware architecture, (5) Linux operation.

3. Discussions

Our system can be characterized by multipurpose, high-speed, parallel, and real-time control capabilities. It is also easy to expand into larger imaging system as needed, either by mosaicing a number of CCDs into a single camera or by linking several CCD cameras in parallel. Our module concept applies not only to the control of CCD cameras, but also to the basic data processing. This will enable very efficient data management of modern astronomical observations which involve a very large volume of digital data.

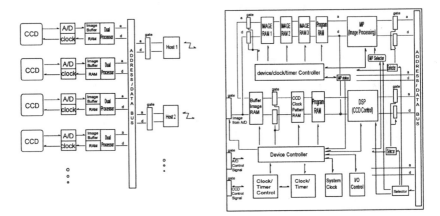

Figure 1.   Schematic diagram of our modulized CCD control system.
Dual processors control the CCD and also perform basic image processing. The RAM contains clock pattern data, which are used to drive
the CCD. The output signal gets converted to digital data and laced
in the Image Buffer. Each module is interfaced with multiple hosts via
the Address/Data Bus.

Figure 2.   (*Left*) CCD operation simulator. It generates imaginary
A/D signals and forwards them to an image buffer. It also takes the
clock pattern from computer and forwards it to the CCD clock driver.
(*Right*) The Multi-CCD controller during development. The top unit
is responsible for DSP and CCD control. We are developing algorithms
for device control signals between different parts based on FPGA. The
bottom unit is for MP image processing development.

Small–Telescope Astronomy on Global Scales
ASP Conference Series, Vol. 246, 2001
W.P. Chen, C. Lemme, B. Paczyński

The Multiple Telescope Telescope, an Inexpensive Fiber Fed Spectroscopic Facility

Reed L. Riddle

Department of Physics & Astronomy, Iowa State University, Ames, IA 50011, riddle@iastate.edu

William G. Bagnuolo, Jr.

Center for High Angular Resolution Astronomy and Department of Physics & Astronomy, Georgia State University, Atlanta, GA 30303-3083, bagnuolo@chara.gsu.edu

Abstract. A unique telescope dedicated to spectroscopy is discussed. This instrument, the Multiple Telescope Telescope, uses many novel approaches to achieve high quality, medium to high resolution spectroscopy at low cost.

The Georgia State University Multiple Telescope Telescope (MTT) is located at Hard Labor Creek Observatory (80 km east of Atlanta, GA), and is a spectroscopic facility designed to gather medium to high resolution spectra of stellar objects, with high efficiency and accuracy, at a fraction of the cost of a traditional instrument. The MTT was originally designed in 1989 as an inexpensive alternative approach to building a telescope for spectroscopic studies (Bagnuolo *et al.*1990; Barry 1995).

The initial motivation for the MTT was to construct an instrument that could support stellar spectroscopic studies, particularly of massive interacting binary systems. This instrument would allow dedicated monitoring of objects without the need to apply for telescope time at other facilities, and allow for studies of interesting but ephemeral objects such as novae and supernovae. The determination to do spectroscopy alone helped to motivate the current design using lightweight structures, fiber optics and a segmented primary mirror. The main reason to use a segmented mirror system is to avoid building a telescope that is structurally monstrous; a solid mirror weighs much more than a segmented mirror of the same size and therefore requires substantially greater structural strength, larger mechanical systems, larger structures to house the telescope and more infrastructure for system construction. A segmented mirror system requires substantially smaller structures for support, leading to a smaller telescope structure and a resultant decrease in costs that can be applied elsewhere.

As can be seen in Figure 1, the MTT primary consists of nine separate mirrors instead of the classical telescope paradigm of a single large mirror. Each mirror injects light into a fiber optic assembly mounted at prime focus on the upper telescope structure. The MTT is not alone in using separate mirrors for its light gathering surface, but it is unique in that each mirror is treated as a separate aperture, gathering an independent spectrum that is then combined

117

Figure 1.   A photograph of the Multiple Telescope Telescope illustrating the important design elements. Note the multiple mirror configuration and the fiber bundles mounted opposite each mirror on the
structure at prime focus.

in post-processing with the spectra from the other apertures to create the final
spectrum. Therefore, each coupled mirror/fiber pair acts as a separate telescope system, from which the "Multiple Telescope" moniker is derived. Each
of the nine mirrors (33.3 cm diameter, f/4.5 parabolic) is controlled by three
stepper motors, which allow kinematic motions in tip, tilt and focus through a
computer interface, and automated software allows for mirror alignment during
observations.

The Newtonian-Ebert Spectrograph was constructed along with the MTT,
and designed to function at a variety of resolutions, concentrating in the medium
resolution regime (i.e. $R = \lambda/\delta\lambda = 10,000$–$30,000$). The basic design is the
Fastie-Ebert (Fastie 1952), which shares the camera and collimator mirror, a
design that is not explored in modern astronomical instrumentation. There are
many advantages in this system (simpler alignment, fewer optics), which are
further explored elsewhere (Barry 1995; Bagnuolo, Riddle, & Barry 2001).

The entire MTT telescope system, including the spectrograph, enclosure,
optics, computers, and everything else, cost about $100K (with the spectro-
graph CCD accounting for 40% of this cost); a comparable 1 m class, single
mirror telescope would cost five to ten times as much today. Two other groups

have constructed telescopes based on the MTT design: the NDAMTT in Japan (Kambe 1999) and the SUMMIT in Australia (Moore 2001). The MTT has participated in numerous scientific projects (Riddle 2000), and has evolved from an experimental project into a robust scientific instrument.

References

Bagnuolo, W. G., Jr., Riddle, R. L., & Barry, D. J. 2001, PASP (in preparation)

Bagnuolo, W. G., Jr., Furenlid, I. K., Gies, D. R., Barry, D. J., Russell, W. H., & Dorsey, J. F. 1990, PASP, 102, 604

Barry, D. J. 1995, Ph.D. Dissertation, Georgia State University

Fastie, W. G. 1952, JOSA, 42, 641

Kambe, E. 1999, private communication

Moore, A. M. 2001, Ph.D. Dissertation, Sydney University

Riddle, R. L. 2000, Ph.D. Dissertation, Georgia State University

There were quiet moments (top from left: Lemme, M. Y. Chou, and Y. S. Li) and wild ones (bottom from left: T. L. Lee and Z. Y. Lin).

Small–Telescope Astronomy on Global Scales
ASP Conference Series, Vol. 246, 2001
W.P. Chen, C. Lemme, B. Paczyński

The Lick Observatory Supernova Search with the Katzman Automatic Imaging Telescope

Alexei V. Filippenko, W. D. Li, R. R. Treffers, and Maryam Modjaz

Department of Astronomy, University of California, Berkeley, CA 94720-3411 USA

e-mail: (alex, wli)@astro.berkeley.edu

Abstract. The Katzman Automatic Imaging Telescope (KAIT) at Lick Observatory is a fully robotic 0.76-m reflector equipped with a CCD imaging camera. Its telescope control system checks the weather, opens the dome, points to the desired objects, finds and acquires guide stars, exposes, stores the data, and manipulates the data without human intervention. There is a 20-slot filter wheel, including $UBVRI$. Five-minute guided exposures yield detections of stars at $R \approx 20$ mag when the seeing is good ($\leq 2''$).

One of our main goals is to discover nearby supernovae (SNe; redshifts generally less than 5000 km s^{-1}), to be used for a variety of studies. Special emphasis is placed on finding them well before maximum brightness. A limit of ~ 19 mag is reached in the 25-sec unfiltered, unguided exposures of our Lick Observatory Supernova Search (LOSS). We can observe over 1200 galaxies in a long night, and we try to cycle back to the same galaxies after 3 to 4 nights. Our software automatically subtracts template images from new observations and identifies supernova candidates that are subsequently examined by student research assistants. LOSS found 20 SNe in 1998, 40 in 1999, and 36 in 2000, making KAIT the world's most successful search engine for nearby SNe. We also find novae in the Local Group, comets, asteroids, and cataclysmic variables. Multifilter follow-up photometry is conducted of the most important SNe, and all objects are monitored in unfiltered mode. A Web page describing LOSS is at http://astro.berkeley.edu/~bait/kait.html .

1. Introduction

KAIT is the third robotic telescope in the Berkeley Automatic Imaging Telescope (BAIT) program. The predecessors to KAIT were two telescopes developed at the Leuschner Observatory, which is located about 10 miles east of the University of California, Berkeley campus (U. C. Berkeley). The first telescope, a 0.50-m Cassegrain system built in 1954, was retrofitted with computers and started gathering data automatically in January 1992. In November 1992, it was joined by a 0.76-m Ritchey-Chrétien telescope. The two BAIT systems were fully automated, and were capable of carrying out various kinds of imaging studies such as searching for and monitoring supernovae (SNe); monitoring other time-variable objects like active galactic nuclei, quasars, and novae; studying solar-system objects such as planets, comets, asteroids, and the Moon; and allowing student access for astronomy laboratory classes. The SN search at that time was called the Leuschner Observatory Supernova Search. More thorough discussions of BAIT can be found in Treffers, Richmond, & Filippenko (1992), Richmond, Treffers, & Filippenko (1993), and Treffers et al. (1995).

Over the years the rising cost to maintain and upgrade the Leuschner facility, which has at best marginal weather for observing (often foggy and with mediocre seeing), prompted us to look for a better site for the BAIT program. We found a home for a new telescope at Lick Observatory on Mt. Hamilton, California, in an old but renovated dome which previously housed a 0.6-m telescope. Lick Observatory, located at an elevation of 4200 feet, has better seeing and clearer skies than Leuschner, has a crew of mountain engineers and telescope operators to respond to unforeseen problems, and is only a two-hour drive from the U. C. Berkeley campus. KAIT began taking data around August 1996, and it was officially dedicated on October 25, 1996.

The Lick Observatory Supernova Search (LOSS), the primary project carried out with KAIT, discovered its first SN in 1997 (SN 1997bs — Treffers et al. 1997; Van Dyk et al. 2000). Many improvements were made to the telescope hardware and the system software in 1998, and LOSS discovered 20 SNe that year. 40 SNe were discovered in 1999, and 36 in 2000, making LOSS the world's most successful search for nearby SNe.

In this paper, we report the KAIT system hardware and software in Section 2, the details of the SN search in Section 3, and our conclusions in Section 4.

2. The System Hardware and Software

KAIT possesses a 0.76-m (30-inch) diameter primary with a Ritchey-Chrétien mirror set. The telescope mount, designed by Autoscope Corporation, has a focal ratio of $f/8.2$, which yields a plate scale of $33\rlap{.}''2$ mm^{-1} at the focal plane. At the back end of the telescope, the primary beam goes through an open aperture in a diagonal mirror, passes through the filter wheel assembly, and gets imaged onto a CCD. For guiding, the peripheral light from the main beam that strikes the diagonal mirror gets reflected into a guider CCD camera, which is mounted on an X-Y translation stage that can move to "pick-off" light from any part of the diagonal mirror.

Before July 1998, we had been using a Photometrics CCD camera with a Thompson TH 7895 chip as the main detector for KAIT. This is a front-illuminated chip and has rather poor blue response. A new detector, an Apogee

AP7 CCD camera with a SITe 512 × 512 pixel back-illuminated chip (pixel size 24 μm), was installed in KAIT in July 1998. The new CCD camera has much better quantum efficiency than the old one, and its field of view is 6'.8 × 6'.8 (0''.8 pixel^{-1}). The camera is cooled thermoelectrically by forced air to about 60° C below the ambient temperature.

KAIT has a filter wheel with 20 slots, including a set of standard Johnson UBV and Cousins RI filters. Five-minute guided R-band exposures yield detections of stars at $\sim$ 20 mag when the seeing is good ($\leq$ 2'').

A weather station is crucial for robotic observations. The KAIT weather station has an array of sensors monitoring the outside temperature, the relative humidity, wind speed, rain, and clouds. The cloud sensor is an infrared detector with a 12–13 μm filter that points directly overhead. If a single drop of water hits the rain sensor, it closes the dome circuit that directly initiates the closing of the dome slit, bypassing all the control software. As an added safety feature, during normal operation a "keep open" command is issued every two minutes by the control software to the dome slit, without which it automatically closes itself.

There are one workstation and three personal computers (PCs) that control the operation of the system. One PC is responsible for the telescope, the dome, the slit, and the weather station; another PC controls the Apogee imaging camera; the third PC controls the autoguider camera and the filter wheel. These PCs serve as "slaves" that "listen" to the commands given by the "master" computer, the workstation. Upon receipt of a command from the master computer, the PCs go off and execute the orders; meanwhile they cannot receive any more commands. The master computer coordinating the entire operation has to perform several tasks: (1) scheduling observations, (2) forwarding commands to appropriate PCs to perform observations, (3) logging data and error reports, (4) accepting remote logins to change schedules or inspect telescope operation, and (5) allowing manual operations of the system if necessary.

A typical workday for KAIT proceeds as follows. The system starts up at 3:00 pm. The status of all the hardware is checked and initialized, and the observations for the night are scheduled according to all the active request files. When the Sun is 10° above the horizon (usually about one hour before sunset), the dome slit is opened and the CCD camera is cooled. When the Sun is 5° above the horizon, trial flatfield images are taken for the filters that will be used during the night, but usually it is too early and the CCD is saturated in very short exposures. The system then sleeps for a short time and tries to image again. Useful flatfields are taken when the images have reasonable counts. The exposure time is gradually increased to compensate for the darkening of the sky. The final flatfield images are combined for each filter to improve the signal-to-noise ratio.

After flatfield observations are completed, a focus-detecting procedure is executed three times to find the best focus for the telescope. The procedure consists of taking a series of 36 exposures of a bright star in a single image, slightly changing the focus and offsetting the telescope between consecutive exposures. A program then analyzes the image and determines which focus value gave the smallest full-width at half maximum (FWHM) for the star. The entire procedure takes about 5 minutes to run. It is repeated after the first 20 minutes of observations, again after the next 40 minutes, and subsequently every 80 minutes; this is done because the focus of KAIT is not stable.

The arranged targets are observed when the Sun is 8° below the horizon. Each target can be observed and processed in a different manner. For example, photometric observations require the images to be bias subtracted, dark subtracted, and flatfielded; SN search images, on the other hand, are automatically transferred to a computer at the U. C. Berkeley campus to be processed. Observations may use different filters, different exposure times, and may or may not require autoguiding.

During bad weather such as high winds, high humidity, rain, or totally overcast skies, the slit is closed automatically, and the system takes a "nap" for 10 minutes. It tries to do observations again after the nap, and keeps trying every 10 minutes thereafter.

At the end of the night the system is shut down automatically: the dome slit is closed, the telescope is placed at a specific position, and the tracking is turned off. E-mail messages are sent out to users who have requested observations. Many bias and dark-current images are obtained, averaged, and saved for the next night. The system then goes to "sleep" until it wakes up again at 3:00 pm.

3. The Lick Observatory Supernova Search (LOSS)

The primary science project carried out with KAIT is LOSS, whose sample consists of about 5,000 galaxies. The majority of the LOSS galaxies come from the Third Reference Catalogue of Bright Galaxies (de Vaucouleurs 1991), and the others are from the Uppsala General Catalog of Galaxies (Nilson 1973). We could have selected a larger number of galaxies to increase the number of SN discoveries, but we chose to use the restricted sample (at least for our initial search) in order to find most of the SNe well before maximum brightness.

The galaxies are observed with 25-second unfiltered and unguided exposures, and the typical limiting magnitude of the images is ~ 19. Much effort has been made to improve the efficiency of the system, which rose to about 100 images per hour by early 2001. KAIT can observe over 1200 images in a clear winter night. Because of this high efficiency, we can cycle back to the same galaxies after 3 to 4 nights.

The program to automatically process images and detect SNe is an important part of LOSS because with many images observed each night, it is very difficult to visually compare the images to find new objects — it is time-consuming, and one can easily miss SNe in complicated regions. The LOSS image processing software consists of many programs written in C, Fortran, and

IRAF scripts, which combine to perform all necessary tasks for image process-ing: making (and replacing) templates, image subtraction, candidate detection, and checking of results.

The template-making program studies each image and stores in several files properties such as the positions of the stars, the set of stars used for point-spread-function (PSF) matching and intensity transformation, and the locations of the saturated stars.

The image-subtraction routine is one of the most important programs for LOSS. It first detects the stars in the new image, then uses their positions to compute the shift and rotation between the new and the template images, and subsequently aligns them. The PSF-matching stars are used to make the two images have the same PSF. An intensity-transformation region, which is usually an area that includes one of the PSF-matching stars, is studied in both images to derive the intensity ratio between them. The intensity of the new image is then transformed to that of the template. After these transformations, the new and the template images have the same intensity level, the same PSF, and are properly aligned. The subtracted image is then obtained by a simple arithmetic subtraction of the template from the new image.

The candidate detection program finds all new objects in the subtracted images, and classifies them into one of the following categories: (1) residual from imperfect subtraction, if the candidate is within a certain radius of a bright star; (2) cosmic ray, if the candidate has a very small FWHM; and (3) real SN candidate.

Each day, a research assistant (from a group of several undergraduate stu-dents, and sometimes graduate students) uses the result-checking program to examine the results of the image processing. The new image, the template, and the subtracted image are all displayed on the screen, with the SN candidates marked with red circles and the cosmic rays with green circles. Promising SN candidates are then reobserved, and the confirmed SNe are reported to the Cen-tral Bureau of Astronomical Telegrams, where the International Astronomical Union Circulars are issued. The undergraduate assistants also sometimes iden-tify SNe that were missed by the detection program (e.g., those near poorly subtracted stars, or those that are fainter than the automatic detection limit).

LOSS found one SN in 1997 (SN 1997bs), 20 SNe in 1998, 40 SNe in 1999, and 36 in 2000. Tables 1 and 2 list the SNe discovered in 1998 and 1999, respectively. Because of the small interval between LOSS observations of a given galaxy, most of the LOSS SNe were discovered well before their maximum brightness. For this reason, LOSS provides excellent SNe for individual detailed studies. The LOSS SN sample is also ideal for statistical studies of SNe, such as the SN rates in galaxies of different Hubble types.

LOSS also discovered 4 novae in nearby galaxies (e.g., M31) out of the 7 discovered worldwide in 1998, and 7 out of 11 in 1999. Two cataclysmic variable stars were discovered in 1998 and one was discovered in 1999.

Table 1.   LOSS SN discoveries in 1998

SN name	Host galaxy	Date of discovery	mag at discovery	SN type
1998W	NGC 3075	Mar 16	17.3	II
1998Y	NGC 2415	Mar 16	18.3	II
1998bm	IC 2458	Apr 21	17.6	II
1998bn	NGC 4462	Apr 17	17.4	Ia
1998cc	NGC 5172	May 15	18.1	Ib
1998cu	IC 1525	Jun 29	18.4	II
1998de	NGC 252	Jul 23	18.4	Ia-pec
1998dh	NGC 7541	Jul 20	16.8	Ia
1998dj	NGC 788	Aug 08	16.1	Ia
1998dk	UGC 139	Aug 19	17.6	Ia
1998dl	NGC 1084	Aug 20	16.0	II
1998dm	MCG -01-4-44	Aug 22	16.8	Ia
1998dt	NGC 945	Sep 01	17.7	Ib
1998dx	UGC 11149	Sep 10	18.3	Ia
1998eb	NGC 1961	Sep 17	17.8	Ia
1998ef	UGC 646	Oct 18	15.2	Ia
1998en	UGC 3645	Oct 30	18.4	II
1998es	NGC 632	Nov 13	14.6	Ia-pec
1998fa	UGC 3513	Dec 25	18.2	IIb
1998fe	NGC 6027D	Jul 19	18.0	...

Note: See Filippenko (1997) for a discussion of SN types.

Table 2.   LOSS SN discoveries in 1999

SN name	Host galaxy	Date of dis.	mag at discovery	SN type
1999A	NGC 5874	Jan 10	18.3	II
1999ac	NGC 6063	Feb 26	15.2	Ia-pec
1999bg	IC 758	Mar 28	15.5	II
1999bh	NGC 3435	Mar 29	16.8	Ia-pec
1999br	NGC 4900	Apr 12	17.5	II
1999bu	NGC 3786	Apr 16	17.5	Ic
1999bw	NGC 3198	Apr 20	17.8	IIn
1999bx	Anon.	Apr 26	16.5	II
1999by	NGC 2841	Apr 30	15.0	Ia-pec
1999bz	UGC 8959	May 01	17.6	Ic
1999cd	NGC 3646	May 14	17.9	II
1999ce	Anon.	May 16	18.3	Ia
1999cl	NGC 4501	May 29	16.4	Ia
1999co	Anon.	Jun 18	17.4	II
1999cp	NGC 5468	Jun 18	18.2	Ia
1999cq	UGC 11268	Jun 25	15.8	Ib/c
1999cw	MCG-01-02-00	Jun 28	14.2	Ia-pec
1999da	NGC 6411	Jul 05	17.0	Ia-pec
1999dg	UGC 9758	Jul 23	16.7	Ia-pec
1999dh	IC 211	Jul 23	15.4	II
1999dk	UGC 1087	Aug 12	16.7	Ia
1999do	MRK 929	Aug 20	17.4	Ia
1999dp	UGC 3046	Sep 02	18.2	II
1999dq	NGC 976	Sep 02	16.3	Ia-pec
1999eb	NGC 664	Oct 02	16.2	IIn
1999ec	NGC 2207	Oct 03	17.9	Ib
1999ed	UGC 3555	Oct 05	17.8	II
1999ej	NGC 495	Oct 18	18.1	Ia
1999ek	UGC 3329	Oct 20	18.1	Ia
1999em	NGC 1637	Oct 29	13.5	II
1999ew	NGC 3677	Nov 13	16.5	II
1999gb	NGC 2532	Nov 22	16.2	IIn
1999gd	NGC 2623	Nov 24	16.6	Ia
1999ge	NGC 309	Nov 27	15.6	II
1999gf	UGC 5515	Nov 24	18.2	Ia
1999gm	PGC 24106	Dec 15	17.3	Ia
1999go	NGC 1376	Dec 23	15.5	II
1999gp	UGC 1993	Dec 23	17.3	Ia-pec
1999gq	NGC 4523	Dec 23	14.5	II
1999gs	NGC 4725	Dec 28	19.3	...

Note: See Filippenko (1997) for a discussion of SN types.

The KAIT CCD camera has quite a small field of view, yet two comets were caught in the course of LOSS. Comet C/1998Y2 (LI) was discovered on December 25, 1998 (Li 1998), and Comet 1999E1 (LI) was discovered on March 13, 1999 (Li & Modjaz 1999).

Follow-up observations for the discovered SNe are emphasized during the course of LOSS. Our goal is to build up a multicolor photometric database for nearby SNe. Because of the early discoveries of most LOSS SNe, our light curves usually have good coverage from pre-maximum brightening to post-maximum decline. Moreover, all SNe discovered by LOSS are automatically monitored in unfiltered mode as a byproduct of our search. Examples of our light curves of SNe are presented in Figure 1. Journal papers that summarize part of the LOSS data include the study of SN 1997br (Ia; Li et al. 1999), SN 1997bs (IIn; Van Dyk et al. 2000), SN 1998bu (Ia; Jha et al. 1999), SN 1998de (Ia; Modjaz et al. 2001), SN 1999cq (Ic; Matheson et al. 2000), the rise time of SNe Ia (Riess et al. 1999a, 1999b), the peculiarity rate of SNe Ia (Li et al. 2001), and observations of SNe Ib/Ic (Matheson et al. 2001). Several more papers are now in preparation.

4. Conclusions

In this paper we reported the hardware and software control system of KAIT and the details of LOSS. KAIT is a fully robotic telescope that operates by itself without human intervention. LOSS discovered a total of 96 SNe in 1998–2000, making KAIT the world's most prolific search engine for nearby supernovae.

The successful operation of KAIT demonstrates the obvious benefits of robotic telescopes: they are far more efficient than humans at performing repetitive tasks, and do not make mistakes because of physical fatigue. They eliminate the need for astronomers to travel great distances for observing, therefore saving them both time and money, especially on marginal observing nights/days. Robotic telescopes need little human support, which frees telescope operators to perform other duties. Complete automation permits a telescope to be operated at a site that is ideal for science but not for human habitat. Finally, the astronomical community can greatly benefit by coordinating automated telescopes at different places in order to observe targets of opportunity if one site is down. It also prevents redundant observing of targets, thereby increasing the overall operation and the science efficiency.

As telescopes with increasingly larger collecting areas dominate the astronomical scene, many small telescopes below 2-m aperture are slowly being phased out due to the cost of operation. But as witnessed with KAIT and other similar systems (numerous examples of which are described in these *Proceedings*), small automated telescopes unveil new realms of observations that were once thought to be unfeasible because of the demand on the time and stamina of an observer. Moreover, robotic telescopes provide invaluable learning experiences for astronomy students, and offer exciting exercises for laboratory curricula. Automation of small telescopes will allow them to continue occupying a unique role in astronomy for many years.

Acknowledgments. We are grateful to all of the undergraduate and graduate assistants who have helped us with KAIT and LOSS, especially Michael Richmond in the early stages. Our ongoing work on KAIT and LOSS is supported by National Science Foundation grant AST-9987438, as well as by the Sylvia and Jim Katzman Foundation. We acknowledge Sun Microsystems, Inc., the Hewlett-Packard Company, AutoScope Corporation, Lick Observatory, the National Science Foundation, the University of California, and the Sylvia and Jim Katzman Foundation for donations that made KAIT possible. A.V.F. thanks the Guggenheim Foundation for a Fellowship. His travel to this symposium was financed by its organizing committee and by the U. C. Berkeley Committee on Research.

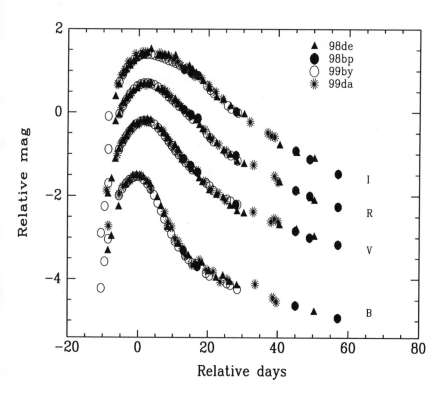

Figure 1.   Examples of SN Ia light curves obtained during the course of LOSS.

References

de Vaucouleurs, G. 1991, in *Third Reference Catalogue of Bright Galaxies* (New York: Springer-Verlag)

Filippenko, A. V. 1997, ARA&A, 35, 309

Jha, S., et al. 1999, ApJS, 125, 73

Li, W. D. 1998, IAU Circ. No. 7075

Li, W. D., & Modjaz, M. 1999, IAU Circ. No. 7126

Li, W. D., et al. 1999, AJ, 117, 2709

Li, W. D., et al. 2001, ApJ, 546, 734

Matheson, T., et al. 2000, AJ, 119, 2303

Matheson, T., et al. 2001, AJ, 121, 1648

Modjaz, M., et al. 2001, PASP, 113, 308

Nilson, P. 1973, Uppsala General Catalogue of Galaxies, Upps. Astro. Obs. Ann., Vol. 6

Richmond, M. W., Treffers, R. R., & Filippenko, A. V. 1993, PASP, 105, 1164

Riess, A. G., et al. 1999a, AJ, 118, 2668

Riess, A. G., et al. 1999b, AJ, 118, 2675

Treffers, R. R., Richmond, M. W., & Filippenko, A. V. 1992, in *Robotic Telescopes in the 1990s*, ed. A. V. Filippenko (San Francisco: ASP, Conf. Ser. Vol. 34), 115

Treffers, R. R., et al. 1995, in *Robotic Telescopes*, ed. G. W. Henry & J. A. Eaton (San Francisco: ASP, Conf. Ser. Vol. 79), 86

Treffers, R. R., et al. 1997, IAU Circ. No. 6627

Van Dyk, S. D., et al. 2000, PASP, 112, 1532

Small-Telescope Astronomy on Global Scales
ASP Conference Series, Vol. 246, 2001
W.P. Chen, C. Lemme, B. Paczyński

The Beijing Astronomical Observatory Supernova Survey

Yulei Qiu & Jingyao Hu

Beijing Astronomical Observatory, Chinese Academy of Sciences A20, Datun Rd., Chaoyang District, Beijing 100012, China
qiuyl@nova.bao.ac.cn, hjy@class1.bao.ac.cn

Weidong Li

Department of Astronomy, University of California, Berkeley, CA 94720-3411 weidong@urania.berkeley.edu

Abstract. We present here the Beijing Astronomical Observatory Supernova Survey (BAOSS). The motivation and the strategy of the survey are briefly discussed, and the hardware and the software of the system are described. BAOSS started in 1996 and has discovered 40 supernovae (SNe) in four years, thus playing an important role in the search for nearby supernovae. The results of several well-observed SNe discovered by BAOSS, SNe 1996W, 1996cb, 1997br, and 1998S, are reported.

1. Introduction

After the explosions of SN 1987A in the Large Magellanic Cloud and SN 1993J in M81, there has been an increasing interest in the observation and study of SNe. The wide-spread use of CCD detectors makes finding SNe relatively easy and the total number of SNe discovered has doubled in the last decade. A great number of photometric and spectroscopic observations have been published. As a result, the physics of supernovae is better understood than ever before. It is now widely accepted that SNe can be divided into several types, i.e., Type Ia, II, Ib, and Ic. The Type Ia SNe (SNe Ia), due to their relative homogeneous brightness, are used as distance indicators for cosmology, and exciting results have been obtained on the expansion history of the Universe from studies of high red-shift SNe Ia (e.g., Schmidt et al 1998; Perlmutter et al 1999). Some SNe may be correlated with Gamma Ray Bursts (Wheeler et al 2000).

The idea of conducting a supernova survey with the 60cm telescope at the Xinglong Station of Beijing Astronomical Observatory (BAO) was first proposed by one of us (Hu) in 1994, after the successful observations and analysis of SN 1993J (Wang & Hu 1994) at BAO. The motivation was to systematically discover supernovae in nearby galaxies before their optical maxima. The Xinglong Station, at a latitude of 40.5 degrees, is about 900m in altitude and 150 kilometer northeast of Beijing with no city-light pollution. The best observing seasons are winter and spring, which is ideal for SN searches because of the long nights and more nearby galaxies observable than other seasons. The 2.16m telescope of BAO is conveniently located on the same mountain and can be used to get spectroscopy for SNe.

Modifications to the 60cm telescope began in 1995 and continued until mid 1996. The optics was modified so that a CCD camera could be installed at the primary focus of the telescope. Software was compiled to automatically control the telescope and the CCD camera and to process the observations. Meanwhile a sample of about 5,000 nearby galaxies was constructed from the Third Reference Catalogue of Bright Galaxies (RC3, De Vaucouleurs 1991), the New General Catalogue of Galaxies (NGC, Huchra 1993), and the Uppsala General Catalogue of Galaxies (UGC, Warren 1993). Our strategy was to find as many SNe as possible, as early as possible. The system can observe about 500 galaxies in one night and recycle the sample in about one week.

BAOSS discovered the first SN, SN 1996W, in April of 1996, which is also the first SN discovered by Chinese astronomers with a telescope. Six SNe were discovered during the first year of operation (1996) and 17 were discovered in 1997. In four years, a total of 40 SNe were discovered by BAOSS, most of which well before their optical maxima. Some of these SNe were well observed and studied, both by BAO and by other astronomers worldwide. The success of this system proves that a small dedicated automatic telescope is the best way to do SN searches.

More modifications to the system were done in 2000. A CCD camera with a better quantum efficiency than the old one was purchased, and the optics of the telescope was modified so that the camera could be installed at the Cassegrain focus. These changes improved the efficiency for both SN searches and SN photometry. A small auto-guiding telescope was also installed which enabled us to take long exposures. More emphasis was put on finding SNe really early (e.g., two weeks before maxima), so the sample of galaxies was reduced to 2,500. With the current high efficiency each galaxy in the sample can be revisited every 3 to 4 days.

This paper is organized as follows. The hardware of the system is described in Section 2 and the software in Section 3. The discoveries are reported in Section 4. Section 5 discusses several well-observed SNe in BAOSS.

2.   The Telescope and the Detector

The 60cm telescope used by BAOSS has an f/15 Cassegrain system. The focal ratio at the prime focus is f/4.3. The telescope is automatically controlled by a PC.

The original CCD camera mounted at the primary focus was a TI 215 with 1024x1024 pixels. The pixel size was 12 μ, yielding a scale of about 1"/pixel. The field of view (FOV) of the camera was 17' X 17'. The camera was thermo-electrically cooled to about - 50^0C, so the dark current was relatively high. In 2000, we purchased a new CCD camera manufactured by Princeton Instruments, which is liquid nitrogen cooled and has a peak quantum efficiency of over 90%. The new camera has 1340x1300 pixels, each 20 μm in size. The FOV is reduced to 10' X 10' but it is still big enough for SN search and photometry observations.

3. The Software

The software for BAOSS keeps evolving over time, but the main purpose remains the same: to schedule observations, to process the images automatically to find SNe, and to automatically control the hardware.

3.1. The Scheduling Program

This program has two functions. One is to select objects from the master sample of galaxies to be observed for the night, and the other is to arrange for their observation at the most appropriate time. The weather condition and the moon phase are considered by the program in order to optimize the selection of the objects. During nights that are partly cloudy, the observer can opt to choose bright nearby galaxies.

The major selection criteria for the galaxies are their priorities, defined as follows:

$$Priority = (D1 - D0)/D$$

where $D1$ is the current date, $D0$ is the date of the last observation, and D is the control time.

The choice of the control time D reflects our survey strategy. For galaxies between 20Mpc - 100Mpc, we choose D to be as big as possible, as long as no SNe will be missed by the search. For galaxies less than 20Mpc, we choose D to be the duration of a normal SNe Ia between the limiting magnitude of our system and its peak magnitude in the corresponding galaxy. Assuming ideal weather and no competition, these choices of D would enable our search to find all the nearby SNe before their maxima, and to find all the more distant SNe.

Once the priorities for all the galaxies are calculated, the program selects the galaxies from higher to lower priorities, with additional consideration of hour angles of the targets and the moon phases. This scheduling program is very useful for BAOSS. During seasons with few available observing nights, it is essential to use the scheduling program to make good use of the limited observation time.

3.2. The Image Processing Program

The principle of finding supernovae is to find new objects in the current observations by comparing them to the existing template images. A template image for a galaxy is the one with the highest quality, but if the new observation is found to be better than the template, the new image becomes the new template. Since SNe are likely to occur in the optical disks of the host galaxies, it is often difficult to find them visually because of the light contamination from the host galaxies. The goal of the image processing is thus to remove the complicated galaxy background and to reveal any changes in the images.

Since the new observation and the template are taken under different conditions (pointing, seeing, moon phase, transparency), various steps have to be taken before a final subtraction can be achieved. First, the two images are aligned to the same position, then convolved so the stars in the images have a similar Point-Spread-Function (PSF). Finally, the intensity levels of the images are matched. New objects can then be detected in the resultant subtraction and listed as SN candidates. These candidates are reobserved and if any of them are confirmed, reports are sent to the Central Bureau of Astronomical Telegrams (CBAT). If possible, spectroscopy is also attempted by the 2.16m telescope.

The first version of the image processing program was compiled in 1995, based on programs provided by Brian Schmidt. It was heavily modified in the summer of 2000 using new tasks then available to IRAF (the Image Reduction and Analysis Facility) in the package "images/immatch".

An example is shown in Fig. 1 to explain how the program works to find SN 2000dw in UGC 11955. Four images are displayed: A) the new observation; B) the template image; C) the new observation after image alignment, PSF matching, and intensity matching; and D) the subtracted image with SN 2000dw standing out. It is obviously much easier to find SN 2000dw in image "D" than in image "A".

3.3. The Automatic Observation Programs

The automatic observation programs have two parts: one part controls the CCD camera and is written in Visual Basic Language; the other controls the telescope and is written in C Language. The two programs communicate with each other via a Linux network. The automatic observations are taken in the following sequences. The telescope is first pointed to the new target and if at position, the CCD camera is directed to start an exposure. When the exposure is finished the telescope moves to the next object and the cycle continues.

4.   The Discoveries

The first supernova discovered by BAOSS was SN 1996W (Li et al. 1996). From 1996 April to 2001 March, 40 SNe, 5 cataclysmic variables, and 2 extra-galactic novae have been discovered by BAOSS. A brief listing of the SNe statistics discovered by BAOSS is presented in Table 1. Table 2 lists more details of these SNe.

Table 1.   Statistics of the SNe discovered by BAOSS

Total	Ia(normal)	Ia(peculiar)	II(normal)	II(peculiar)	Ib &Ic
40	10	4	18	3	5

In Table 1, "Ia(normal)" are those SNe Ia that show normal spectra like SN 1986I. "Ia(peculiar)" are those SNe Ia that are similar to either 1991T (overluminous) or 1991bg (subluminous). "II(normal)" are those Type IIP and IIL SNe. "II(peculiar)" are those Type IIn and IIb SNe.

In the BAOSS sample, SNe II represent about 50 percent of the total, SNe Ia 35 percent, and Ib/Ic about 12 percent. This frequency distribution is different from that found by Cappellaro et al (1999). In particular, the percentage of SNe Ia in the BAOSS sample seems to be lower than theirs. One possible explanation is that only SNe Ia occur in both spiral and elliptical galaxies, and a possible bias against elliptical galaxies in the BAOSS galaxy sample may have resulted

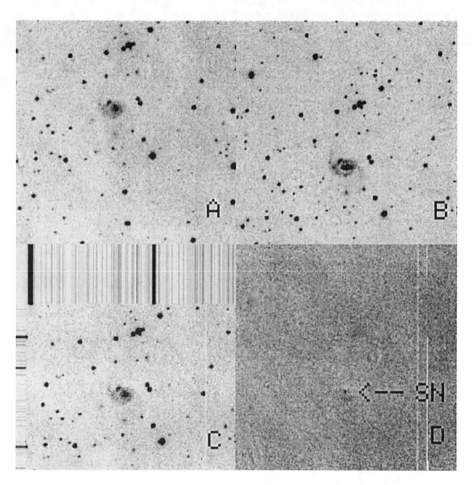

Figure 1. Example images of the supernova finding program. A: the new observation; B: the template image; C: the new observation after image alignment, PSF matching, and intensity matching; D: the subtracted image with SN 2000dw identified.

in a smaller percentage of SNe Ia. The low SN Ia frequency could, however, be caused by the small number statistics.

One interesting result of the statistics from our discoveries is the high occurence rate of peculiar SNe Ia. Combining our discoveries with those of the Lick Observatory Supernova Survey (LOSS)(Li et al 2001), we found the rate of peculiar SNe Ia to be about 36%±9%, much higher than earlier estimates (< 17%, Branch et al, 1993).

5. Some Well Studied SNe Discovered by BAOSS

One important goal of BAOSS is to do photometric and spectroscopic follow-up observations of bright SNe. Here we report on the observations of several SNe discovered in the course of BAOSS. The photometric observations were taken with the 60cm telescope and the spectroscopic date were obtained with the OMR low resolution spectrograph of the 2.16m BAO telescope.

5.1. SN 1996W in NGC4027

SN 1996W was discovered on April 10, 1996 (Li et al. 1996) and was the first SN discovered by BAOSS. Fig. 2 shows the BVR light curves and some spectra of SN 1996W. The spectra show prominent hydrogen Balmer lines, indicating that SN 1996W is a Type II SN.

The B band light curve of SN 1996W is consistent with that of the template Type IIP SN. The V band light curve also shows a prominent plateau. The spectra show typical Type IIP spectral features similar to those of SN 1986I. In particular, the Hα shows a conspicuous P Cygni profile. These observational features are consistent with the suggestion made by Schlegel (1996), that there is a correlation between the plateau shape of light curves and the P Cygni profile of the spectra of Type IIP supernovae.

5.2. SN 1996cb in NGC3150

SN 1996cb was discovered by Japanese astronomers and by BAOSS (Qiao et al 1996) independently. We confirmed the discovery spectroscopically and classified it as a Type II SN. In Fig. 3, we present the BVR light curves and some spectra of SN 1996cb.

The first few spectra of SN 1996cb showed that it was a typical type II, with prominent hydrogen Balmer lines. But as the supernova evolved, the spectra deviated from typical type II SNe and became similar to those of type Ib supernovae. These kind of supernovae are defined as Type IIb SNe (Wheeler & Filippenko 1996, Filippenko 1996). SN 1996cb turned out to be the third type IIb SN ever discovered, the other two being SN 1987K and SN 1993J.

The light curves of SN 1996cb are very similar to those of SN 1993J. SN 1996cb resembles SN 1993J in its spectral evolution except for some subtle differences (Qiu et al 1999), from which we concluded that the progenitor of SN 1996cb was smaller than that of SN 1993J but has a thicker hydrogen envelop.

Table 2.   Supernovae Discovered By the BAO Supernova Survey

SN	Galaxy	Coordinates of SN	Mag	SN Type	Discovery Date
1996 W	NGC4027	115928.98 -191521.9	V16.0	II	1996 Apr 10
1996 bo	NGC673	014822.86 +113115.1	V16.5	Ia	Oct 18
1996 bv	UGC3432	061613.04 +570308.4	V16.6	Ia	Nov 3
1996 bw	NGC664	014344.51 +041319.0	17.5	II	Nov 30
1996 by	UGC3379	055824.96 +682712.1	16.5	Ia	Dec 14
1996 cb	NGC3510	110341.98 +285413.7	16.5	IIb	Dec 15
1997 C	NGC3160	101356.18 +384900.5	17.5	Ic	1997 Jan 13
1997 Y	NGC4675	124531.40 +544417.0	14.8	Ia	Feb 2
1997 aa	IC2102	045153.60 -045736.0	17.0	II	Mar 1
1997 bn	UGC4329	081902.20 +211101.7	16.4	II	Apr 3
1997 bo	A121952+0743	121952.66 +074349.7	16.1	II	Apr 3
1997 br	ESO576-G40	132042.40 -220212.3	14.5	Iapec	Apr 10
1997 cn	NGC5490	140957.76 +173232.3	17.2	Iapec	May 19
1997 co	NGC5125	132401.18 +094222.9	18.2	II	May 29
1997 cw	NGC105	002517.27 +125306.2	16.5	Iapec	Jul 10
1997 dc	NGC7678	232828.41 +222523.0	18.3	Ib	Aug 5
1997 dg	A234014+2612	234014.21 +261211.8	16.7	Ia	Sep 27
1997 di	UGC4015	074744.45 +621939.2	18.1	II	Oct 27
1997 do	UGC3845	072642.50 +470536.0	15.8	Ia	Oct 31
1997 ds	M-01-57-007	222411.51 -032910.5	16.2	II	Nov 17
1997 dt	NGC7448	230002.93 +155850.9	15.4	Ia	Nov 21
1998 C	UGC3825	072334.63 +412604.2	18.1	II	1998 Jan 21
1998 D	NGC5440	140259.28 +344454.3	15.5	Ia	Jan 29
1998 S	NGC3877	114606.11 +472855.8	V16.0	IIpec	Mar 2
1998 T	NGC3690	112833.16 +583343.7	14.5	Ib	Mar 3
1998 ab	NGC4704	124847.24 +415528.3	B16.6	Iapec	Apr 1
1998 ar	NGC2916	093458.86 +214257.5	18.4	II	Apr 14
1998 dn	NGC337A	010127.08 -073636.7	15.8	II	Aug 22
1998 ec	UGC3576	065306.11 +500222.1	16.9	Ia	Sept 26
1999 D	NGC3690	112829.50 +583340.6	15.6	II	1999 Jan 16
1999 aa	NGC2595	082741.11 +212915.5	16.3	Ia	Feb 11
1999 an	IC755	120110.5 +14061	15.0	II	Mar 7
1999 dn	NGC7714	233614.70 +020908.8	16.0	Ib	Aug 19
1999 ed	UGC3555	065000.95 +253754.5	17.8	II	OCt 5
1999 el	NGC6591	203717.83 +660611.5	15.4	IIn	Oct 19
2000 dw	UGC11955	221349.06 +391415.1	17.5	II	2000 Oct 17
2001 B	IC 391	045719.24 + 781116.5	15.5	Ib	2001 Jan 3
2001 K	IC 677	111356.09 121806.6	17.8	II	Jan 15
2001 T	MCG-02-37-6	143253.17 -125843.6	16.8	II	Feb 6
2001 X	NGC 5921	152155.45 +050342.1	17.0	II	Feb 27

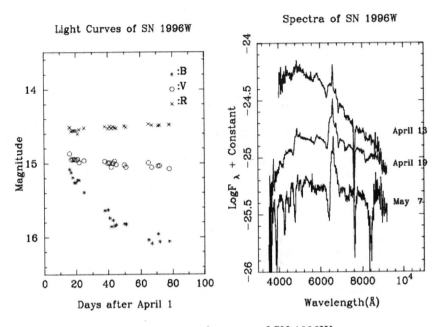

Figure 2.   The light curves and spectra of SN 1996W

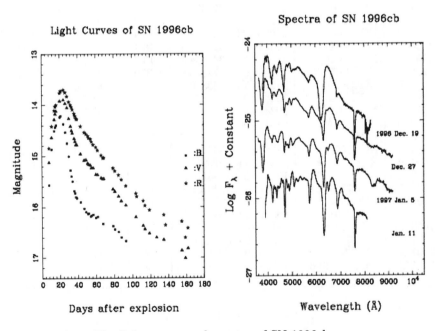

Figure 3.   The light curves and spectra of SN 1996cb

Light Curves of SN 1997br **Spectra of SN 1997br**

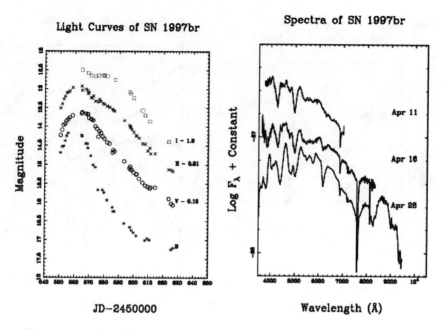

JD-2450000 Wavelength (Å)

Figure 4.   The light curves and spectra of SN 1997br

5.3. SN 1997br in ESO 576-G40

SN 1997br was discovered by BAOSS on Apr. 10.6 UT, 1997 (Li et al 1997). In Fig. 4, we show the UBVI light curves and some spectra of SN 1997br.

SN 1997br is a peculiar SN Ia whose prototype is SN 1991T (Filippenko et al 1992). The spectra of this kind of SNe Ia before maxima are dominated by high excitation Fe III lines . As the SN evolves, it develops lines of the intermediate-mass elements that are typical in normal SNe Ia.

The light curves of SN 1997br are similar to those of SN 1991T, an overluminous Ia SN, with a slow decline rate after the B maximum. The spectra of SN 1997br also resemble those of SN 1991T with subtle differences. Spectra of SN 1997br seem to indicate an earlier transition to the dominant phase of Fe-peak elements after the B maximum (Li et al 1999).

5.4. SN 1998S in NGC3877

SN 1998S was discovered on Mar 2, 1998. It is a peculiar type II SN, showing narrow Balmer lines in the spectra. Schlegel (1990) defined this kind of SNe as IIn SNe, where "n" stands for narrow lines. In Fig. 5, we present the BVR light curves and some spectra of SN 1998S.

SN 1998S is a very bright SN reaching a magnitude of about 12 at maximum. It was discovered at a very early phase when it was only 16 mag in the R band. The absolute magnitude in the B band at the maximum for SN 1998S is -18.6, about 2 magnitude brighter than for a normal type II supernovae. Due to its brightness and peculiarity, SN 1998S has been extensively observed worldwide, including several visits by the Hubble Space Telescope. The spectral evolution

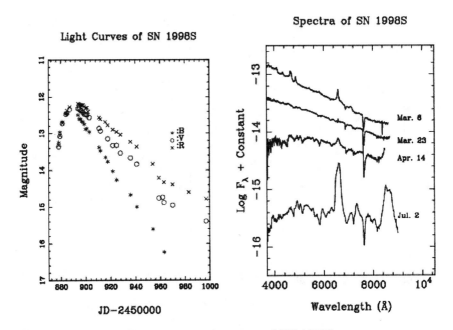

Figure 5.   The light curves and spectra of SN 1998S

of SN 1998S (Liu et al 2000) is similar to that of SN 1979C, a typical supernova of SNe II-L. The spectra before maximum showed high-ionization N III and He II emission lines superimposed on a strong blue continuum, which is the Wolf-Rayet star feature. The light curves showed a steep and linear decline, consistent with those of SN 1979C. Observations by HST and in near-infrared wavelenghts show that there is a strong interaction between the ejecta and the circumstellar materials of the SN. SN 1998S thus provides a good opportunity to study the environment of SNe.

Acknowledgments.   BAOSS is supported by the National Science Foundation of China. We would like to thank the support from one sub-project of the 973 Project, "NKBRSF G 19990754".

References

Branch, D., Fisher, A., Nugent, P. 1993, AJ, 106, 2338

Cappellaro, E., Evans, R., Turatto, M. 1999, A&A, 351, 459

De Vaucouleurs, A., Corwin, H. G. 1991, Third Reference Catalogue of Bright Galaxies (New York: Springer-Verlag)

Filippenko, A. V., Richmond, M., Matheson, T. et al. 1992, ApJ, 384, L15

Filippenko, A. V. 1997, in Thermonuclear Supernovae, ed. R. Canal, O. Ruiz-Lapuente, & I. Isern (Dorfrecht: Kluwer), 1

Huchra, J. P. 1993, Redshifts Catalogue, Electronic version, Washington D. C.: NASA,

Li, W. D., Qiu, Y. L., Zhu, X. H.; Hu, J. Y. 1999, AJ, 117, 2709

Li, W., Qiao, Q., Qiu, Y., Hu, J. 1996, IAUC 6497

Li, W. & Li C. 1998, IAUC 6829

Li, W., Filippenko, Treffers, R., Riess, A., Hu, J., Qiu, Y. 2001, ApJ, 546, 734

Liu, Q. Z., Hu, J. Y., Hang, H. R.; Qiu, Y. L., Zhu, Z. X., Qiao, Q. Y. 2000 A&AS, 144, 219

Perlmutter, S., Aldering, G., Goldhaber, G. et al 1999, ApJ, 517, 565

Qiao, Q., Li, W., Qiu, Y., Hu, J. 1996, IAUC 6527

Qiu, Y., Li, W., Qiu, Q., Hu, J. 1999, AJ, 117, 736

Schlegel, E. M., 1990, MNRAS, 244, 269

Schlegel, E. M. 1996, AJ, 111, 269

Schmidt, B., Suntzeff, N., Phillips, M. et al 1998, ApJ, 507, 46

Warren, W., H., Uppsala Catalogue of Galaxies, Electronic version, Washington D. C.: NASA, 1993

Wang, L.F., & Hu, J. Y. 1994, Nature, 369, 380

Wheeler J. C., & Filippenko, A. V. 1996, in IAU Colloq. 145, Supernovae and Supernova Remnants, Ed. R. McCray & Z. Wang(New York: Cambridge Univ. Press), 241

Wheeler, J. C., Yi, I., Hoflich, P., & Wang, L. 2000, ApJ, 537, 810

Trip to Kenting Observatory

Small–Telescope Astronomy on Global Scales
ASP Conference Series, Vol. 246, 2001
W.P. Chen, C. Lemme, B. Paczyński

Optical Identification of Gamma-Ray Bursts at Kenting Observatory

Wei-Hsin Sun

National Central University, Chung-Li, Taiwan 320, ROC

Shun-Tang Tseng

National Central University, Chung-Li, Taiwan 320, ROC

Abstract. A remotely-controlled observing station, the Kenting Observatory, has been established in the Kenting National Park, at the southern tip of Taiwan. There are four observing systems at the Observatory: (1) a Meade 16-inch reflector with an SBIG spectrograph plus an SBIG STV; (2) an RC-16inch reflector with an FLI CCD; (3) a Celestron 14-inch with an ST-7E CCD; and (4) a Binocular system of two Canon F1.8 telephoto lens with two FLI CCDs. One of the scientific tasks of this observatory is to provide fast response to satellite alerts in the hopes of identifying the optical counterparts of Gamma-Ray Bursts (GRBs) via the global effort of GRB Coordinated Network (GCN). In this conference proceedings, we describe the instruments and the science projects being carried out in the Observatory.

1. Introduction

The Kenting Observatory (hereafter KTO) is located in the National Museum of Marine Biology and Aquarium (NMMBA). The Museum is situated in Cher-Cheng Village in the Heng-Chun Peninsula of Ping-Dong County, about 18km north to the Kenting area. The location of NMMBA belongs to the Kenting National Park, which governs a large area of the Southern tip of Taiwan.

The Observatory is established for the purpose of providing the universities and high schools in Taiwan a major observing facility that could be accessed remotely through Internet, in view of the many advantages that remote observation could offer. However, in addition to the educational and promotional purposes, KTO also provides a platform for many scientifically significant research projects. Because the Observatory is located in a National Museum at sea-level, the night sky transparency and darkness are not superb. Nevertheless, these factors do not prevent the Observatory from doing projects of discovery type, such as the search for optical counterparts of Gamma-Ray Bursts (GRBs). Below we describe briefly the telescopes and associated equipments of the Observatory, and then present the scientific attempts.

墾丁中央大學遠距遙控天文台

Figure 1.   The AstroHaven clam-shell dome at the Kenting Observatory. The two halves open to the North and the South, respectively, and can stop at desired angle, providing effective wind-shielding.

2.   The Observatory and the Equipments

The remote observing project is funded by the Ministry of Education (MOE), the National Science Council (NSC), and the National Central University (NCU). The NMMBA was very generous and willing to support KTO with an office space, power, and network connection. We thus decided to set the Observatory in NMMBA, to save the funding and effort of developing a new site.

The first observing instrument of the Observatory, the Meade 16-inch telescope, was erected on May 4th, 2000, which was protected by simple tent before the actual dome was constructed. The dome arrived in late June. It is a clam-shell type enclosure and is made by AstroHaven in Canada (Figure 1). It has two halves and can open one side or both sides to any desired opening angles. This provides effective wind shielding, which is especially important in the winter when the gusty North-Eastern seasonal wind prevails in Heng-Chun Peninsula. We have implemented a number of modifications before this dome is in full operation and behaves satisfactorily. These modifications include tighter rain-proof sealing, more reliable wire-control for the open and close functions of the shells instead of wireless-control, revised latch-stop, and air-conditioning.

This clam-shell type of dome, which does not use shutter mechanism, offers a great convenience in terms of the equipment it could accommodate. More than one telescope can be placed in the dome, as long as they do not interfere

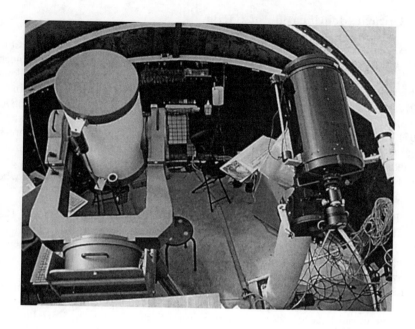

Figure 2.   There are two telescopes side by side in the dome, a Meade 16-inch on a custom-made equatorial mount (left) and a Celestron 14-inch on a Paramount GT-1100 (right).

with each other. Currently there are two telescopes in our dome side by side, a Meade LX200 16-inch and a Celestron 14-inch (Figure 2).

The Meade 16-inch is equipped with an SBIG low resolution spectrograph. We have removed the slit so the system can perform slitless spectroscopy. The Celestron 14-inch has an ST-7E attached, with the built-in color filter wheel using standard broad-band filters. Each telescope/CCD system is controlled by a Pentium II level computer system. Aside from the control computers, there is one more computer in the dome dedicated to receiving satellite alerts from the GCN.

In addition to the above instruments, there is another telescope of Ritchey-Chretien type with an aperture of 16-inch. This telescope will be mounted on a GT-1100ME equatorial mount and may replace the C-14 once the RC-16 is ready. We also set up a binocular type of system which consists of two Canon 200mm F1.8 telephoto lens. This system will be used to carry out a narrow band imaging survey of Northern sky in several wavelength ranges of major emission lines.

Both remote control and satellite linkage need fast and reliable network connection. Chinese Telecom provides the Observatory with a dedicated T1 line from NMMBA to Ping-Dong University of Science and Technology, where the T1 line joins the TANET, the academic backbone supported by MOE.

3. Optical Identification of GRBs

Since the discovery of GRBs in the early 70's and the subsequent confirmation of their cosmological distribution in the 90's, the nature and the emission mechanisms of these powerful celestial objects have become one of the major challenges in high energy astrophysics (van Paradijs et al. 2000 and references therein). In 1997, ground-based observations prompted by satellite alerts (e.g. BeppoSAX) revealed for the first time the optical counterpart of a GRB (van Paradijs et al. 1997). This is an important step forward in understanding how and where the large amount of energy is generated. However, due to the extreme difficulty of immediate response to satellite alerts and the fast fading of the optical afterglows (OAs), there have only been a handful (about two dozens) of GRBs whose optical afterglows were recorded.

The success of ROTSE observations of GRB990123 (Akerlof et al. 1999) has stimulated the inauguration of many small (aperture~50cm) telescopes that are connected to the Gamma- Ray Bursts Coordinated Network (GCN, see the web site: http://gcn.gsfc.nasa.gov/gcn/ for further references) for satellite alert and related information. However, the best example in the search of optical counterpart, the GRB990123, has put forth stringent requirements on the systems which hope to acquire optical images of GRBs. The total of 6 white light images acquired between 22 sec and 10 min by ROTSE after the initial gamma-ray event showed that the estimated V magnitude of the OA dropped from 8.9 to 14.3. This fast fading nature asks for prompt acquisition and exposure within a few minutes of the burst, or the OA will fade and not be detected by small telescopes.

Thus to cope with this rigorous requirement, we have arranged that the telescopes at Kenting Observatory, while performing other observational projects, are always at standby mode for the response to the satellite alerts. Currently we receive GCN notices in the email mode, which means that the alerts are in ordinary email format (time delay: 5 − 30 sec). In the near future, we will collaborate with HETE-2 project and convert the receiving mode to socket control (time delay: 0.1 − 2.0 sec) .

Fast response is essential for it will deliver the information on the temporal behavior at the early stage of the phenomenon, but it is not enough. Spectral evolution, on the other hand, provides much more information on how the post-burst shock evolves (Sari et al. 1998). We thus set up the GRAB project – Gamma-Ray Alerted Binoculars at KTO for these scientifically challenging tasks. The scientific goals of the GRAB project are two-folded: to obtain multi-wavelength as well as multi-epoch observations of GRBs.

The GRAB project at first included two F1.8 telephoto lens, as in the ROTSE project, to cope with the large positional uncertainty of alerts from CGRO. However, as the CGRO was de-commissioned and as HETE-2 comes in as replacement, the positional accuracy has been improved significantly, from degrees down to arcminutes suitable for target acquisition with ordinary long focal length telescopes. We thus reassign the scientific missions for the F1.8 binoculars, and bring in Meade-16inch and C-14inch (with RC-16inch later) for the GRAB project. In short, the Binoculars have developed from two 11-cm to

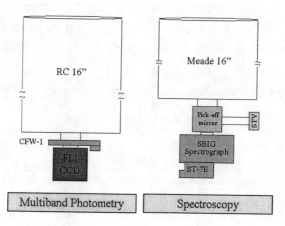

Figure 3. This schematic diagram shows the observing algorithm of the GRAB project. The RC-16 (now C-14) takes multi-color photometry of the target field while the Mead-16 obtains the slitless spectroscopy.

two 40-cm tubes. The C-14 (eventually RC-16) is equipped with direct imaging CCD plus standard broad-band filters. While the Meade-16 will take the slitless spectra of the target field (Figure 3). Practises and drills show that we could obtain the first 10-sec exposure of the HETE-2 alert field in 2.5 minutes.

4. An Example of Other Science Projects at KTO – Asteroid Occultation

KTO, with observing staff present all the time, offers a high probability of coverage of astronomical events. Asteroid occultations are scientifically important events that happen all year round. Right before this meeting, in the evening of January 3rd, 2001, we observed an event of the asteroid *12Victoria*, with a diameter of 117km, occulting a background star *SAO92745*, from UT 11:52 to 13:39. The star is bright, at about 8th magnitude. To our surprise, the asteroid is only 3 magnitude less and can be seen directly in the sequence of images depicting the progress of the event (Figure 4). We will provide the observations to the IOTA for record and will perform analysis on the following issues: the brightness variation of the asteroid, the exact time of occultation, and the orbit determination. These information will allow us to learn more about the rotation of the asteroid and help refine the orbit prediction models.

5. Conclusions

With the help of MOE, NSC, NCU, and NMMBA, we have set up a remotely controlled observatory in the Kenting area. One of the major scientific goals is to provide fast response to the satellite alerts in order to obtain optical observations,

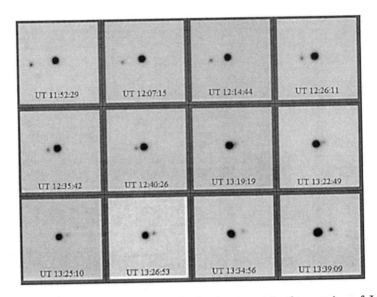

Figure 4.   KTO observed an occultation event in the evening of January 3rd, 2001. From UT11:52 to 13:39, the asteroid *12Victoria* moved from left to right, and occulted the background star *SAO92745* for a brief moment. The star is at about 8.3 magnitude while the asteroid is 3 magnitudes fainter. Both objects are clearly seen in the images.

both temporal and spectral, for the optical counterparts of GRBs. The observing system has been completed and has been in standby mode for satellite alerts. Additional projects, such as asteroid occultation, are also being carried out with preliminary results obtained and being analysed.

Acknowledgments. The authors thank the support of NMMBA for the logistic support and thank Chinese Telecom for the dedicated T1 line. We thank Dr. Claudia Lemme for providing vital information of asteroid occultation events. We also thank Prof. Jing-Yao Hu, Drs. Xiao-Jun Jiang, Yu-Lei Qiu, and Jan-Yan Wei (BAO) for the help in the observation of the occultation event. This project is supported by the grants of MOE 89070250, NSC 89-2515-S-008-003, NSC 89-2515-S-008-005, and NSC 89-2515-S-008-006.

References

Akerlof et al. 1999, *Nature*, 398, 400.

Sari 1998, ApJ, 494, L49.

van Paradijs et al. 1997, *Nature*, 386, 686.

van Paradijs et al. 2000, *Ann.Rev.Astro.&Astrophys.*, Vol.38, 379.

Small–Telescope Astronomy on Global Scales
ASP Conference Series, Vol. 246, 2001
W.P. Chen, C. Lemme, B. Paczyński

Early Results from HETE-2

N. Kawai[1], A. Yoshida[1], T. Tamagawa[1], M. Matsuoka[2], Y. Shirasaki[2],
G. Ricker[3], G. Crew[3], J.P. Doty[3], A. Levine[3], R. Vanderspek[3],
J. Villasenor[3], G. Monnelly[3],J.-L. Atteia[4], G. Vedrenne[4], J.-F. Olive[4],
M. Boer[4], E.E. Fenimore[5], M. Galassi[5], J.-L. Issler[6], C. Colongo[7],
K. Hurley[8], J.G. Jernigan[8], D.Q. Lamb[9], C. Graziani[9], G. Pizzichini[10],
S. Woosley[11], K. Takagishi, I. Hatsukade, M. Yamauchi[12], T. Cline[13]

[1] *RIKEN, 2-1 Hirosawa, Wako, Saitama 351-0198, Japan,*
[2] *National Space Development Agency of Japan, Tsukuba, Japan,*
[3] *MIT Center for Space Research, Cambridge, USA,*
[4] *Centre d'Etude Spatiale des Rayonnements, Toulouse, France,*
[5] *Los Alamos National Laboratory, Los Alamos, USA,*
[6] *Centre Nationale d'Etudes Spatiales, Toulouse, France,*
[7] *Sup'Aero, Toulouse, France,* [8] *UC-Berkeley, USA,*
[9] *University of Chicago, USA,* [10] *TESRE/CNR, Bologna, Italy,*
[11] *UC-Santa Cruz, USA,* [12] *Miyazaki University, Japan,*
[13] *NASA GSFC, USA*

Abstract. The High Energy Transient Explorer 2 is a small scientific satellite designed to detect and localize gamma-ray bursts (GRBs). The coordinates of GRBs detected by HETE-2 will be distributed to interested ground-based observers within seconds of burst detection, thereby allowing detailed observations of the initial phases of GRBs. HETE-2 was launched successfully on October 9, 2000. The GRB positions will start to be delivered after a few months of the complete testing and calibration of the spacecraft system and the science instruments.

1. Introduction

HETE-2 is a mission to provide big possibilities for small telescopes to study the extreme physics occurring at the cosmological distances.

Gamma-ray bursts have been one of the most profound mysteries in high-energy astrophysics since their discovery in the late 1960's. Attempts to associate GRB sources with any of the known categories of celestial objects had long been unsuccessful, partly because the GRBs are short transient events, typically lasting only tens of seconds. Furthermore determination of the source position in the gamma-ray band is intrinsically difficult. In order to understand the origin of the gamma-ray bursts, the most critical observation should be the accurate determination and prompt dissemination of the position of the source. With such information, detailed studies of the counterparts in optical and other wavebands should provide the clue linking GRBs to known objects. This was the consensus reached at the "High Energy Transients" workshop at Santa Cruz (1983), and

149

High Energy Transient Explorer (HETE) was conceived as the mission to achieve this goal (Woosley et al 1984).

The original HETE mission was selected as an unsolicited NASA pilot mission in 1990 with G. Ricker (MIT) as the principal investigator, joined by international partners from Japan (RIKEN) and France (CESR). The satellite was launched in November 1996, but the mission was lost due to failure of the launching rocket. The importance of the study of GRBs and the HETE mission was recognized in all the participating countries, and the funding to rebuild the HETE satellite was approved in US, Japan, and France in 1997.

A breakthrough in the study of GRBs was brought about in 1997 by the BeppoSAX X-ray astronomy satellite. Its wide-field camera provided GRB positions with sufficient accuracy ($\sim$several arcmin) and relatively small delay ($\sim$hours), and the discovery of X-ray afterglows of GRBs with its narrow-field instruments enabled detailed follow-up observations of afterglows in X-ray, optical, and radio bands (Costa et al. 1997, van Paradijs et al. 1997, Frail et al. 1997). These observations led to identification of GRBs with sources at cosmological distances, and provided information about their distance, host galaxies, and the environment (*e.g.* Metzger et al. 1997). Timely localizations also started to be provided by other space missions such as RXTE ASM (Takeshima et al. 1998) and a network of Interplanetary spacecrafts (IPN) (Cline et al. 1999).

Some of the original goals of HETE were achieved by these missions, but it was also made clear that very rapid GRB position information with sufficient accuracy is very important. It was soon found that the afterglows fade rapidly following a power-law with time. Several hours after the burst, the afterglows are typically 19th magnitude or dimmer, and detailed spectroscopy is only possible with the 8–10 meter class large telescopes. The host galaxies are typically 23 mag or dimmer, and the red-shift measurements are also possible only with the world's largest telescopes (*e.g.* Sahu et al. 1997, Kulkarni et al. 1998). This constraint has limited the progress in the study of the GRB population significantly. If the position information is provided 100 times earlier, *i.e.* within minutes of the burst, the optical afterglow could be detected at 10-15th magnitude, and 1-meter class (or even smaller) telescopes could contribute spectroscopic observations. The detection of the 9th magnitude prompt optical emission from GRB990123 by the ROTSE experiment (Akerof et al. 1999) also imply that very rapid ($\lesssim$10 seconds) positioning can lead to observations of very bright optical counterparts, which could be spectroscopically studied with small telescopes.

The current limitation of the delay in obtaining the GRB position information is due to the delay in downlinking of the satellite data, and the time required for the analysis on the ground. HETE is designed to overcome these limitations by localizing GRBs with its on-board processors, and the positions will be immediately broadcasted to the burst-alert stations distributed around the world to provide almost continuous real-time contact.

2. Mission Overview

The primary goals of the HETE mission are the multiwavelength observation of gamma-ray bursts (GRBs) and the prompt distribution of precise GRB coordi-

Table 1. Scientific Instruments of HETE-2

	FREGATE	WXM	SXC
Built by	CESR	RIKEN, LANL	MIT CSR
Instrument type	NaI(TI); cleaved Scintillator	1-dim PSPC (Xe 1.4 atm)	CCD (15μm pixel)
Energy Range	6 – 400 keV	2 – 25 keV	0.5 – 10 keV
Resolution			
Timing	10 μs	1 ms	1.2 s
Spectral	25% (20 keV) 9% (662 keV)	22% (8 keV)	46 eV (525 eV) 129 eV (5.9 keV)
Angular	–	±11′ (at 8 keV)	±33″
Effective Area[a]	120 cm^2	350 cm^2	72 cm^2
Coded Mask	N. A.	1/3 open	1/5 open
Sensitivity (10 σ) (erg cm^{-2}s^{-1})	3 × 10^{-8} (8 keV–1 MeV)	8 × 10^{-9} (2 keV–10 keV)	3 × 10^{-8} (1-10 keV)
Field of View	3 str	1.6 str	0.9 str

[a] Total effective area of the detector surface.

nates to the astronomical community for immediate follow-up observations. To achieve these goals, HETE-2 is equipped with one gamma-ray and two X-ray detectors, which share a common field of view of ~1.5 steradians, and, together, are sensitive to photons in the energy range of ~0.5 keV to over 400 keV. The two X-ray detectors are coded-aperture imagers, allowing HETE to determine the location of a GRB to a precision of 10 arc-minutes (typical) to as low as 10 arc-seconds (Table 1). Sophisticated on-board processing software allows the location to be calculated on board in real time, and ground post-burst analysis will provide refined localizations

In orbit, the HETE spacecraft will always point in the anti-solar direction for optimal exposure of the solar panels to the Sun. As a result, ground observers will always know approximately where HETE is observing. In addition, all bursts detected by HETE will be at least 120 degrees from the Sun and, therefore, in a prime position for observations by ground-based optical observers. The scientific instruments operate during orbit twilight and night, when the Earth is not blocking their view. .

3. Ground Stations

The HETE-2 spacecraft communicates with the ground over either of two radio links.

- Command uplink and high-data-rate downlink occurs over the S-band, at 2.1-2.3 GHz. The downlink data rate over this link is 250 kbit/s. S-band communications are done using the three Primary Ground Stations: Kwajalein Atoll (Marshall Islands), Cayenne (French Guiana), and Singapore.

- Burst alert messages and spacecraft housekeeping data are transmitted in the VHF band, at 138 MHz. The data rate over this link is 300 bits/s. VHF data reception is via the Burst Alert Station Network. There are about a dozen burst-alert stations, distributed evenly around the equator to provide continuous contact with the satellite.

All operations of the ground stations and the satellite are controlled by the MIT Command and Control Center (MCC). Commands for the spacecraft are sent to the primary ground stations from the MCC, and data received from either the primary or burst alert stations are forwarded to the MCC.

4. Burst Alert Network

One of the key features of the HETE-2 satellite is its ability to calculate precise localizations of GRBs on board within seconds of burst onset, and then to transmit the burst localizations to the ground as soon as they have been calculated. The HETE-2 satellite utilizes a low-rate VHF transmitter to continuously broadcast the burst information; on the ground, an array of listen-only burst alert stations (BAS) receive the data and transmit them to the MIT Control Center. Once received at MIT, the burst information is immediately relayed to the GRB Coordinate Distribution Network (GCN) at the Goddard Space Flight Center for distribution to interested ground observers.

4.1. Burst Information Downlinked by HETE

Information about a GRB will come to the ground in two ways:

- The results of real-time analysis performed on the spacecraft are transmitted via the VHF to the burst alert stations. These results will be of moderate quality, as the spacecraft processors are computationally limited.

- The raw data taken by all science instruments are transmitted via the S-band data link. With these data, sophisticated ground analysis can refine the burst position.

4.2. The GCN messages

The GCN messages devoted to HETE data are the following:

- The HETE_ALERT message indicates that a burst has been detected. Preliminary brightness and trigger time information are included in this message, but no localization information will be available. This message is meant as a warning of an impending localization message.

- HETE_FLIGHT messages contain updates of burst intensities and durations and, most importantly, burst localizations. There can be multiple versions of this message sent during a burst: as new or better localization information becomes available, it will be relayed to the GCN. These messages are likely to be of most use to those GCN sites with automated instrumentation.

- The HETE_FINAL message contains a summary of the best information available from analysis done on the satellite. This message should be of general interest to all.

- HETE_GND_ANALYSIS messages will be sent out after ground analysis of the raw data have been completed. More than one such message could be distributed. These messages will give useful a posteriori updates on the burst position and intensity.

5. Mission Status

5.1. Launch and Orbit

HETE-2 was launched successfully on October 9, 2000 0538(UT) with a Pegasus Rocket from the ocean near the Kwajalein Atoll of the republic of Marshall Islands at the center of the Pacific Ocean. The orbit is a semicircular low-earth orbit with an apogee of 635 km, a perigee of 595 km, and an inclination of 1.95°. The orbit life is sufficiently long to support its planned mission life of two years with possible extension of extra few years.

5.2. Crab Calibration

The Crab Nebula was observed by HETE-2 in November 2000 – January 2001. These data are used to calibrate spectral, timing, and astrometric performance of the detectors and the spacecraft system. The timing performance for WXM and FREGATE has been verified through the successful reconstruction of the Crab Pulsar light curve. The shadow patterns of the coded aperture were successfully recorded for hundreds of different positions of the Crab Nebula in the fields of view of WXM and SXC, which provided a data base to calibrate the alignment between the X-ray detector systems and the optical aspect cameras. All four optical cameras are fully operational. Data taken with the cameras can be used to measure the spacecraft aspect to a precision of 10–20″ each second of orbit night. The relative positions of the cameras have been measured and thermal variations in their pointing can be measured with a precision of a few arcseconds.

5.3. Burst Detections

As of February 2001, in limited operations, the HETE French Gamma Telescope (FREGATE) has detected 6 cosmic GRBs (001102, 001105, 001106B, 001225, 010107, and 010126) that have been confirmed by other spacecraft, and 10 likely GRBs that have not been confirmed (001027, 001115A, 001115B, 001124, 001222, 001226, 010111, 010204, 010213, and 010225). The 010126 event is particularly notable as a milestone for HETE, since the event triggered the on-board burst detection software, initiating a successful VHF alert through the Burst Alert Network.

On 10 January, an unconfirmed burst event was detected by the WXM during the Crab calibration. Because of the state of the instrument calibration at the time of the burst, localization in only one direction was possible. During the FREGATE-triggered burst events of 010126 and 010204, the WXM responded by contributing appropriate data formats to the HETE downlink. However, because the 010126 and 010204 GRB events were outside the FOV of the WXM, localizations were not possible. Presently, the WXM is fully set to respond to triggers initiated by FREGATE.

5.4. An Early Result: GRB010213

On 13 February 2001 at 12:35:35 UTC, a soft spectrum, high energy transient at high galactic latitude was detected and localized by HETE. Both the WXM and the FREGATE instruments detected the event. Although the WXM and FREGATE cover extremely broad energy bands (WXM: 2–25 keV; FREGATE: 6–500 keV), the burst event was only detectable in the 2–18 keV band by WXM, and in the 6–10 keV band by FREGATE. Because of the unusual spectrum of the transient, neither instrument triggered.

The preliminary coordinates of the burst $(10^h 31^m 36^s, +5°30'39'')$ (J2000), were derived by combining data from the WXM and Boresighted Optical Cameras. The statistical error radius in the WXM localization is $3.5'$ (95% confidence) In addition, we estimate a systematic error radius at present of $30'$ about this location. The spacecraft aspect was known to an accuracy of $\pm 30''$ (95% confidence) from the optical cameras, and will be improved.

The burst exhibited a double-peaked structure and lasted about 30 seconds. The incident flux measured with the WXM (2–18 keV) at the first peak was about 0.7 Crab, and at the second peak was about 2.4 Crab. The spectrum in the 2–18 keV range was harder than that of the Crab nebula at the first peak, and softer than that of the Crab at the second peak. The peak flux seen with FREGATE (6–10 keV) was ~2 Crab. Assuming a Crab-like spectrum, the peak energy flux was ~1.5 10^{-8}erg cm^{-2} s^{-1} in the 6–10 keV range.

The high galactic latitude of the source, well away from the Galactic Bulge, and the shape of its light curve suggest that it is a gamma-ray burst with an unusual spectrum. Conceivably, it could instead be a nearby X-ray burst source. A preliminary catalog search of the WXM error circle revealed no correspondence with known globular sources, cataclysmic variables, low mass X-ray binaries, or flare stars. (GCN Circular 934)

References

Akerlof, C. et al. 1999, 398, 400

Cline, T. L. 1999, A&AS, 138, 557

Costa, E. et al. 1997, Nature, 387, 783

Frail, D. A. et al. 1997, Nature 389, 261

Kulkarni, S. R. et al. 1998 Nature, 393 35

Metzger, M. R. et al. 1997, Nature, 387, 878

Sahu, K. C. et al. 1997 Nature, 387, 476

Takeshima, T. et al. 1998, in AIP Conf. Proc. 428, Gamma-Ray Bursts: 4th Huntsville Symp., eds. Meegan, Preece, & Koshut. (New York: AIP), 414

van Paradijs, J. et al. 1997, Nature, 386, 686

Woosley, S. E. et al. 1984, in AIP Conf. Proc 115, High Energy Transients in Astrophysics., Woosley, S.E., ed. Santa Cruz, 1983. (New York: AIP)

Small–Telescope Astronomy on Global Scales
ASP Conference Series, Vol. 246, 2001
W.P. Chen, C. Lemme, B. Paczyński

RIBOTS: An Automatic Telescope System for Gamma-Ray Burst Follow-Up Observations

Yuji Urata, Nobuyuki Kawai, Atsumasa Yoshida, Mitsuhiro Kohama

The Institute of Physical and Chemical Research (RIKEN),Wako,Saitama 351-0198, Japan

Tetsuya Kawabata, Kazuya Ayani

Bisei Astronomical Observatory, Ohukura 1723-70, Bisei, Oda, Okayama 714-1411, Japan

Abstract. We are constructing a fully automatic observation system named RIBOTS (RIken-Bisei Optical Transient Seeker). We aim to detect optical flashes and early afterglows of Gamma-ray bursts (GRB) with RIBOTS. We are constructing RIBOTS with a small telescope because a quick pointing to the burst is essential for our purpose. RIBOTS is linked to the GRB alert system provided by the HETE-2 satellite.

1. Introduction

Gamma-ray bursts are one of the most elusive phenomena for high-energy astrophysicists. It is well known that an important clue for resolving the puzzle is the detection of transient optical emission associated with the bursts. This can be achieved by a small optical telescope system capable of rapidly moving towards the position of the event.

2. RIBOTS

RIBOTS is sited at Bisei Astronomical Observatory (Okayama prefecture, Japan). This system points fully automatically and starts observing after receiving a HETE-2 alert with position information. We constructed the system with a small aperture telescope and a CCD camera (Table 1). The mount slews at 6 degrees per second (max). RIBOTS will track the center of the field of view of HETE-2 (60×60deg), so that any GRB position reported by HETE-2 can be reached within 5 seconds after the position is available. Since HETE-2 is expected to deliver a GRB position within 10 seconds after the burst triggers, RIBOTS can start observing GRBs as early as 15 seconds from their onset.

With RIBOTS, a new optical transient will be quickly picked up in an image, and its position with 5 arcseconds accuracy will be reported automatically in 3 minutes including the time required to download the image to the computer. This refined position information will be very useful for spectroscopic observations with other, larger telescopes.

We control the system with two PCs. One PC (using Windows) controls the CCD camera while the other (using Linux) manages the whole system including weather monitoring, telescope control, time sequencing, communication with GCN and quick analysis.

Table 1.   Specifications of RIBOTS hardware

Telescope Meade LX200-30		CCD camera SBIG ST-8E	
Aperture	300mm	Pixel	$510\times340(3\times3)$
Focal length	1000mm (with 0.33 reducer)	Resolution	5.6 arcsec(3×3)
FOV	47.2'×31.4'	Read out time	12 sec (3×3)
Slew speed	6 deg/sec	Wavelength	350-925nm(QE $\geq$20%)

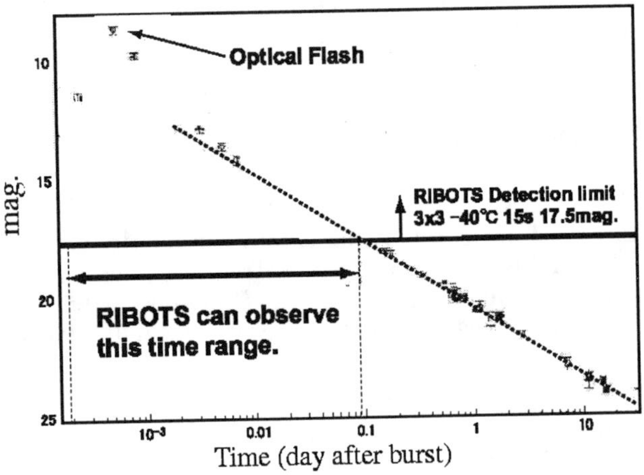

Figure 1.   Comparison of the GRB990123 light curve with the RI-BOTS detection limit. RIBOTS can cover the early phase of the GRB flash/afterglow (2-3 hours after the burst).

3. Feasibility Study

We evaluated the limiting magnitude of RIBOTS by correlating its images with the USNO-A2.0 catalog, and found it to be 17.5-18.0 mag (non-filter, 15 seconds exposure, 3×3 binning mode, dark sky at Bisei). Fig.1 shows a comparison of the GRB990123 light curve and the RIBOTS detection limit. RIBOTS has enough sensitivity to catch the optical flash and early afterglow. RIBOTS can cover the time period up to 2-3 hours after the burst, so we can obtain the light curve of the early phase afterglow phenomenon.

Considering RIBOTS's 15 sec quick pointing capability and sensitivity, RI-BOTS can detect much dimmer optical flashes than that of GRB990123 (9th mag).

References

T.J.Galama, et al., Nature 398, 394 (1999)

N.Kawai, et al., this volume

T.Kawabata, et al., this volume

Small–Telescope Astronomy on Global Scales
ASP Conference Series, Vol. 246, 2001
W.P. Chen, C. Lemme, B. Paczyński

A Spectrograph for Prompt Observations of Gamma-Ray Bursts with a 1-m Telescope

Tetsuya Kawabata and Kazuya Ayani

Bisei Astronomical Observatory, Ohkura 1723-70, Bisei, Oda, Okayama 714-1411, Japan

Mitsugu Fujii

Fujii Bisei Observatory, 4500 Kurosaki, Tamashima, Okayama 713-8126, Japan

Yuji Urata, Nobuyuki Kawai and Noboru Ebizuka

RIKEN, Saitama 351-0198, Japan

Abstract. We are developing a new slit-less spectrograph for gamma-ray bursts (GRBs) at Bisei Astronomical Observatory. We can quickly point the 1-m telescope to the GRB coordinates provided by HETE-2 via the GRB Coordinates Network. The pointing is readjusted when the position is refined to 5" accuracy by the small optical telescope RIBOTS using the optical image of the GRB, and we then take an exposure immediately with the slit-less spectrograph.

1. Introduction

GRBs were discovered more than thirty years ago. They generally have durations between 0.1 s and 100 s and appear at random times from unpredictable positions on the sky. Recent observations of GRB afterglows support that the bursts originate in other galaxies at cosmological distances (e.g. Metzger et al. 1997). The physical mechanism which produces these brief flashes of high-energy radiation has remained a mystery. In order to investigate the physical processes of GRBs we need the spectra in early phase. The networking between satellites and ground-based telescopes permits small telescopes to observe the GRB spectra before they decline. We are therefore developing a new slit-less spectrograph and control system for prompt observations of GRBs at Bisei Astronomical Observatory (BAO).

2. Spectrograph and Telescope Control

The Bisei Imaging Spectrograph (BIS) is attached to the cassegrain focus of the BAO 1-m telescope and is always standing by for prompt observations. BIS is a very low-resolution spectrograph, consisting of a removable slit, the optical elements and the CCD camera ST6. Figure 1 shows the observed sample image

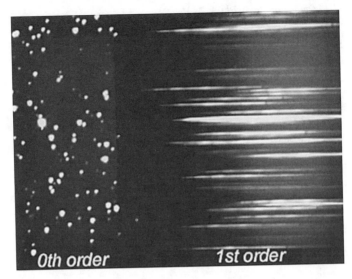

Figure 1.   An observed BIS image of the open cluster M37 is shown.
The left and right side is 0th and 1st order image, respectively.

of M37. Both the 0th and 1st order spectra in the field of 6' x 2'.5 are taken
simultaneously. The resolution is 7nm with a seeing size of 3" and the wavelength
range is from 400nm to 800nm. The limiting magnitude is 14 with 5 minutes of
exposure time. After an observation the slit is inserted into the focal plane and
adjusted to the position of the GRB counterpart in order to take the flat fields
and the comparison frames for wavelength calibration.

HETE-2 (Kawai et al. 2001) detects GRBs and gives their positions to the
GRB Coordinates Network (GCN) whoose accuracy is around 10'. The small
optical telescope RIBOTS (Urata et al. 2001) and the 1-m telescope move
quickly to the GRB coordinates provided by HETE-2 via GCN. RIBOTS takes
the optical images for refining the accurate position to within 5" and sends it
to the 1-m telescope. The pointing of the 1-m telescope is then readjusted and
BIS immediately takes an exposure.

3.   Discussion

HETE-2 shall localize around 50 GRB events per year. The opportunity for
spectroscopic observations is a few times per year because of geographic and
weather conditions.

References

Metzger, M. R., et al. 1997, Nature, 387, 878
Kawai et al. 2001, thess proceedings
Urata et al. 2001, thess proceedings

Small–Telescope Astronomy on Global Scales
ASP Conference Series, Vol. 246, 2001
W.P. Chen, C. Lemme, B. Paczyński

Cataclysmic Variables: A 'SWOT' Analysis

Brian Warner

Department of Astronomy, University of Cape Town, Rondebosch 7700, South Africa

Abstract. A brief review of the nature and subtypes of cataclysmic variables (CVs) is given. The catalogue of CVs is still very incomplete. All-sky surveys should add large numbers of CVs, which will improve knowledge of the space density of these systems. It is pointed out that the nova-like variables, which are the most difficult to discover, are the subtype having the highest space density. Their discovery is therefore the highest priority – they fix the frequency of CVs, which is important in population syntheses of binary stars.

1. Introduction

The cataclysmic variables (CVs) will feature prominently in any All-Sky monitoring programs, both as transient events and as periodic phenomena. The amplitudes of the largest nova eruptions rival those of supernovae but reach much brighter apparent magnitudes (e.g., Nova Cygni 1975 which had been fainter than 21st magnitude until a week before its eruption, when it rose to 13.5 mag, then erupted to 2.2 mag, and has settled back at 16.3 mag). Dwarf novae erupt on time scales from days to years, with most being in the region of months, and ranges of two to eight magnitudes. Thanks largely to the sterling efforts of amateur astronomers over the past century, continuous light curves are available for a few dozen dwarf novae. Ideally, many more should be available in order to assist theoreticians to understand the accretion disc instabilities that cause the dwarf nova outbursts. Ironically, the most frequent outbursters, exemplified by V1193 Ori and ER UMa, have been recognised only in recent years, and so far only about 5 are known; this was a result of undersampling in earlier All-Sky photographic surveys. Many CVs do not have outbursts, but many of these non-eruptive CVs have irregularly spaced low states, often many magnitudes deep. More challenging to find are the nova-like variables without either outbursts or low states; these CVs in stable states nevertheless may have brightness variations of a few tenths of a magnitude and should be readily discoverable from their photometric unsteadiness alone in repeated All-Sky surveys. As with all unreddened CVs, they should also have UV excesses or distinctive colours.

All CVs are close binaries, transferring mass from a companion to a white dwarf, with orbital periods from tens of minutes to hundreds of days. Only 7 of the shortest period systems, the AM CVn stars, have been so far found, but it is likely that very many more brighter than 17th magnitude remain to be discovered; more than a hundred CVs with orbital periods in the range 1.5 hours to a few days are known; but only a few with giant secondaries and periods of hundreds of days have been found, and there are hundreds of CVs and suspected CVs that still have no certain classification or known orbital periods. Many CVs have brightness modulations at the orbital period - with amplitudes ranging from a fraction of a magnitude to eclipses many magnitudes deep. The SU UMa class of dwarf novae during superoutbursts have superhumps with slightly varying periods that are a few percent longer than the orbital periods. There is extensive evidence for precession of the accretion discs in many CVs, with periods of days. On shorter time scales, there are Dwarf Nova Oscillations and Quasi Periodic Oscillations in many dwarf novae during outbursts and in nova-like variables, on time scales of seconds to tens of minutes. These vary their periods rapidly and await really comprehensive monitoring, with round-the-Earth coverage. The problem has been the unpredictability of their appearance in any chosen CV, so collaborative monitoring has been difficult (for example, despite having been known for nearly 30 years, only recently have Dwarf Nova Oscillations in the optical and X-ray been observed simultaneously - in SS Cyg - and that was achieved by arranging Target of Opportunity time).

Turning to the 'SWOT' aspects. The strengths of CVs are that they are some of the most rapidly varying close binaries - several orbits of data may be collected in a night. They are also the objects in which accretion discs can best be studied - with a wide range of mass transfer rates $(\dot{M})$ and transitions between them. Accretion discs appear around young stars, in close binaries containing neutron stars and black holes, and in Active Galactic Nuclei, but it is in the CVs that much of the physics can most readily be seen in action. There are many other valuable aspects to CVs - a list is given in Warner (2000). The Weaknesses and Threats aspects probably can be combined into one category - that, CVs being stars, it is becoming more difficult to get observing time on large telescopes, and to get grant funding, to follow up the discoveries made with small telescopes: the extragalactic and extra-terrestrial planet bandwagons have strong lobbyists.

It is the Opportunities provided by All-Sky monitoring and by global co-operation among smaller telescopes that can provide a major step forward in understanding of CVs. Many of the brighter CVs are still being overlooked or are inadequately studied. The asymmetry between the numbers of CVs known in the northern and southern celestial hemispheres, and the large number of quite bright CVs for which orbital periods are unknown, illustrates this clearly. Even novae that reach apparent magnitudes 4 to 6 are being missed (Duerbeck 1990). The nova-like CVs, which do not have large outbursts to draw attention to themselves, are the most difficult to discover. They have been found mostly in surveys of ultraviolet-rich stars at higher galactic latitudes - these are still incomplete, especially in the southern hemisphere. Large numbers of nova-likes should be found in the all-sky photometric surveys (e.g. the Sloan Digital Survey) that are underway or planned. They have distinctive colours, but large amounts of small telescope time will be needed to classify all the CVs found this way. Detailed studies of the brighter nova-likes will provide the opportunity to have worked on a CV prior to its becoming a nova - many of the novae of the twentieth century were found (retrospectively from archived photographic plates) to have developed from relatively bright nova-likes that had been previously overlooked (e.g., V603 Aql 1918, pre-eruption magnitude 11.4; RR Pic

1925, 12.0; HR Del 1967, 11.9). Sometime soon a known nova-like is going to become a nova. Only by having studied all the bright nova-likes comprehensively will we be able to make a comparison of before and after (e.g., a small orbital period change, which shows how much mass and angular momentum are carried away by a nova eruption).

The opportunity for discovering CVs in All-Sky surveys is of considerable importance. A full inventory of CVs to some well-defined magnitude limit is necessary if we are to have reliable statistics on space densities of the various subtypes. This is important for population and evolution studies of binary systems in general. The CVs and their detached precursors are a significant channel of evolution for the shorter period (less than about 1 year) binaries. Getting agreement between observation and theory will involve better understanding of common envelope evolution and of the processes of angular momentum loss from binary orbits. This knowledge will spill over into other important studies, including, for example, the precursors of Type Ia supernovae.

Below I give an illustration of approximate space densities of some of the more common subtypes of CVs. Missing from any such estimates are the CVs that have temporarily stopped transferring mass, i.e., that are currently detached systems, being as a result only as bright as the sum of the luminosities of the stellar components. These, and systems with extremely low mass transfer rates that are both intrinsically faint and are very infrequent outbursters, may appear in large numbers at twentieth magnitude or fainter. To complete our picture of CV evolution the numbers of such objects must be known, and their orbital period distribution determined.

2. The Space Density of Cataclysmic Variables

Eventually, with a complete inventory of CVs and their distances, it will be possible to make accurate determinations of their space density. This is likely to progress in several stages. At present we are probably missing a large fraction of CVs - especially those currently in a low or zero state of mass transfer - so we can only provide lower limits to the density. Current or near future surveys should find all the active CVs to at least 20th magnitude. At present we must continue to use statistical methods to determine space densities. Here I will apply a method that I used 27 years ago (Warner 1974) which uses the concept of statistical robustness of the tenth brightest object in a group. This technique, introduced by Allen (1954) to deal with All-Sky Statistics of stars of selected types, in principle gives a good estimate of space density, but requires knowledge of the absolute magnitude and dispersion of the group, and involves allowance for interstellar absorption.

Knowledge of the absolute magnitudes of CVs has improved greatly in the past decade. There are period–absolute magnitude relations for dwarf novae at maximum, and better absolute magnitude–rate of decline relationships for erupting novae (e.g. Warner 1987, 1995). There are, however, selection effects in operation that must be evaluated with some care. First, I briefly describe the requirements for Allen's method. The space density D is given by

$$\log_{10} D = -2.62 - 0.6 m_v(10) + 0.6 M_v + 1.38(q - 0.6)\sigma^2 - \log_{10} S \quad (1)$$

where $m_v(10)$ is the (smoothed) magnitude of the tenth brightest object, M_v is the mean absolute magnitude of the class, q is the slope of the accumulative distribution (the graph of $\log N$ versus m_v, where N is the rank of the object), σ is the dispersion about M_v, and S is the interstellar extinction correction (obtainable from Allen's tables and graphs). The latter depends on the average galactic latitude and the average distance modulus (with allowance for dispersion) of the brightest stars of the group.

The values of $m_v(10)$ and q are found by plotting $\log N$ versus m_v for the brightest 20 or so objects and then fitting a straight line whose slope provides q and whose intercept with $m_v = 10$ gives the smoothed value for $m_v(10)$. For a sample uniformly distributed in transparent space, q takes the value 0.6.

In performing the analysis I have used the catalogue of CVs prepared by Downes, Webbink and Shara (1997; hereafter DWS). I have segregated CVs into the various subclasses, and have treated CVs at maximum and in quiescence separately. These are slightly different samples because, e.g., the ten dwarf novae that are brightest at maximum are not exactly the same as the ten that are brightest at minimum. For the most populous of the subclasses I discuss below the names of the stars selected. The state of CV classification and observation is such that even the choice of the brightest ten members of a subclass has some subjectivity in it.

Some comments on selection effects that influence the choice of sigma are required. At quiescence the accretion disc dominates the luminosity of almost all CVs. For all high $\dot{M}$ CVs (i.e., nova-likes and dwarf novae in outburst) the accretion disc certainly dominates. But the brightness of a disc depends strongly on the angle at which it is viewed, and the absolute magnitude–orbital period relation is provided only for discs viewed at a standard orbital inclination of 57° (Warner 1987). A disc viewed face on is 1.0 mag brighter than at 57°, and a disc viewed at 74° is 1.0 magnitude fainter than at 57°. There is therefore a strong selection effect against larger inclinations, which have much fainter magnitudes. This is seen in the lists of the ten brightest objects given below: only one of the nova-likes is an eclipsing system and only two of the U Gem stars and two of the SU UMa stars are eclipsing. Therefore, although high inclination is statistically more probable than low inclination, this is offset by the much greater brightness of the low inclination systems. By keeping U Gem stars and SU UMa stars separate the intrinsic spread of M_v caused by different disc radii (at different orbital periods) is greatly reduced. I assume, therefore, that the range in M_v for high $\dot{M}$ disc systems is probably quite small - a total range of M_v of 1.5 mag, or a sigma of 0.4 mag seems appropriate. At quiescence, however, all CVs show considerable spread in M_v (Warner 1995) and twice the dispersion at maximum is more appropriate. For novae at maximum there is a spread in magnitude caused by the grouping together of novae of different speed classes, and a sigma of 1.5 mag is adopted (Warner 1974).

The absolute magnitudes have been taken from the evidence presented in Warner (1995 - Figures 3.5, 3.10, 4.16, 5.3), and the assumption that the brightest magnitudes for SU UMa stars are those for superoutbursts, which are on average 1.0 mag brighter than normal outbursts. Care has to be taken with the nova-likes: the absolute magnitudes given in Warner (1987) and Warner (1995) are for average high state brightnesses, whereas the brightest magnitudes in the

DWS catalogue (and listed below for the top ten) are extrema. This only affects the adopted absolute magnitude by a few tenths. Before presenting the results of the application of Allen's technique I will discuss the choice of the brightest CVs in the three subclasses that contribute most to the space density.

3. The Brightest U Gem Stars

There are about 20 U Gem stars that reach brighter than magnitude 11.0 at maximum of outburst. Not unexpectedly, almost all of these have been known for half a century or more. It is possible, however, that a few infrequent outbursters may remain to be found. The most recent suspected bright candidates are Lib4 (in the DWS Catalogue), 10.0 at maximum (Debehogne 1990), which has been shown not to be a CV (Liu, Hu, Li & Cao, 1999) and there are two candidates in the Catalogue of Suspected variables: NSV 11280 (Aql), 10.4 at maximum, and NSV 7956 (UMi), 9: at maximum. Neither of the latter has been confirmed to be a U Gem star, even though discovered several decades ago.

Omitting these uncertain candidates, the ten brightest U Gem stars are:
SS Cyg (8.2 at max), U Gem (8.2), RU Peg (9.0), WW Cet (9.3), HL CMa (10.0), HP And (10.5), SS Aur (10.5), Z Cam (10.5), BV Cen (10.5) and YY Dra (10.6).

Of these, only HL CMa, which is very close to Sirius, has been discovered in relatively recent times.

4. The Brightest SU UMa Stars

As with the U Gem stars, we may expect that the majority of the brightest SU UMa stars are known, except for a few with WZ Sge characteristics - i.e. superoutbursts on time scales of decades. The ten brightest SU UMa types are:
WZ Sge (7.1 at max), VW Hyi (8.5), SW UMa (9.4), WX Cet (9.5), T Leo (10.0:), VY Aqr (10.3), EK TrA (10.4), YZ Cnc (10.5), BZ UMa (10.5) and CU Vel (10.7).

5. The Brightest Nova-likes.

Among the ten brightest nova-likes in the DWS catalogue are three from the Catalogue of Suspected Variables - namely NSV 2872 (mag 11.2 at maximum), NSV 3432 (mag 10.6), and NSV 13022 (mag 10.5). Of these, only one, NSV 2872, has been inspected spectroscopically and has been found not to be a nova-like (Liu & Hu 2000). I am reluctant to admit the other two NSV stars as nova-likes on the basis of their poorly recorded photometric behaviour alone. HM Aur, which is in the DWS catalogue as a nova-like, has been shown by Liu & Hu not to be a CV. This leaves the following as the brightest currently known:
IX Vel (9.1, 1985), TT Ari (9.5, 1975), UU Aqr (9.6, 1986), V3885 Sgr (9.6, 1972), RW Sex (10.4, 1972), QU Car (11.1, 1972), KR Aur (11.3, 1980:), RZ Gru (11.5, 1981), EC 04224-2014 (11.5, 1997) and V747 Cyg (11.7, 1993:).

The dates in brackets show the year in which the star was recognised as being a thick disc CV. As can be seen, many of the brightest nova-likes have been found in the past 20 years. No doubt there are many more, brighter than 12th magnitude, that will be found in ongoing and forthcoming All-Sky surveys. The magnitude of the tenth brightest nova-like was 12.5 in 1974, but is now 11.7 in the DWS catalogue; as pointed out above, these are extrema magnitudes, although this doesn't affect the ranking of the nova-likes.

We will see below that U Gem stars, SU UMa stars and nova-likes contribute about equally, and are the dominant suppliers of CV space density. Unlike the dwarf novae, therefore, further discovery of nova-likes is the area that will probably contribute to higher CV space densities. It is interesting that although many nova-likes (the intermediate polars) have been found from X-ray surveys, and are bright in the X-ray sky, none of them make it into the optical top ten; so it is unlikely that further bright nova-likes will be found from future X-ray surveys.

6. Results

The results of the analysis of the brightest members of the most populous CV subclasses are given in Table 1. Dwarf novae of the U Gem and SU UMa subclasses contribute about equally, with the U Gem class only slightly outnumbering the SU UMa class. The latter subclass is probably less completely sampled than the former - all of the dwarf novae with very long outburst intervals are SU UMa stars, and it is conceivable that a few more that reach 11th magnitude or brighter are still to be found. The nova-likes are about as populous as the U Gem stars and, as pointed out above, are probably still quite poorly sampled. It is with the nova-likes that we can expect the estimate of space density to be pushed up further. The space densities found here for the U Gem and SU UMa stars are about twice those given previously (Warner 1995); that for the nova-likes is similar to what was found before. The current estimate of the space density of optically active CVs, from Table 1, is $\sim 1.9 \times 10^{-6}$ pc^{-3}. To this must be added the space density of the strongly magnetic CVs (polars) found largely from X-ray surveys, which is $\sim 5 \times 10^{-7}$ pc^{-3} (Beuermann & Burwitz 1995). The other, minor contributors (novae, and the recurrent novae and intermediate polars not considered here), add at most about 20 percent to this value, so the total amounts to $\sim 3 \times 10^{-6}$ pc^{-3}.

This may be a considerable underestimate of the true CV space density, as has long been known (e.g., Patterson 1984). Although densities from the most comprehensive optical survey (the Palomar-Green Survey) also arrive at $3-6 \times 10^{-6}$ pc^{-3}, X-ray All-Sky surveys, which are particularly good at detecting the hard X-rays from systems of low M, give densities $\sim 1 \times 10^{-5}$ pc^{-3} (Patterson 1984, 1998) or even a factor of two higher (Warner 1995). Patterson (1998) discusses this problem, and the further one that for short period CVs there is an order of magnitude disagreement with the predictions of population synthesis. He concludes that there is a need for an intensive search for the faint CVs predicted by population synthesis with orbital periods around 80 - 100 minutes that have passed through the 'orbital period minimum' at about 80 min and have increasing orbital periods. If large numbers of these can be found (and these would be among the low M systems detected by X-ray surveys) it would go some way to reconciling the observed and predicted space densities and frequency of short orbital period systems. Such a program can only be carried out with the kinds of small telescope collaborations under consideration in this symposium.

Table 1.   Space Densities D for Cataclysmic Variable Subtypes

Class	M_v	q	σ	$m_v(10)$	b	$\log S$	$\log D$
Dwarf Novae (max)	4.4	0.36	0.4	10.6	27	-0.4	-6.0
Dwarf Novae (min)	8.0	0.36	0.8	14.2	25	-0.4	-6.2
SU UMa(max)	4.2	0.57	0.4	10.8	28	-0.3	-6.2
SU UMa(min)	9.2	0.35	0.8	16.2	28	-0.5	-6.5
Nova-likes	4.5	0.32	0.4	11.6	26	-0.7	-6.2
Na (max)	-8.8	0.25	1.5	2.8	10	-2.0:	-8.7:
Na (min)	4.3	0.27	0.4	15.6	10	-2.0:	-7.5:
Nb (max)	-6.0:	0.19	1.5	7.4	7	-2.5:	-9.4:
Nb (min)	4.3	0.18	0.4	17.6	7	-2.5:	-8.2:

7.   Concluding remarks

The Cataclysmic Variables provide exciting and extensive opportunities for All-Sky surveys and for small telescope collaborations on a variety of time scales. Large telescopes will not be used for these studies, which are more time consuming than photon limited. The importance of CVs for the investigation of the shorter period binary stars, and for stellar evolution in general, is enormous. The understanding that is emerging from the observation and modelling of mass transfer, accretion discs and their instabilities, accretion onto the primary, and nuclear runaways, has miscellaneous applications in areas as wide ranging as black hole and neutron star binaries, supernova explosions, and active galactic nuclei.

References

Allen, C.W. 1954, MNRAS, 114, 387

Beuermann, K. & Burwitz, V. 1995, ASP Conf.Ser. 85, 99

Debehogne, H. 1990, IAU Circ.No. 5131

Downes, R, Webbink, R.F. & Shara, M.M. 1997, PASP, 109, 345

Duerbeck, H.W. 1990, IAU Colloq.No. 122, p. 134

Liu, W. & Hu, J.Y. 2000, ApJS, 128, 387

Liu, W., Hu, J.Y., Li, Z.Y. & Cao, L. 1999, ApJS, 122, 257

Patterson, J. 1984, ApJS, 54, 443

Patterson, J. 1998, PASP, 110, 1132

Warner, B. 1974, MNASSA, 33, 21

Warner, B. 1987, MNRAS, 227, 23

Warner, B. 1995, Cataclysmic Variable Stars, Cambridge University Press

Warner, B. 2000, PASP, 112, 1523

(Top, from left) Lai, Meiszel, M. Y. Tsai, and S. I. Lin;
(bottom, from left) Gaustad, Jassur, and Saucedo

Small–Telescope Astronomy on Global Scales
ASP Conference Series, Vol. 246, 2001
W.P. Chen, C. Lemme, B. Paczyński

Optical and Near-IR Monitoring of Symbiotic Binary Systems

Joanna Mikołajewska

N. Copernicus Astronomical Center, Bartycka 18, 00716 Warsaw, Poland, e-mail: mikolaj@camk.edu.pl

Abstract. Symbiotic stars are long-period interacting binary systems in which an evolved red giant star transfers material to its much hotter compact companion. Such a composition places them among the most variable stars. In addition to periodic variations due to the binary motion, they often show irregular changes due to nova-like eruptions of the hot component. In some systems the cool giant is a pulsating Mira-type star usually surrounded by a variable dust shell. Here, I present results of optical and IR monitoring of symbiotic systems as well as future prospects for such studies.

1. Introduction

Most stars in the Universe are binaries. Among them, symbiotic stars are interacting binaries in which an evolved giant transfers material to a hot and compact companion. In a typical configuration, a symbiotic binary contains an M III giant and a white dwarf accreting material lost in the cool giant wind. The wind is ionised by the hotter of the binary components giving rise to symbiotic nebula (cf. Mikołajewska 1997).

Based on the near-IR colours, two distinct classes of symbiotic stars have been defined (Allen 1983): the S-type (stellar) with normal red giants, and the D-type (dusty) with Mira primaries surrounded by a warm dust shell. The distinction between the S and D types seems to be one of orbital separation: the binary must have enough space for the red giant, and yet allow it to transfer sufficient mass to its companion. In fact, all symbiotic systems with known orbital periods – of the order of a few years – belong to the S-type, while the orbital periods for the D-type systems are generally not known, probably because they are longer than periods covered by existing observations (cf. Belczyński et al. 2000).

Symbiotic stars are thus interacting binaries with the longest orbital periods and the largest component separations, and their study is essential to understand the evolution and interactions of detached and semi-detached binary stars. They are also among the brightest (intrinsically) stars, which makes them excellent observational targets both in our Galaxy as well as in nearby galaxies even for relatively small telescopes.

In the following, I will present results of optical and infrared monitoring of symbiotic stars as well as future prospects for such studies.

2. Variable Phenomena in Symbiotic Stars

The composition of a typical symbiotic binary, specifically the presence of an evolved giant and its accreting companion, places symbiotic stars among the most variable stars. They can fluctuate in several different ways, which can be revealed and studied by patient monitoring of their light curves and radial velocity changes. Namely, binary motion can be manifested by eclipses of the hot component by the giant, by modulation of the giant's light due to reflection effect (with orbital period) and due to tidal distortion (with $P_{orb}/2$), as well as radial velocity changes. The cool giant can also show intrinsic variability, in particular, radial pulsations (all D-type and some S-type systems) and semi-regular variations (S-type) with timescales on the order of months and years as well as long-term light variations due to variable obscuration by circumstellar dust (most D-type systems), solar-type cycles, spots, etc. The effects of mass accretion onto the hot component also involve different variable phenomena. The hot component in the vast majority of symbiotic systems seems in fact to be a luminous ($\sim 1000\,L_\odot$) and hot ($\sim 10^5$ K) white dwarf powered by thermonuclear burning of the material accreted from its companion's wind. Depending on the accretion rate, these systems can be either in a steady burning configuration or undergo hydrogen shell flashes. In many cases such flashes can last for decades due to the low mass of the white dwarf (Mikołajewska 1997). In addition, the hot components in many systems show activity with timescales of a few years which cannot be simply accounted for by the thermonuclear models. A possible and promising explanation involves fluctuations in mass transfer and/or accretion instabilities.

Below, I present examples of light curves for well-studied, though not yet completely understood, symbiotic binaries: RX Pup, CI Cyg and CH Cyg, which are representative for the wealth of variable phenomena observed in these systems.

2.1. RX Puppis: a Possible Recurrent Nova with a Symbiotic Mira Companion

RX Pup is a symbiotic binary composed of a long-period Mira variable pulsating with $P \approx 578$ days, surrounded by a thick dust shell, and a hot $\sim 0.8\,M_\odot$, white dwarf companion. The binary separation could be as large as $a \geq 50$ a.u. (corresponding to $P_{orb} \geq 200$ yr) as suggested by the permanent presence of a dust shell around the Mira component (Mikołajewska et al. 1999). In particular, the Mira is never stripped of its dust envelope, and even during relatively unobscured phases the star resembles the high-mass loss galactic Miras with thick dust shells. In general, the binary component separations in D-type systems must be larger than the dust formation radius. Assuming a typical dust formation radius of $\gtrsim 5 \times R_{Mira}$, and a Mira radius of $R_{Mira} \sim 1 - 3\,au$ (e.g. Haniff, Scholz & Tuthill 1995), the minimum binary separation is $a \gtrsim 20\,au$, and the corresponding binary period is $P_{orb} \gtrsim 50\,yr$, for *any* D-type system.

Recent analysis of multifrequency observations shows that most, if not all, photometric and spectroscopic activity of RX Pup in the UV, optical and radio range is due to activity of the hot component, while the Mira variable and its circumstellar environment is responsible for practically all changes in the infrared

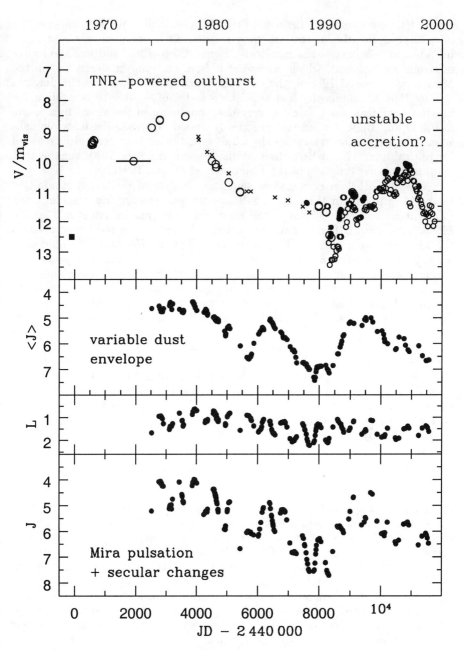

Figure 1. Optical and IR light curves of RX Pup from Mikołajewska et al. (1999). In the V/m_{vis} light curve, small open circles represent observations from RASNZ; large open circles and dots published V magnitudes; crosses FES magnitudes. The optical light is dominated by the hot component activity whereas the IR light curves are dominated by the Mira pulsation and variable dust obscuration of the Mira by circumstellar dust.

range (Mikołajewska et al. 1999, and Fig. 1). In particular, RX Pup underwent a nova-like eruption during the last three decades. The hot component contracted in radius at nearly constant luminosity from 1975 to 1986, and was the source of strong stellar wind, which prevented it from accreting material lost in the Mira wind. Around 1988/9, the hot component turned over the HR diagram and by 1991 its luminosity had faded by a factor of ~ 30 with respect to the maximum plateau value (see the very deep minimum in the visual light curve in Fig.1) and the hot wind had practically ceased. By 1995 the nova remnant started to accrete material from the wind, as indicated by a general increase of the optical flux. The earliest observational records from the 1890s suggest that another nova-like eruption of RX Pup occurred around 1894.

The near-IR light curves show significant long-term variations in addition to the Mira pulsation (Fig. 1). The long-term changes are best visible in the $\langle J \rangle$ light curve after removal of the Mira pulsation (middle panel in Fig. 1). Mikołajewska et al. (1999) have found large changes in the reddening towards the Mira accompanied by fading of the near IR flux. However, the reddening towards the hot component and the emission line regions remained practically constant and was generally less than that towards the Mira. These changes do not seem to be related to the orbital configuration nor to the hot component activity. Similar dust obscuration events seem to occur in many well covered symbiotic Miras (e.g. Whitelock 1998), as well as in single Miras (e.g. Mattei 1997, Whitelock 1998), and they are best explained as intrinsic changes in the circumstellar environment of the Mira variable, possibly due to intensive and variable mass loss. The last increase in extinction towards the Mira in RX Pup has been accompanied by large changes in the degree of polarization in the optical and red spectral ranges. This confirms that these long-term variations are driven by changes in the properties of the dust grains, such as variable quantity of dust and variable particle size distribution, due to dust grain formation and growth (Mikołajewska et al. 2001).

2.2. CI Cygni: a Tidally Distorted Giant with a Disc-accreting Secondary

Although most symbiotic binaries seem to interact by wind-driven mass loss, a few of them may contain a Roche-lobe filling giant. They also show activity with time scales on the order of years that can be related to the presence of accretion discs. Among them, CI Cyg is one of the best studied. Kenyon et al. (1991) demonstrated that it consists of an M5 II asymptotic branch giant, $M_g \sim 1.5 \, M_\odot$, and a $\sim 0.5 \, M_\odot$ hot companion separated by 2.2 AU. They also argued that the hot companion is a disc-accreting main sequence star. However, quiescent IUE data from the early 1990s can also be accounted for by a hot and luminous stellar source powered by thermonuclear burning which makes the case for CI Cyg as an accreting MS star less clear.

The outburst light curve of CI Cyg (Fig. 2) in addition to deep eclipses of the hot component by the red giant shows a $0.5 - 1.0$ mag oscillations with a period of $\sim 0.9 \, P_{orb}$. Some other S-type systems show similar secondary periodicities best visible in their outburst light curves, in all cases $10 - 15 \, \%$ shorter than the orbital periods. The nature of this secondary periodicity is unknown. Recently, Galis et al. (1999) suggested that in the case of AG Dra, it is due

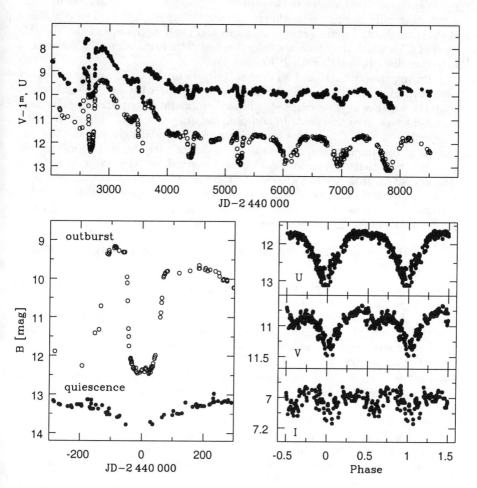

Figure 2.   Optical and near infrared light curves of CI Cyg. The *UBVI* data are from Belyakina (1979, 1984, 1991, 1992), Khudyakova (1989) and Meuninger (1981). Deep eclipses and the large outburst which started in 1975 are the most prominent features of the optical light curves (upper panel). During the outburst and its decline eclipses were narrow with well-defined eclipse contacts whereas the quiescent light curves show very broad minima and almost continuous sinusoidal variation (left panel). The ellipsoidal variability of the red giant is visible only in quiescent visual and red light curves (right panel).

to radial pulsations of the giant, and the outbursts are driven by resonances between the pulsations and binary motion. On the other hand, the secondary periodicities are best visible in the optical light where the contribution from the giant, especially during the outburst, is very low or negligible. There is also a striking similarity between these variations and the superhumps of the SU UMa class of CVs which may indicate that they are rather related to the presence of accretion discs (Mikołajewska & Kenyon 1992).

During the outburst and its decline, eclipses in the UBV continua and optical H II, He I and He II emission lines were narrow with well-defined eclipse contacts whereas at quiescence very broad minima and continuous nearly sinusoidal changes were observed. In addition, the quiescent VRI light curves show a modulation with $P_{orb}/2$ as expected for an ellipsoidal light curve. The amplitude of this modulation, $\Delta I \sim 0.15$ is consistent with the system inclination, $i \sim 73°$, and the mass ratio, $M_g/M_h \sim 3$ derived by Kenyon et al. (1991). The transition from narrow eclipses to sinusoidal variations was accompanied by large spectral changes and appearance of a radio emission with a spectral distribution that cannot be simply accounted for by any of the popular models for symbiotic stars (Mikołajewska & Ivison 2001).

2.3. CH Cygni: Triple or Binary System with a Magnetic White Dwarf

The record for the complexity of variable phenomena found in a single symbiotic object may be held by CH Cyg, the symbiotic system with the longest ($P_{orb} \sim 15.5$ yr) measured orbital period (Mikołajewski, Tomov, & Mikołajewska 1987; Hinkle et al. 1993), whoose light curves are presented in Fig. 3. Both the light curves and the radial velocity curves show multiple periodicities: a $\sim 100^d$ photometric period, best visible in the VRI light curves, which has been attributed to radial pulsation of the giant (Mikołajewski, Mikołajewska, & Khudyakova 1992), while the nature of the secondary period of $\sim 756^d$ also present in the radial velocity curve is not clear (Hinkle et al. 1993; Munari et al. 1996). There is a controversy about whether the system is triple or binary, and whether the symbiotic pair is the inner binary or the white dwarf is on the longer orbit. The near-IR light curves also show long-term variations similar to the dust obscuration phenomena found in symbiotic Miras (cf. Munari et al. 1996).

The hot component also shows very spectacular activity. In particular, we deal with irregular outbursts accompanied by fast, massive outflows and jets, rapid brightness variations with a time scale on the order of minutes, and other peculiarities which cannot be explained in the frame of the classical models proposed for symbiotic stars. Mikołajewski et al. (1990) proposed that this peculiar activity is powered by unstable accretion onto a magnetic white dwarf secondary.

3. Present State-of-the-art and Future Prospects

The recently published catalogue of symbiotic stars includes 188 symbiotic stars as well as 30 objects suspected of being symbiotic (Belczyński et al. 2000), Among them, 173 are in our Galaxy, 14 in the Magellanic Clouds and 1 in

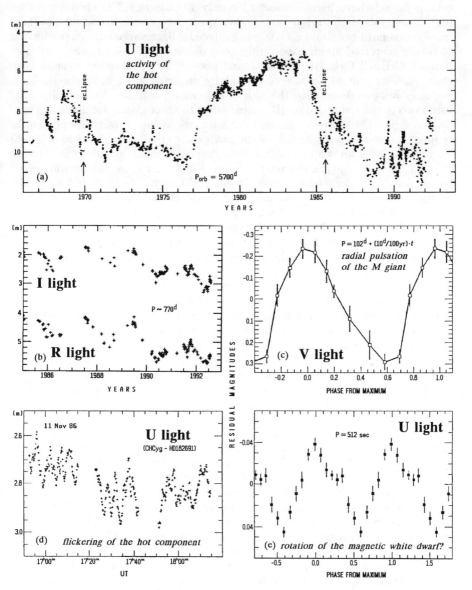

Figure 3. Variable phenomena in CH Cyg: (a) U light curve from the recent series of outbursts. Arrows indicate times of eclipses of the hot component. (b) R and I light curves of the M giant (c) Mean light curve for residuals of the V magnitude binned and folded with the 102^d-pulsation period of the M giant (Mikołajewski et al. 1992). (d) Rapid variability in U light. (e) Residual normal points of the data plotted in panel (d) folded with the 512-sec period (Mikołajewski et al. 1990).

Draco-1. They are excellent targets for small telescopes, especially for long-term monitoring of their complex photometric and spectroscopic variability. Although we have ~ 120 S-type symbiotic systems with $V \lesssim 15^m$, photometric orbital periods have been measured for only 30 objects (18 of then are eclipsing). Twentyone systems have also known spectroscopic orbits and for 8 of them mass ratios have been also estimated. Ellipsoidal light variations, characteristic of tidally distorted stars, have rarely been observed. Thus far, only four systems, T CrB, CI Cyg, BD$-21°3873$, and possibly EG And, seem to show such changes. The general absence of the tidally distorted giant in symbiotic binaries may however be due to the lack of systematic searches for the ellipsoidal variations in the red and near-IR range, where the cool giants dominate the continuum light. On the other hand, tidal interactions are certainly important in symbiotic systems as suggested by the practically circular orbits of most (~ 80 %) systems with known orbital solutions, and specifically of all those showing the multiple CI Cyg-type outburst activity. We do not know the orbital period for any of the extragalactic symbiotic stars, although 8 of them belong to the S-type, and with $V \sim 15 - 17$ mag, are bright enough for optical monitoring even with a relatively small telescope.

Similarly, among 33 galactic D-type systems ($K \lesssim 8^m$), pulsation periods have been observed – and thus the Mira presence confirmed – for only 12 systems. Pulsation periods are also unknown for the few extragalactic D-type systems ($K \sim 10 - 13^m$).

Optical and near-IR monitoring of symbiotic stars is essential not only to the understanding of variable phenomena in symbiotic stars and more generally long-period interacting binaries, but also to studying such phenomena in several other astrophysical environments (giant stars, planetary nebulae, novae, supernovae, supersoft X-ray sources, hot stars and even AGNs).

Studies of the symbiotic Miras are important for understanding the evolution and interaction of detached low-mass binaries. For example, there is ample observational evidence for systematic differences between the symbiotic Miras and average single galactic Miras. In particular, their average pulsation periods are longer, the colours redder and the mass-loss rates higher than typical periods, colours, and mass-loss rates for single Miras (cf. Mikołajewska 1999). It is interesting which and how these differences are related to the binary nature of symbiotic Miras. The symbiotic Miras are often associated with extended radio and/or optically resolved nebulae. These nebulae have usually very complex structure, often with bipolar lobes and jet-like features.

There are also many important questions posed by the active S-type systems. What powers the multiple outburst activity in CI Cyg and other similar systems? How many of these contain tidally distorted giants? What is the nature of the secondary periodicity, $\lesssim 0.8 - 0.9\,P_{orb}$, visible at outburst in some of them? Can the secondary periodicities be considered as evidence for the presence of an accretion disc? Such periodicities have not been found in any symbiotic nova during either optical maximum or constant luminosity phase (the plateau portion of white dwarf cooling tracks), including the best studied case – AG Peg (Kenyon et al. 1993), and their presence indicates that the outbursts in CI Cyg and other similar systems are not powered by thermonuclear reactions. The timescales and relative amplitudes for these eruptions are very similar to the timescales and amplitudes of the hot component luminosity variations (high and low states) in symbiotic recurrent novae (e.g. T CrB, RS Oph, RX Pup) between their nova eruptions (Anupama & Mikołajewska 1999; Mikołajewska et al. 1999) and in other accretion-powered systems (CH Cyg, Mira A+B). It is possible that the main difference between CI Cyg, Z And, AX Per, and other related symbiotic systems with multiple eruption activity and the activity of accretion-powered systems (symbiotic recurrent novae and CH Cyg) is that the hot component in the former burns more or less stably the accreted hydrogen whereas not in the latter. Systematic optical and near infrared monitoring with small telescopes can provide an answer to these and many other questions.

Finally, in 1990s several new symbiotic stars with $V \lesssim 15$ mag have been found. This means that systematic searches for symbiotic stars may in principle reveal many new objects with, say, $V \lesssim 20$ mag. Any red giant with blue/ultraviolet excess may be a good candidate. However, spectroscopic observations are essential to confirm symbiotic nature of such candidates.

Acknowledgments. I gratefully acknowledge Maciej Mikołajewski and Toma Tomov for providing Figure 3. I would also like to thank the LOC for their support. This research was partly funded by KBN Research Grant No. 5 P03D 019 20.

References

Allen, D.A. 1983, MNRAS, 204, 113

Anupama, G.C., Mikołajewska, J. 1999, A&A, 344, 177

Belczyński, K., Mikołajewska, J., Munari, U., Ivison, R.J., Friedjung, M. 2000, A&AS, 146, 407

Belyakina, T.S. 1979, Izv. KAO, 59, 133

Belyakina, T.S. 1984, Izv. KAO, 68, 108

Belyakina, T.S. 1991, Izv. KAO, 83, 118

Belyakina, T.S. 1992, Izv. KAO, 84, 45

Haniff, C.A., Scholtz, M., Tuthill, P.G. 1995, MNRAS, 276, 640

Hinkle, K.H., Fekel, F.C., Johnson, D.S., Scharlach, W.W.G. 1993, AJ, 105, 1074

Galis, R., Hric, L., Friedjung, M., Petrik, K. 1999, A&A, 348, 533

Kenyon, S.J., Oliversen, N.A., Mikołajewska, J., Mikołajewski, M., Stencel, R.E., Garcia, M.R., Anderson, C.M. 1991, AJ, 101, 637

Kenyon, S.J., Mikołajewska, J., Mikołajewski, M., Polidan, R.S., Slovak, M.H. 1993, AJ, 106, 1573

Khudyakova, T.N. 1989, PhD Thesis, Leningrad University

Mattei, J.A. 1997, JAAVSO, 25, 57

Meinunger, L. 1981, MVS, 9, 67

Mikołajewska, J. (ed.) 1997, Physical Processes in Symbiotic Binaries and Related Systems, Copernicus Foundation for Polish Astronomy, Warsaw

Mikołajewska, J. 1999, in Stecklum, B., Guenther, E., Klose, S., eds, Optical and Infrared Spectroscopy of Circumstellar Matter, ASP Conf. Ser., vol. 188, 291

Mikołajewska, J., Ivison, R.J. 2001, MNRAS, 324, 1023

Mikołajewska, J., Kenyon, S.J. 1992, AJ, 103, 579

Mikołajewska, J., Brandi, E., Hack, W., Whitelock, P.A., Barba, R., Garcia, L., Marang, F. 1999, MNRAS, 305, 190

Mikołajewska, J., Brandi, Garcia, L., Ferrer, O., W., Whitelock, P.A., Marang, F. 2001, in Szczerba R. et al., eds, Post-AGB Objects as a Phase of Stellar Evolution, Kluwer, in press, astro-ph/0103495

Mikołajewski, M., Mikołajewska, J., Khudyakova, T.N. 1992, A&A, 254, 127

Mikołajewski, M., Tomov, T., Mikołajewska, J. 1987, Ap&SS, 131, 733

Mikołajewski, M., Mikołajewska, J., Tomov, T., Kulesza, B., Szczerba, R., Wikierski, B. 1990, Acta Astr., 40, 129

Munari, U., Yudin, B.F., Kholotilov, E., Tomov, T. 1996, A&A, 311, 484

Whitelock, P.A. 1998, in Takeuri, M., Sasselov, D., eds, Pulsating Stars – Recent Developments in Theory and Observation, Universal Academy Press, Tokyo, 31

Small–Telescope Astronomy on Global Scales
ASP Conference Series, Vol. 246, 2001
W.P. Chen, C. Lemme, B. Paczyński

The Rotation and Variability of T Tauri Stars: Results of Two Decades of Monitoring at Van Vleck Observatory

W. Herbst

Astronomy Dept., Wesleyan University, Middletown, CT 06459 USA

Abstract. A brief history of photometric monitoring of T Tauri stars with a 0.6 m telescope at Wesleyan University is given. This is followed by discussion of three recent results: 1) the mass dependence of stellar rotation rates in the Orion Nebula Cluster, 2) a check on stellar radii estimates of pre-main sequence stars, and 3) the nature of a very peculiar object in NGC 2264.

1. Introduction and Brief History

T Tauri stars (TTS) are low mass (M < 2 M$_\odot$) pre-main sequence (PMS) stars with ages in the range of 0.1 to 10 My. They were originally identified as an interesting class of irregular variable stars by Joy (1942). The nature and cause of their variations remained largely unknown until the 1980's and is still mysterious in some aspects. Our photometric monitoring program, employing the 0.6 m (Perkin) telescope of Van Vleck Observatory, situated on the campus of Wesleyan University, began in 1981 with two principal goals. First, we wished to elucidate the nature of T Tauri variables which meant observing them on a regular basis. Second, we wanted to involve students in research and train them in some methods of observational astronomy. The program began with visual observations of objects in the Orion Nebula region and quickly moved to UBVRI and Hα photoelectric photometry of bright T Tauri and Herbig Ae/Be (HAEBE) stars after construction of an appropriate instrument, with the help of a grant from Research Corporation. Our first papers (Herbst, Holtzmann & Phelps 1982; Herbst, Holtzmann & Klasky 1983; Herbst & Stine 1984) established distinctions between what are now known as Type II (hot spot) variability and Type III (UXor) variability including discovery of the color "turnaround" at faint magnitudes which is characteristic of the latter.

In the early 1980's it was discovered that some T Tauri stars were not, in fact, irregular variables but at least some of the time showed distinctly periodic behavior (Schaefer 1983; Rydgren & Vrba 1983). The original interpretation proposed, that we were witnessing the rotation of a spotted star, has been verified by subsequent studies and is now well established. In most cases, the spots are cooler than the photosphere, although a few examples of hot spot periodicity have been found, including one star (DN Tau) which showed hot and cool spot periodicity simultaneously at one epoch (Vrba et al. 1986). Photometric variations caused by the rotation of a star with an asymmetric distribution of cool spots on its surface was dubbed "Type I" variability by Herbst et al. (1994)

and is now recognized as the most common cause of variations in PMS stars. Carpenter et al. (2001) estimate that 80% of the near-IR variations they detect in stars in the Orion Nebula Cluster (ONC) and surroundings can be accounted for by this mechanism. It appears to be the ONLY mechanism of variability operating in weak TTS (WTTS) - those lacking evidence for accretion disks - except for an occasional flare analogous to solar flares. The best studied star exhibiting this kind of variations is V410 Tau, which was a target of intensive study at Wesleyan. We showed that the star had two large, cool, long-lived spots in opposite hemispheres which drifted in longitude relative to each other, but persisted for at least seven years and maybe longer (Vrba, Herbst & Booth 1988; Herbst 1989). These results were basically confirmed by Doppler imaging studies in the early 1990's (Joncour, Bertout & Menard 1994; Rice & Streissmauer, 1996).

The recognition that it was possible to obtain rotation periods for many PMS stars by photometric monitoring led to a new class of studies which has persisted until today and continues to grow. Bouvier and collaborators applied the method to study rotation of WTTS and CTTS in associations such as Taurus using PMT's (Bouvier et al. 1986; Bouvier et al. 1993). We organized campaigns on some stars involving many observers (including some amateur astronomers) at a wide range of longitudes around the world. An early success of this program was the discovery of a rotation period for T Tauri itself (Herbst et al. 1986). However, it soon became clear that most of the bright CTTS and HAEBE stars were not going to yield periodicities easily, if at all. Unlike the WTTS, where dogged monitoring is usually rewarded with a period (e.g. Grankin 2001), the CTTS and HAEBE stars only rarely show significant periods (Herbst et al. 1987). Claims to the contrary by Bouvier et al. (1993) were based on a too-optimistic interpretation of noise peaks in periodograms (Herbst & Wittenmyer 1996; Rebull 2001). When reasonably priced, off-the-shelf CCD systems became available in the late 1980's we were able to switch from PMTs to these area detectors and began focusing attention at Wesleyan on young clusters for obvious reasons of efficiency.

A pilot study of the very center of the ONC, the so-called Trapezium cluster, with our original chip, a 5Kx5K device about 1 cm in size, immediately yielded the first (8) rotation periods for PMS stars in the ONC (Mandel & Herbst 1991). This was followed by a more extensive study in which we monitored 11 fields within the ONC using the small chip and discovered 35 rotation periods. Their frequency distribution was strikingly non-uniform and distinctly bimodal, with peaks near 2 days and 8 days and a clear gap at 4 days (Attridge & Herbst 1992). Further study confirmed this result, expanding the number of rotation periods to 75. We proposed an explanation for the bimodal period distribution in terms of the "disk-locking" mechanism of Königl (1991) and Ostriker & Shu (1995) (Choi & Herbst 1996; see also Edwards et al. 1993). It was a surprise, therefore, when Stassun et al. (1999) claimed, based on their wider field and deeper study of the ONC, that the period distribution was not bimodal. The issue was resolved by Herbst et al. (2000a) who showed that, in fact, only the brighter, more massive stars in the ONC - the ones included in the studies at VVO, which employed only the 0.6 m telescope - showed a bimodal distribution. Fainter, lower mass stars, which dominate the Stassun et al. (1999) sample have, in fact, quite different rotation properties as a group! The issue will be updated in the next section after a brief retreat from rotation issues to the general question of TTS and HAEBE star variability.

Probably the most active young star monitoring program in the world over the last decade and a half was run at Maidanak Observatory in Uzbekistan by V. Shevchenko of the Tashkent Astronomical Institute. Shevchenko and his colleagues, many of whom were students, obtained photoelectric photometry (UBVR) of hundreds of PMS stars each year using primarily 0.4 m telescopes. The excellent conditions at Mt. Maidanak (Ehgambendlev et al., 2000), including long strings of clear weather, aided this work. Supported by a grant from the Civilian Research and Development Corporation for States of the Former Soviet Union, we were able to collaborate with Shevchenko to monitor common objects at different longitudes and to make the Maidanak data easily available to the world through an ftp sight on the World Wide Web (ftp://ftp.astro.wesleyan.edu/pub/ttauri/). Analysis of these data has contributed to many papers on TTS variability, including some of our own (Herbst et al. 1994; Shevchenko & Herbst 1998; Herbst & Shevchenko 1999). Hopefully, the PMS monitoring programs at Mt. Maidanak will continue, led by Grankin, Ibraghimov and others, following the untimely death of Shevchenko.

2. Updating the ONC Rotation work

Work on rotation in the ONC, which began with Mandel & Herbst (1991), has expanded dramatically in the last few years (Choi & Herbst 1996; Stassun et al. 1999; Herbst et al. 2000a; Rebull 2001) spurred by new observational programs, wider field CCDs and some scientific controversy. As mentioned previously, results obtained by Stassun et al. (1999) on fainter, lower mass stars in the ONC and its surroundings seemed to be at odds with the primary result of Choi & Herbst, that a bimodal period distribution was present in the ONC. Herbst et al. (2000a) showed that the difference was primarily due to the different mass ranges probed in the studies - the bimodal distribution only applies to stars more massive than 0.25 $M_\odot$. Rebull (2001) surveyed a large area around the ONC (the so-called ''flanking fields'') and also found that the period distribution was not bimodal, although she did not have masses for most of her stars, so could not check on mass dependence issue. However, her work raises the possibility that bimodality is a feature only of the ONC itself and does not apply generally to a more heterogeneous set of PMS stars. This, in fact, is what one might expect from the disk-locking scenario since it requires rather special conditions - a set of very young stars most of which had been released from their disks only a short time (much less than 1 My) ago - to produce a bimodal distribution. A more heterogeneous set of disk release times would quickly eliminate bimodality from the period distribution as the stars contract rapidly and spin up conserving angular momentum on a short timescale (see Herbst et al. 2000a).

It is clearly important to consider stellar mass when discussing rotational evolution and to push the rotation studies in all clusters to the lowest possible masses. We have initiated a program in collaboration with R. Mundt of Max-Planck-Institute for Astronomy in Heidelberg, Germany, to study the ONC with the recently commissioned Wide Field Imager on the 2.2 m telescope at the European Southern Observatory in Chile. Preliminary results were reported by

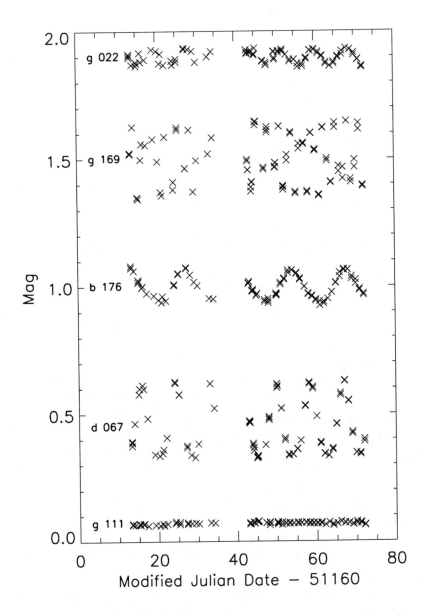

Figure 1.   Some examples of periodic light curves of TTS in the ONC obtained with the 2.2 m telescope at the European Southern Observatory in collaboration with astronomers at the Max-Planck-Institute in Heidelberg, Germany (Herbst et al. 2000b). The bottom light curve is a non-variable star shown as a comparison object.

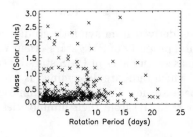 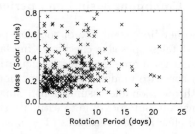

Figure 2.   Rotation Period versus Mass for stars in the ONC. The left hand panel shows the full range of the data, while the right hand panel enlarges the well-populated region of mass less than 0.8 M$_\odot$.

Herbst et al. (2000b) and are briefly updated here. A full discussion is in preparation. Rotation periods were determined for more than 400 stars in the ONC. Comparison with previous studies mentioned above shows that there were 111 stars in common and 99 of these yielded identical periods. In all but one of the remaining cases, the period differences were consistent with either aliasing (with a one day observing interval) or period doubling (caused by the existence of two spots in opposite hemispheres at some epochs). This is a gratifying affirmation of the methods used to identify periodic stars and of the basic interpretation of the periodicity in terms of rotation.

A sample set of light curves for periodic stars (plus one comparison object) is shown in Figure 1. The period distribution as a function of mass is shown in Figure 2, where the left panel shows the full data set and the right panel shows an enlarged view of the important region below 0.8 M$_\odot$. A clear trend can be seen in the sense that the median rotation period decreases with stellar mass among the lower mass stars. The gap around 4 days which exists for stars more massive than 0.25 M$_\odot$ is also readily apparent. Histograms showing these features more clearly are shown by Herbst et al. (2000b).

What can it mean that, within the mass range considered here, the lower mass stars in the ONC rotate more rapidly than those of higher mass? A full discussion is not possible within the space constraints of this review. Indeed, the answer to this question is unknown at present. For the purposes of this discussion, therefore, only one speculative possibility will be mentioned. If disks control the spin-up of stars during the accretion stage then it may be that the faster rotation of the lower mass stars indicates less action by their disks in slowing the spin-up. This, in turn, may simply mean that lower mass stars have lower mass disks which cannot control stellar rotation to the same extent as their higher mass counterparts. The issue is of particular relevance to planet formation scenarios since reasonably long-lived disks of sufficient mass are required for planet formation. If low mass stars (by far the most common in the galaxy) do not have such disks, then the large majority of stars in the galaxy may not have planets. Clearly these issues require a great deal more analysis before they can be settled; our purpose here was to illustrate one important issue to which monitoring programs with small telescopes may speak.

3.   Combining v sin i and period

Another important use of rotation periods derived from synoptic studies with small telescopes is in checking a fundamental property of PMS stars - their radii. Stellar radii are essential for calculating masses and ages of PMS stars by comparison with models. Yet, the method by which radii are determined is fraught with questionable assumptions. Since there are few PMS eclipsing binaries it is usually only possible to estimate stellar radii (R) from their luminosities (L) and effective temperatures (T_{eff}), as

$$ R = \frac{L^{\frac{1}{2}}}{(4\pi\sigma)^{\frac{1}{2}} T_{eff}^2}. $$

A fundamental assumption of this method is that the stars involved radiate isotropically. However, we know this cannot be strictly true for most T Tauri stars, since they exhibit optical and near-IR variations of 10-20% or more as the stars rotate - this is how their rotation periods are determined! Effective temperatures are usually assigned by comparing the spectra of these heavily magnetized, spotted and sometimes veiled TTS with spectra of normal dwarfs and giants leading to a classification on the MK system, although usually without a luminosity class. Then, a spectral type - T_{eff} calibration based on normal stars is invoked. There is debate in the literature about whether it is best to compare TTS with dwarfs or with giants in assigning an effective temperature, a decision which can affect the adopted T_{eff} by up to a few hundred degrees. Distances to the stars and extinction corrections also factor into the inferred radii through the luminosity. Clearly it would be of value to check whether results obtained by this method are reliable.

Fortunately, this can be done if one has a sufficiently large sample of stars which have rotation periods (P) and v sin i measurements. Assuming a random distribution of inclination angles (i) of the rotation axes to the line of sight, one expects that,

$$ < sin i >= \frac{\pi}{4} = < \frac{v sin i}{v_{eq}} > $$

where $v_{eq} = 2\pi R/P$ is the equatorial rotational velocity based on a star's rotation period and radius. Rhode, Herbst & Mathieu (2001) have recently determined v sin i values for about 250 stars in the ONC, of which more than half have measured rotation periods. They find a substantial discrepancy between results obtained by these two independent methods of determining radii. The sense of the discrepancy is that the method based on effective temperatures yields radii that are systematically larger, particularly for stars of spectral class M0-M3. It is impossible to summarize here all of the possible causes and implications of this result - the reader is referred to the original paper. It is simply noted that there may well be a problem with the current method of assigning T_{eff} which is resulting in these stars being considered too cool by up to 600K and, therefore, too low a mass by up to 60%. If true, this also means that age estimates for PMS stars may be somewhat in error, complicating attempts to test various theoretical models.

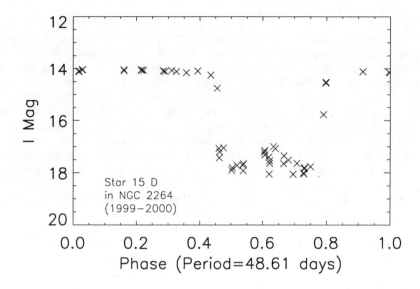

Figure 3.   Light Curve of 15D. The period of 48.61 days applies only
to the data obtained during the 1999/2000 season.

4.   A bizarre and potentially important object

The last point I wish to make in this review, is that monitoring programs with
small telescopes may also turn up rare and potentially important objects which
can be marked for intensive study using larger instruments if necessary. An
example is the star known as 15D in NGC 2264 (Kearns et al. 1997; Kearns &
Herbst 1998). The light curve for this peculiar object during the most recent
observing season is shown in Figure 3. It is clearly an eclipsing system with a
period of 48.35 days but an extremely unusual, in fact, unique (to the author's
knowledge) light curve. During eclipse, the star fades by about 3 magnitudes
and stays in eclipse for about 16 days, or 1/3 of its period! A recently obtained
spectrum of the system during minimum light (Hamilton, 2001) indicates that
the spectrum, which is typical of a K7 star at maximum light, is not significantly different during eclipse. Clearly, the eclipse must be caused by a large
dust clump orbiting the K7 star, either on its own or attached to a secondary
companion, perhaps as a circumstellar disk. In either case, the optical variations
of the system during primary minimum must contain information on the density
distribution of the dust clump as it passes across our line of sight. Detailed study
of this system is under way and promises to reveal, for the first time, information
on the distribution of dust in a circumstellar disk at much higher spatial resolution than has ever been obtained. Without dedicated monitoring programs this
unique and potentially important object might not have been found for many
years to come.

5. Final Comment on the Educational Value of this Monitoring Program

It is sometimes argued that only robotic telescopes should be used for monitoring programs in the future because of their efficiency. I argue that, if this were to be the case, it would be a loss for science, because small telescopes and monitoring programs such as the one at Wesleyan University play an important role in education. The problem with robots for teaching is that they are remote, black boxes to students, who may have no idea about the details of where the data are coming from or how they are obtained. By contrast, when they acquire the data for themselves using semi-automated telescopes on campus, they gain a greater appreciation of what one needs to be careful about (e.g. clouds, bad seeing, proper dome alignment, stray light, the Moon, air mass effects, instrumental problems, telescope tracking or pointing problems, etc.) and how much faith to put in their data and its interpretation. If students only use robotic data, who will build the next generation of robots? I believe that the teaching function of small telescopes on campuses, which can involve students closely in actual observations should not be ignored and, in some cases, outweighs the potential efficiency gains of robotic systems. I hope that there will continue to be a place in astronomy for small telescopes engaged in monitoring programs which involve students for pedagogical purposes.

Acknowledgments. This paper is dedicated to the memory of my colleague and friend, Valery Shevchenko, who led the impressive photometric monitoring efforts of a group based at Mt. Maidanak in Uzbekistan for so many years. His untimely death at such a young age was a real loss for astronomy. I wish to thank the many collaborators who have contributed enormously to the work described here. Chief among them are Fred Vrba (USNO), Lynne Hillenbrand (CalTech), Katherine Rhode (Yale), Eric Williams (Wesleyan), Reinhard Mundt (MPIA-Heidelberg), Coryn Bailer-Jones (MPIA-Heidelberg), Bob Mathieu (Wisconsin), and Mansur Ibraghimov (Ulug Bek Astron. Inst., Tashkent). This work would not have been possible without the energy, enthusiasm and talents of more than 50 Wesleyan students and Keck Northeast Astronomy Consortium summer students who have participated in the research over the years. I would particularly like to thank the following (current affiliations given in parentheses): Jon Holtzmann (NMSU), John Booth (South Pole), John Filhaber, Jody Attridge (MIT Haystack), Peter Stine (Bloomsburg), Nancy Eaton, Philip Choi (UC-Santa Cruz), Lisa Frattare (STScI), Aaron Steinhauer (Indiana), Kristin Kearns, Kathy Rhode (Yale), Gillian Curran (Wellesley), Andrew Rhodes (Wesleyan), Candice Shih (Haverford) and Arianne Donar (Wesleyan). I would also like to acknowledge the financial support over the years from many sources, including Research Corporation, the Perkin Fund, NSF, NASA, USCRDF, Sigma Xi, Dudley Observatory, the W. M. Keck Foundation and Wesleyan University.

References

Attridge, J. A. & Herbst, W. 1992, ApJ 398, L61

Bouvier, J., Bertout, C., Benz, W., & Mayor, M. 1986, A&A 165, 110

Bouvier, J., Cabrit, S., Fernandez, M., Martin, E. L., & Matthews, J. M. 1993, A&A 101, 485

Carpenter, J. M., Hillenbrand, L. A. & Skrutskie, M. F. 2001, preprint

Choi, P. I. & Herbst, W. 1996, AJ 111, 283

Edwards, Suzan, Strom, Stephen E., Hartigan, Patrick, Strom, Karen M., Hillenbrand, Lynne A., Herbst, William, Attridge, Joanne, Merrill, K. M., Probst, Ron, Gatley, Ian 1993, AJ 106, 372

Ehgamberdiev, S. A., Baijumanov, A. K., Ilyasov, S. P., Sarazin, M., Tillayev, Y. A., Tokovinin, A. A., & Ziad, A. 2000 A&A 145, 293

Grankin, K. N. 2001, private communication

Hamilton, C. H. 2001, private communication.

Herbst, W. 1989, AJ 98, 2268

Herbst, W., Bailer-Jones, C. A. L., Mundt, R., Meisenheimer, K. and Wackermann, R. 2000b, A.S.P. Conf. Series, "Planetary Systems in the Universe: Observation, Formation and Evolution", A.J. Penny, P. Artymowicz, A.-M. Lagrange, & S.S. Russell, eds., in press

Herbst, W., Booth, J. F., Chugainov, P. F., Zajtseva, G. V., Barksdale, W., Covino, E., Terranegra, L., Vittone, A. & Vrba, F. 1986, ApJ 310, L71

Herbst, W., Booth, J. F., Koret, D. L., Zajtseva, G. V., Shakhovskaya, H. I., Vrba, F. J., Covino, E., Terranegra, L., Vittone, A., Hoff, D., Kelsey, L., Lines, R., & Barksdale, W. 1987, AJ 94, 137

Herbst, W., Herbst, D. K., Grossman, E. J. & Weinstein, D. 1994, AJ 108, 1906

Herbst, W., Holtzman, J. A. & Phelps, B. E. 1982, AJ 87, 1710

Herbst, W., Holtzman, J. A. & Klasky, R. S. 1983, AJ 88, 1648

Herbst, W. & Shevchenko, V. S. 1999, AJ 118, 1043

Herbst, W. & Stine, P. C. 1984, AJ 89, 1716

Herbst, W., Rhode, K. L., Hillenbrand, L. A., & Curran, G. 2000a, AJ 119, 261

Herbst, W. & Wittenmyer, R. 1996, B.A.A.S. 189, 4908

Joncour, I., Bertout, C., Menard, F 1994, A&A 285, L25

Joy, A. H. 1942, PASP 54, 15

Kearns, K. E., Eaton, N. L., Herbst, W. & Mazzurco, C. J. 1997, AJ 114, 1098

Kearns, K. E. & Herbst, W. 1998, AJ 116, 261

Königl, A. 1991, ApJ 370, 39

Mandel, G. N. & Herbst, W. 1991, ApJ 383, L75

Ostriker, E. C. & Shu, F. H. 1995, ApJ 447, 813

Rebull, L. M. 2001, in press

Rhode, K. L., Herbst, W. & Mathieu, R. 2001, AJ in preparation

Rydgren, A. E. & Vrba, F. J. 1983, ApJ 267, 191

Schaefer, B. E. 1983, ApJ 266, 458

Shevchenko, V. S. & Herbst, W. 1998, AJ 116, 1419

Stassun, K. G., Mathieu, R. D., Mazeh, T., & Vrba, F. J. 1999, AJ 117, 2941

Rice, J. B. & Streissmauer, K. G. 1996, A&A 316, 164

Vrba, F. J., Herbst, W. & Booth, J. F. 1988, AJ 96, 1032

Vrba, F. J., Rydgren, A. E., Chugainov, P. F., Shakovskaia, N. I., & Zak, D. S. 1986. ApJ 306, 199

Small–Telescope Astronomy on Global Scales
ASP Conference Series, Vol. 246, 2001
W.P. Chen, C. Lemme, B. Paczyński

High-Speed Photometry of Bright roAp Stars With Small Telescopes

D. W. Kurtz

*Centre for Astrophysics, University of Central Lancashire, Preston
PR1 2HE, UK,
Department of Astronomy, University of Cape Town, Rondebosch 7701,
South Africa, and
Laboratoire d'Astrophysique, Observatoire Midi Pyrenees,
14 Av. Eduoard Belin, 31400 Toulouse, France*

Abstract. The rapidly oscillating Ap stars are magnetic peculiar A stars which pulsate in multiple p modes with periods in the range of about 6 to 16 minutes with their oscillation axes aligned with the oblique magnetic axes of the stars. Some of these stars have the richest frequency spectra of any non-degenerate stars other than the sun. This paper shows how photometric observations using small telescopes can be used to work on several astrophysically interesting problems posed by these stars. An example of high precision photometry is shown. The proof of oblique dipole pulsation, the distortion of pulsation modes (probably by the magnetic field), and the determination of asteroseismic luminosities are all discussed. The latter, especially when combined with new theoretical developments concerning magnetic field-pulsation interaction, suggests that Ap stars have lower effective temperatures and/or smaller radii than has been previously thought. It is pointed out that this may be related to the recently discovered extreme discrepancy in effective temperature determined from the wings and cores of the $H\alpha$ line.

1. Introduction

Small telescopes have one great advantage over large telescopes: large amounts of time are available on them. This means that research projects which demand extended time coverage can in practice *only* be done with small telescopes. Small telescopes have a further advantage over large telescopes in that extended coverage campaigns – observing runs which need multisite longitude coverage to avoid temporal gaps in the data – are much more easily organised with small telescopes where the demands for time are more relaxed. Papers in this volume demonstrate both of these advantages in a variety of ways: for all-sky monitoring, for temporal coverage of variable objects, for supernova searches and gamma-ray-burst optical identification; for the search for extra-solar planets; for the study of near-earth objects and Kuiper-belt objects; for microlensing observations; and, more and more importantly, for education and public understanding of science.

187

Many of these new uses of small telescopes are revolutionary, such as the tens of thousands of variable stars discovered and studied in detail by the MACHO (Massive Compact Halo Object) survey. But one of the main, traditional uses of small telescopes is alive and well – the photometric study of pulsating variable stars. Here are some examples: There are now hundreds of Slowly Pulsating B (SPB) stars known (see Balona 2000; Dziembowski 1998; Aerts, Waelkens & de Cat 1998) and nearly as many of the recently discovered γ Dor stars (Kaye et al. 2000). Both of these classes are pulsating in g modes which are particularly useful for probing conditions in the stellar cores. Helioseismologists have long sought even a single g mode in the sun (unsuccessfully), so the prospect of the asteroseismic study of g modes in SPB and γ Dor stars is exciting and, with multiple pulsation periods in the range of 1–3 days typically, only small telescopes can provide the long-term photometric data needed to determine their frequency spectra. Similarly, asteroseismological studies of g modes in white dwarfs (Kawaler 1998) and p modes in δ Sct stars (Handler 2000; see also papers in the proceedings edited by Breger & Montgomery 2000) are the preserve of small-telescope astronomy.

My own research interest is primarily in rapidly oscillating Ap (roAp) stars which, like other examples just given, are stars which can be studied intensively photometrically with small telescopes. In the next section I will give some brief background material on these stars and why they are interesting, and give some examples of research on them that has been done with small telescopes.

2. roAp Stars With Small Telescopes

There are many types of peculiar stars on the upper main sequence. I have discussed the nomenclature for these stars in detail (Kurtz 2000) and recommend that paper for readers who wish to know more about the various classes and subclasses of peculiar stars, and the convoluted terminology associated with them. In this paper I will discuss the rapidly oscillating Ap (roAp) stars which are mid-A to early-F main sequence Ap SrCrEu stars which pulsate in high-overtone p modes with periods in the range 6–16 min and peak-to-peak amplitudes $\Delta B \leq$ 0.016 mag – generally much less. They mostly lie within the δ Sct instability strip, but a few of them are cooler. Although they largely overlap with the δ Sct stars, it now seems probable that the driving mechanism for their pulsation is H I ionisation (Dziembowski & Goode 1996; Cunha et al. 2001), rather than He II ionisation as in the δ Sct stars.

The roAp are oblique pulsators: they pulsate in non-radial p modes with their pulsation axes aligned to their magnetic axes which are themselves inclined to the rotation axes of the stars. This is a unique characteristic not known in any other pulsator, and it allows us to see the non-radial mode from a varying viewing angle. That provides more information about the mode type – its degree ℓ and order m – than can be obtained for other types of pulsating stars. For much more detail about the roAp stars and their interpretation, see the extensive recent review by Kurtz & Martinez (2000), or earlier reviews and discussions by Weiss (1986), Shibahashi (1987, 1990), Kurtz (1990), Matthews (1991), Martinez, Kurtz & Kaufmann (1991), Martinez (1993), and Martinez & Kurtz (1994a,b; 1995).

2.1. High-Precision, High-Speed Photometry of roAp Stars

Some of the highest precision photometric observations ever made have been on the roAp stars. The technique used to observe these stars is high-speed photometry which uses continuous integrations without interruptions for comparison star observations. This is not possible for observing lower-frequency variable stars because of variations in the transparency of the Earth's atmosphere, but at the frequencies of the pulsations of the roAp stars (equivalent to periods in the 6 to 16-min range) the source of noise is predominantly atmospheric scintillation which has a flat frequency spectrum in this range (when the observations are made under excellent photometric conditions). The roAp stars are also bright, so photon statistical noise is less than the scintillation noise.

Fig. 1 shows an example of the very high precision that is possible using small telescopes. This is a continuous 12.5-hr light curve of the roAp star HR 3831 using the Cerro Tololo Interamerican Observatory (CTIO) 0.6-m telescope and the South African Astronomical Observatory (SAAO) 0.75-m telescope. It is typical of the quality of data obtainable from good photometric sites using very small telescopes. Low-frequency sky transparency variations have been filtered from the data; this is why I use the term *high precision* to describe the data, not *high accuracy*. Although the relative variations shown are precisely known, the absolute brightness of the star is poorly known because of the lack of the use of comparison and standard stars.

The light curve shown in Fig. 1 was part of a 17-day, two-site campaign on HR 3831 by Kurtz, Kanaan & Martinez (1993 – KKM) from CTIO and SAAO. Their data set could only be obtained on small telescopes, given the length of the campaign. A low-resolution Fourier Transform of their data is shown in Fig. 2. There the fundamental frequencies (there actually 7 of them – see KKM) and their harmonics can be seen. The level of the *highest* noise peaks is less than 0.1 mmag and the formal error on the amplitudes of the pulsation modes is formally 9 μmag (KKM). This high precision has been used to show that the pulsation mode of HR 3831 cannot be described as a single normal mode, and that, in turn, has been the impetus for many theoretical investigations of the way in which the magnetic field and the pulsation modes interact in roAp stars (Shibahashi 1986, 1987, 1990; Shibahashi & Takata 1993; Takata & Shibahashi 1995; Dziembowski & Goode 1996; Cunha & Gough 2000; Bigot et al. 2000).

2.2. A Distorted Dipole Mode

HR 3831 pulsates in a single mode which is a distorted oblique dipole mode – probably distorted by the magnetic field, although that is not yet understood. Fig. 3 shows the way the pulsation amplitude and phase vary with the 2.851976-d rotation period of the star (Kurtz et al. 1997). Each point in Fig. 3 represents, typically, one hour of observation covering about 5 pulsation cycles. Kurtz et al. (1997) had a long-term monitoring program for HR 3831 where they observed it as often as possible throughout its observing season for years with the SAAO 0.5-m telescope. Thus the astrophysical information gleaned from Fig. 3 was obtained from observations with a very small telescope indeed.

From a broad-band linear polarisation study of the magnetic field in HR 3831 we know that the rotational inclination, i, is near 90° (Bagnulo, private communication; see Landolfi et al. 1997 for a discussion of the technique for constraining

HR3831 JD2448313 B 40

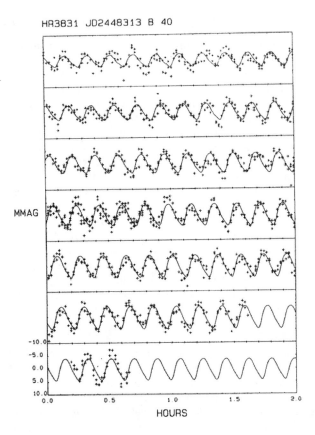

MMAG

HOURS

Figure 1.   A light curve of HR 3831 taken on JD2448313 with the
CTIO 0.6-m and SAAO 0.75-m telescopes. Each panel is two hours
long; the entire light curve spans about 12.5 hr. Notice in the fourth
panel that the data from the two observatories overlap and agree well.
Each data point represents 40-s integration, an average over four of
the original 10-s integrations used at the telescope. This is a typical
procedure: using 10-s integrations means that less data is lost when bad
points are discarded, but averaging to 40-s integrations filters some of
the very high frequency noise to give a better presentation. It also
reduces computing time when analysing the data. Low frequency noise
caused by sky transparency variations has been filtered so there is no
long-term variation in the light curve, but this has no effect on the
time scale of the variations seen here. It does help with the visual
comparison of the frequency solution (solid line) to the actual data
points. These data are typical of the kind of data that can be obtained
on roAp stars with small telescopes under good conditions and with
well-functioning photometers. (from KKM 1993)

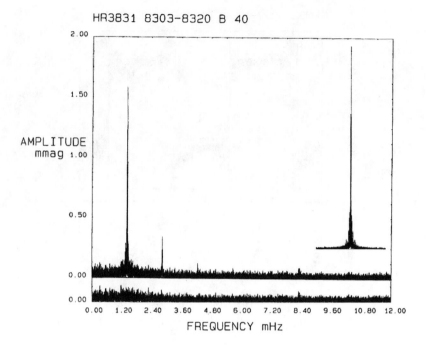

Figure 2.   This amplitude spectrum shows that an excellent signal-to-noise ratio can be obtained for pulsation amplitudes less than a mmag with small telescopes at the frequencies of pulsation of the roAp stars. The noise level of the highest noise peaks in this figure are about 0.1 mmag and decrease with increasing frequency. This high precision, which is obtainable with small telescopes for bright rapidly oscillating stars (not just roAp stars), means that there are many astrophysically interesting observations to be done with small telescopes and photoelectric photometers. (from KKM 1993)

the magnetic field geometry), and its magnetic obliquity, β, is small – Bagnulo estimates $\beta \sim 8°$, whereas I argue for a value nearer 30-40° from consideration of the pulsation amplitude spectrum. This disagreement is yet to be resolved. Whatever the value of β, with i near to 90° we see both magnetic poles as the star goes through its rotation cycle. This means we see both pulsation poles, too, since the magnetic and pulsation poles are aligned. With that in mind, Fig. 3 clearly demonstrates that the pulsation mode is a dipole mode. Rotation phase 0.0 is defined to be the time of pulsation maximum. At that time the pulsation semi-amplitude is about 4 mmag, as can be seen in the bottom half of Fig. 3. As HR 3831 rotates, the amplitude of the pulsation drops until the star is seen in quadrature where the amplitude is either zero, or very close to it. This is also the phase of magnetic crossover, and it occurs at rotation phase 0.25 because $i \approx 90°$. As the opposite magnetic and pulsation pole comes into view, the amplitude again increases, and the pulsation phase reverses by 180°,

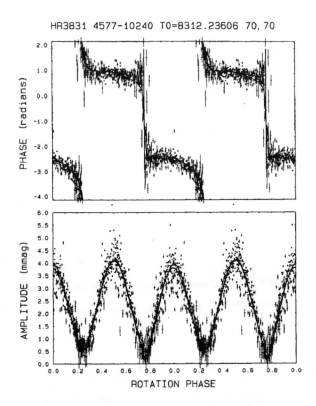

Figure 3.   This diagram plots the pulsation amplitude (bottom) and phase (top) for HR 3831 over its 2.851976-d rotation period. Two full rotation cycles are shown. Each point represents typically one hr, or 5 pulsation cycles, of data. This diagram clearly demonstrates that HR 3831 pulsates in a distorted dipole mode. (from Kurtz et al. 1997)

as can be seen in the upper panel of Fig. 3. This form of amplitude modulation with the 180° phase reversal is a clear signature of a dipole pulsation mode.

    From a study by Pekeris (1938) it was long thought that dipole pulsation modes were unphysical for pulsating stars because the geometric centre of the star undergoes an oscillation in space, apparently in disagreement with simple Newtonian physics.   Christensen-Dalsgaard (1976) showed that the centre of mass of a stellar dipole oscillator is not displaced, hence such modes are possible for stars (although not for an incompressible sphere such as the Earth).   We now know that the sun, some white dwarfs, and roAp stars do pulsate in dipole modes. Fig. 3 is the clearest demonstration of that, and it is largely the result of work with an 0.5-m telescope.

    The mode in HR 3831 is not purely dipolar, however.   It cannot be described by a single spherical harmonic; that is, it is not a normal mode. Kurtz (1992) demonstrated that the mode can be decomposed into a spherical harmonic series with $\ell = 0, 1, 2, 3$ components. Shibahashi & Takata (1993) and

Takata & Shibahashi (1995) used that as a basis to examine first the effects of the distortion of the pulsation eigenfunction by the magnetic field, and then the effects on the pulsation mode of a quadrupolar component to the magnetic field. Their results came much closer to matching the observations that previous theory had been able to, but complete agreement was not obtained. This possibly means that the magnetic field is not itself describable as a simple decentred dipole (which is what using a dipole plus quadrupole gives, to first order). Hence, again, these observations obtained with an 0.5-m telescope are initiating and constraining our understanding of both the magnetic field and pulsation geometry and the complex interaction between them. Many more detailed studies of other roAp stars are needed, and only with small telescopes will the large amounts of observing time needed be available.

2.3. Asteroseismic Luminosities

The pulsation modes of the roAp stars are high-overtone p modes that fit an asymptotic relation which describes the frequency separations as a function of the radial overtone, n, and the spherical harmonic degree, ℓ:

$$\nu_{n\ell} = \Delta\nu_0 \left(n + \tfrac{\ell}{2} + \varepsilon\right) + \delta\nu \tag{1}$$

where

$$\Delta\nu_0 = \left(2 \int_0^R \frac{dr}{c(r)}\right)^{-1} \tag{2}$$

(Tassoul 1980, 1990).

In eq. 1 $\nu_{n\ell}$ is the frequency of a mode with radial overtone n and spherical degree ℓ, ε is a constant of order 1 depending on the stellar structure, and $\delta\nu$ is a second-order term which depends on the central condensation of the star. The "large spacing", $\Delta\nu_0$, is the inverse of the sound travel time across the star; for the roAp stars it can be used to determine an asteroseismic luminosity. This is done by comparing the observed value of $\Delta\nu_0$ with values computed from standard A-star models (e.g. see Heller & Kawaler 1987) for which lines of constant $\Delta\nu_0$ are similar to lines of constant radius. The more luminous the star, the lower the value of $\Delta\nu_0$, since higher luminosity means lower sound speed (because of lower density) and longer sound travel times (because of larger radii).

Matthews, Kurtz & Martinez (1999) compared parallaxes computed from asteroseismic luminosities for roAp stars with true parallaxes determined by HIPPARCOS. The agreement was excellent, as is shown in Fig. 4. The luminosities of the Ap stars are notoriously difficult to determine by spectroscopic and photometric means because of their extreme spectral peculiarities. The photometric techniques – particularly the use of the Strömgren δc_1 – fail spectacularly. For those roAp stars which are too far away to have HIPPARCOS luminosities, the asteroseismic luminosities are now the best available.

However, in Fig. 4 it can be seen that there is a systematic shift in the sense that the derived asteroseismic parallaxes are slightly smaller than the true HIPPARCOS parallaxes, meaning that the asteroseismic luminosities are overestimated. Matthews, Kurtz & Martinez (1999) suggested that this indicates

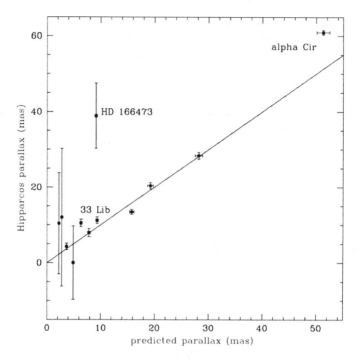

Figure 4.   This diagram shows the asteroseismic parallaxes calculated from standard A-star models for 12 roAp stars compared to the HIP-PARCOS parallaxes. The agreement is very good, demonstrating that for roAp stars that do not have HIPPARCOS parallaxes, the asteroseis-mic technique is the most reliable method to get their luminosities – a quantity difficult to determine spectroscopically or photometrically for the peculiar A stars. There is a small systematic shift in the sense that the asteroseismic parallaxes are too small. This suggests that the effective temperatures and/or the radii of Ap stars are systemati-cally smaller than those determined using normal A-star models. (from Matthews, Kurtz & Martinez 1999)

that the roAp stars probably have lower effective temperatures or smaller radii than those of the standard A star models that were used to calculate the as-teroseismic luminosities. New theoretical work by Cunha & Gough (2000) and Bigot et al. (2000) both indicate that one affect of the magnetic field on the large spacing, $\Delta\nu_0$, expected from eq. 1 is to increase the large spacing. This, in turn, increases the systematic shift in Fig. 4, suggesting even lower effective temperatures, or even smaller radii than previously thought. It is well-known that the atmospheres of the Ap stars are so peculiar that no atmospheric models are yet able to describe them completely. This is very well demonstrated by the "core-wing anomaly" in the Hα line recently discussed by Cowley et al. (2001). For many of the most peculiar of the Ap SrCrEu stars the wings of the Hα line match effective temperatures about 1000 K hotter than the very narrow Hα line cores.

New, more sophisticated models of Ap star atmospheres are needed to understand something so basic as the Hα line, amongst many other outstanding problems for these stars; those better models will also need solve the systematically high luminosity estimates from the asteroseismic $\Delta\nu_0$, thus providing an additional constraint on the models. Here is yet another example of astrophysical understanding resulting from observations which in practice can only be obtained with small telescopes.

References

Aerts, C., Waelkens, C, de Cat, P. 1998, in New Eyes to See Inside the Sun and Stars, eds. F.-L. Deubner, J. Christensen-Dalsgaard, D.W. Kurtz, IAU Symposium 185, Kluwer, Dordrecht, p. 295

Balona, L.A. 2000, in *The Impact of Large-Scale Surveys on Pulsating Star Research*, L. Szabados and D. Kurtz, eds., ASP Conf. Ser., Vol. 203, Astron. Soc. Pac., San Francisco, p. 401

Bigot, L., Provost, J., Berthomieu, G., Dziembowski, W.A., Goode, P.R. 2000, A&A, 356, 218

Breger, M. and Montgomery, M. 2000, in *Delta Scuti and Related Stars, Reference Handbook and Proceedings of the 6th Vienna Workshop in Astrophysics*, Breger, M. and Montgomery, M, eds., ASP Conf. Ser., Vol. 210, Astron. Soc. Pac., San Francisco

Christensen-Dalsgaard, J. 1976, MNRAS, 174, 87

Cowley, C.R., Hubrig, S., Ryabchikova, T.A., Mathys, G., Piskunov, N., Mittermayer, P. 2001, A&A, in press

Cunha, M.S., Gough, D. 2000, MNRAS, 319, 1020

Cunha, M.S., Dolez, N., Vauclair, S., Gough, D.O., Balmforth, N. 2001, A&A, in press

Dziembowski W., Goode P. R. 1996, ApJ, 458, 338

Dziembowski, W. 1998, in New Eyes to See Inside the Sun and Stars, eds. F.-L. Deubner, J. Christensen-Dalsgaard, D.W. Kurtz, IAU Symposium 185, Kluwer, Dordrecht, p. 355

Handler, G. 2000, in *The Impact of Large-Scale Surveys on Pulsating Star Research*, L. Szabados and D. Kurtz, eds., ASP Conf. Ser., Vol. 203, Astron. Soc. Pac., San Francisco, p. 408

Heller C. H., Kawaler S. D. 1988, ApJ, 329, L43

Kawaler, S. 1998, in New Eyes to See Inside the Sun and Stars, eds. F.-L. Deubner, J. Christensen-Dalsgaard, D.W. Kurtz, IAU Symposium 185, Kluwer, Dordrecht, p. 261

Kaye, A.B., Handler, G., Krisciunas, K., Poretti, E. Zerbi, F.M. 2000, in *The Impact of Large-Scale Surveys on Pulsating Star Research*, L. Szabados and D. Kurtz, eds., ASP Conf. Ser., Vol. 203, Astron. Soc. Pac., San Francisco, p. 426

Kurtz D.W. 1982, MNRAS, 200, 807

Kurtz D.W. 1992, MNRAS, 259, 701

Kurtz D.W. 1990, ARA&A, 28, 607

Kurtz, D.W. 2000, in *Delta Scuti and Related Stars, Reference Handbook and Proceedings of the 6th Vienna Workshop in Astrophysics*, Breger, M. and Montgomery, M, eds., ASP Conf. Ser., Vol. 210, Astron. Soc. Pac., San Francisco, p. 287

Kurtz, D.W., Kanaan, A., Martinez, P. 1993, MNRAS, 260, 343

Kurtz, D.W., Martinez, P. 2000, Baltic Ast., 9, 253 (KKM)

Kurtz, D.W., van Wyk, F., Roberts, G., Marang, F., Handler, G., Medupe, R., Kilkenny, D. 1997, MNRAS, 287, 69

Landolfi, M., Bagnulo, S., Landi degl'Innocenti, M., Landi degl'Innocenti, E., Leroy, J. L. 1997, A&A, 322, 197

Martinez P. 1993, PhD thesis, Univ. Cape Town

Martinez P., Kurtz D.W. 1994a, MNRAS, 271, 118

Martinez P., Kurtz D.W. 1994b, MNRAS, 271, 129

Martinez P., Kurtz D.W. 1995, Ap&SS, 230, 29

Martinez P., Kurtz D.W., Kaufmann, G.M. 1991, MNRAS, 250, 666

Matthews J. 1991, PASP, 103, 5

Matthews J.M., Kurtz D.W., Martinez P. 1999, ApJ, 511, 422

Pekeris, C.L. 1938, ApJ, 88, 189

Shibahashi H. 1986, in Hydrodynamic and Magnetohydrodynamic Problems in the Sun and Stars, ed. Osaki Y., Univ. Tokyo, Tokyo, p. 195

Shibahashi H. 1987, Lect. Notes Phys., 274, 112

Shibahashi H. 1990, Lect. Notes Phys., 388, 393

Shibahashi H., Takata M. 1993, PASJ, 45, 617

Takata M., Shibahashi H. 1995, PASJ, 47, 219

Tassoul M. 1980, ApJS, 43, 469

Tassoul M. 1990, ApJ, 358, 313

Weiss W. W. 1986, in *Upper Main Sequence Star with Anomalous Abundances*, ed. C. R. Cowley, M. M. Dworetsky, and C. Mégessier, IAU coll. 90, D. Reidel Publ. Co., Dordrecht, Holland, p. 219

Small–Telescope Astronomy on Global Scales
ASP Conference Series, Vol. 246, 2001
W.P. Chen, C. Lemme, B. Paczyński

The Moscow Long-Term Program of Cepheid Radial Velocities

Nikolai Samus

Institute of Astronomy, Russian Acad. Sci., 48, Pyatnitskaya Str., Moscow 109017, Russia and Sternberg Astronomical Institute

Natalia Gorynya

Institute of Astronomy, Russian Acad. Sci., 48, Pyatnitskaya Str., Moscow 109017, Russia

Abstract. A correlation spectrometer has been used for 15 years by Moscow observers for different programs at several small telescopes and has proven itself to be very effective. Here we present some of the results in the field of Cepheid studies acquired with the spectrometer.

1. Introduction

We present some results of our program based upon observations of Galactic Cepheids with a CORAVEL-type correlation spectrometer. This program seems a good example of using the principal advantage of small-telescope programs, namely the possibility to have plenty of observing time over the course of many years, combined with special advantages of this particular instrument, namely its outstanding effectiveness combined with quite modest requirements in terms of size which allows it to be carried from one observing site to another, according to availability of telescopes and to weather conditions.

The spectrometer used, ILS, was designed and built by Tokovinin (1987) in 1986. It used the ideas first suggested by Felgett (1953) and by Griffin (1967) which were then implemented for echelle spectrometers in the CORAVEL machine (Baranne et al. 1979). In a CORAVEL-type spectrometer, an echelle spectrograph forms an image of a star's spectrum. A physical mask is placed in the focal plane. It is actually an image of the spectrum of a "standard" star, in our case, of Arcturus; the positions of the mask that correspond to spectral lines are transparent, while those between them are not. A special plane mirror added into the optical scheme oscillates at a frequency of about 10 Hz so that the observed spectrum moves back and forth along the mask. The flux passing through the mask is minimal when the lines of the observed spectrum coincide with the "lines" of the mask. The light is then collected and measured with a photomultiplier. A special controller (in the earlier configuration of the instrument) or a computer (currently) serves to collect the measurements of light separately for 50 time intervals of each oscillation cycle of the plane mirror. Thus, we obtain a "generalized spectral line", actually corresponding, in our case, to ~ 1500 lines in the spectrum of the program star or of Arcturus. The registered "line" covers a range of approximately ± 25 km/s, but we can search for the radial velocity in a much wider interval, about ± 300 to 500 km/s, using calibrated rotation of

197

the diffraction grating of the spectrometer. The generalized line is then fitted to a relevant profile (usually Gaussian), and the position of its minimum determines the radial velocity. Preliminary reductions are performed already during observations, but more sophisticated software can be used off-line to improve the accuracy. The reductions take into account all the necessary corrections for the motion of the observer; the zero point correction, usual for such instruments, is determined using observations of radial velocity standards. The profile shape gives additional information about abundances or rotation.

The main technical characteristics of the ILS are the following: It can measure radial velocities of main-sequence stars with "normal" chemical abundances in the spectral type range approximately from F5 to M5; somewhat earlier giants can also be measured. The characteristic accuracy for sufficiently bright stars in the middle of the spectral type range is 0.3 to 0.5 km/s. The limiting magnitude for telescopes of the 1-m class is about $12^m.5$; in record cases, stars as faint as 14^m could be measured. The typical exposure time for brighter stars is about 5 min; for faint stars, it seldom exceeds 30 min.

Several instruments of CORAVEL-type design were built in the 1980s or early 1990s in different countries. Now, there are only few survivors. Surely there are advantages in registering the complete spectrum with a CCD and then determining the radial velocity (and numerous other parameters!) by means of purely digital reductions. However, instruments like the ILS have proved to be excellent machines for "the poor", the reductions are very simple, almost no additional technical support is needed, and you can bring your instrument to an isolated observatory and start observing the next night. We are going to continue the use of the ILS for several more years.

2.   Observations and Some Results

During 15 years of active exploitation of the ILS, it has been used by several groups for many scientific programs, among them: orbits of binary and multiple stars; kinematics of the Galaxy; kinematics of stars in open and globular clusters; pulsations of stars (Cepheids and some others). Here we discuss only the last program.

In 1987–2000, we acquired more than 7000 observations of 144 Cepheids using 11 telescopes in 6 countries (Russia, Ukraine, Georgia, Azerbaijan, Uzbekistan, Bulgaria), having 60 cm to 2 m apertures. Table 1 shows some additional information. Primarily, telescopes of the 1-m class were used. Our main instrument was the 1 m telescope of the Simeiz Observatory (Crimea, Ukraine). The observatory is now a department of the Crimean Astrophysical Observatory, but, despite the disintegration of the Soviet Union, the telescope still belongs to the Institute of Astronomy, Moscow. Note that we use, rather effectively, the 70 cm telescope of the Sternberg Astronomical Institute, installed in Moscow, located less than 10 km from the Kremlin!

The specific feature of our program is that we try to obtain a good coverage of the pulsation velocity curve for each Cepheid during each year season. Our database of original accurate radial velocity measurements for Cepheids currently appears to be the world's richest.

Table 1.   Telescopes and Observations of Cepheids

No	Telescope	Years	No. of observ.
1	70 cm, Moscow, Russia	1987–2000	867
2	60 cm, Nauchny, Crimea, Ukraine	1987–1990	47
3	200 cm, Shemakha, Azerbaijan	1988	2
4	122 cm, Abastumani, Georgia	1988	8
5	125 cm, Nauchny, Crimea, Ukraine	1989–1990	64
6	100 cm, Mt. Maidanak, Uzbekistan	1989–1993	283
7	100 cm, Simeiz, Crimea, Ukraine	1990–2000	3792
8	60 cm, Simeiz, Crimea, Ukraine	1990–1998	1902
9	200 cm, Mt. Rozhen, Bulgaria	1990	16
10	60 cm, Mt. Maidanak, Uzbekistan	1991	212
11	60 cm, Zvenigorod, Russia	1997	18

Grand total: 14 years, 1341 nights, 7211 observations of 144 Cepheids

Figure 1 shows the characteristic radial velocity curves for Cepheids, folded with their pulsation periods. Note that the scatter of data points for X Cyg is very low, and that the velocity curve looks not worse than good photoelectric light curves of Cepheids. Thus, if the Baade–Wesselink technique is applied to determine Cepheid radii, the factor that crucially limits the accuracy is no longer the uncertainty in radial velocities.

The other two examples do not look so nice. If the period of the Cepheid is correct and does not vary strongly, there can be two reasons for the increased scatter. The first of them are double mode pulsations. We have observed several double-mode Cepheids and were able to separate the two pulsation modes in their radial velocities. EW Sct is a well-known double-mode Cepheid, we have more than 100 data points for it. Another interesting example of a double mode Cepheid in our program is V458 Sct, recently discovered by Antipin (1997). For this star, the first overtone amplitude is larger than that of the fundamental tone (Antipin et al. 1999); the same effect is revealed by the star's light curve, but the interpretation of the radial velocity curve in terms of energy is more straightforward, and the coexistence of two oscillations, with the first-overtone having a higher energy, poses a problem to the theory of stellar pulsations.

The second reason for the increased scatter is binarity. TX Del is an extreme case (it is not clear if the star is a classical Cepheid or a Population II Cepheid). We independently discovered the binarity of this star, first noted by Harris and Welch (1989). Its pulsation period is $6^d.17$; its orbital period is only 133^d, a very low value for binary Cepheids (remember that classical Cepheids are supergiants). Even its velocity curve based on observations of a single year reveals quite obvious signatures of binarity.

We discovered the binarity of the classical Cepheids BY Cas, VY Cyg, VZ Cyg, and MW Cyg. Our results for more than 20 spectroscopic binary Cepheids, including the determination of their orbital elements, are summarized in Gorynya et al. (2000). The newest discovery is the spectroscopic binarity of V496 Aql (Fig. 2). For this star, we have two possible values for the orbital period, 1447^d and 573^d; we tend to prefer the shorter one. Our observations show that at least 22% of the Cepheids are binaries; we consider the much higher estimates of some other authors, such as 50% or more, to be too high.

As noted above, our data are very advantageous for determinations of Cepheid radii using the Baade–Wesselink technique. From our original radii of 62 Cepheids, we have derived the period–radius relation in the form (Sachkov et al. 1998):

$$\log R = 1.23(\pm 0.03) + 0.62(\pm 0.03)\log P.$$

The radii derived with the Baade–Wesselink technique are very important as a tool to distinguish between different pulsation modes for Cepheids.

A relation between the light curve shape of a Cepheid and its period, the Hertzsprung sequence, is rather well known, but we could confidently reveal it, for the first time, in their radial velocity curves rather than in their light curves (Gorynya 1998).

In the recent years, we have entered into a fruitful collaboration with Dr. P. Moskalik (Warsaw) who uses our data for Fourier decomposition of radial velocity curves. This sensitive tool of research makes it possible to study the Hertzsprung sequence as well as to reveal resonances between different pulsation modes. Currently, Dr. Moskalik is studying our most recent observations and gives us advice on our program of observations, indicating insufficient coverage of velocity curves for stars of interest, hints to binarity, *etc.* We are planning to continue this cooperation. Also, we are trying to apply our instrument to the study of different variability types, but this is outside the scope of this presentation.

Acknowledgments. We are grateful to the astronomers who took part in the observations of our program, in particular, to Drs. S. Antipin, A. Rastorgouev, and M. Sachkov.

Our study was financially supported, in part, by grants from the Russian Foundation for Basic Research, the Program of Support for Leading Scientific Schools of Russia, and the Russian Federal Scientific and Technological Program "Astronomy".

References

Antipin, S.V. 1997, IBVS No. 4485

Antipin, S.V., Gorynya, N.A., Sachkov, M.E. et al. 1999, IBVS No. 4718

Baranne, A., Mayor, M., & Poncet, J.L. 1979, Vistas in Astronomy 23, 279

Felgett, P.B. 1953, Optica Acta 2, 9

Gorynya, N.A. 1998, IBVS No. 4636

Gorynya, N.A., Samus, N.N., Sachkov, M.E. et al. 2000, in: The Impact of Large-Scale Surveys on Pulsating Star Research, L. Szabados and D.W. Kurtz (eds.), ASP Conference Series, 203, 242

Griffin, R.F. 1967, ApJ 148, 465

Harris, H.C. & Welch, D.L. 1989, AJ 98, 981

Sachkov, M.E., Rastorguev, A.S., Samus, N.N., & Gorynya, N.A. 1998, Astronomy Letters 24, 377

Tokovinin, A.A. 1987, Soviet Ast. 31, 98

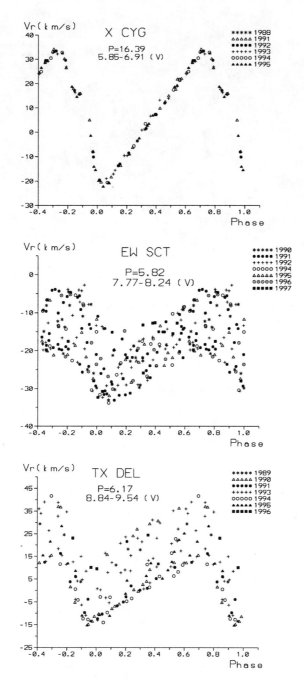

Figure 1.   Radial velocity curves for three Cepheids, folded with their pulsation periods. EW Sct is a double-mode Cepheid, and TX Del is a spectroscopic binary.

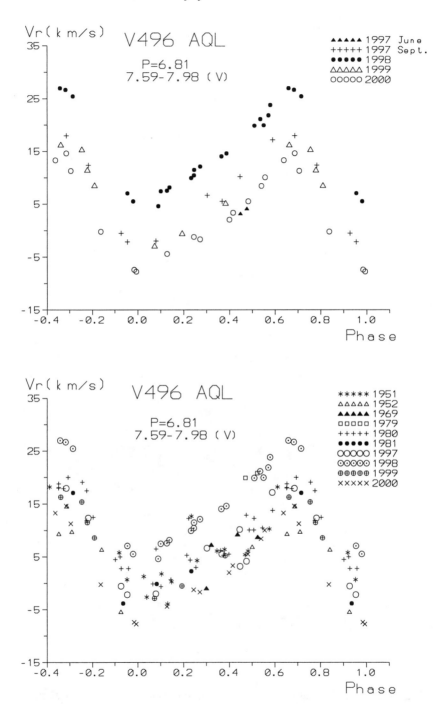

Figure 2.   Radial velocities of V496 Aql, plotted with the period of pulsations.

Small–Telescope Astronomy on Global Scales
ASP Conference Series, Vol. 246, 2001
W.P. Chen, C. Lemme, B. Paczyński

High Speed CCD Photometry of Flare Stars

Sun-Youp Park

Department of Astronomy, Yonsei University, Seoul, 120-749, Korea

Yong-Ik Byun

Department of Astronomy, Yonsei University, Seoul, 120-749, Korea

Abstract. Previously, flare stars have been observed by fast two channel photoelectric photometry. We have started an observing program to monitor flare stars using a normal CCD camera via a method called "trailed mode photometry". For this program, we developed a fully automated trail photometry software.

1. Introduction

Solar and stellar flares are believed to take place because of the heating of the corona by X-ray emission and material evaporation from dense heated region of the chromosphere. There are however several different models for solar and stellar flare activities. While flares in general share some common features, they can be radically different in others. For example, the total energy released during a stellar flare of dMe-type flare stars can be up to 1,000 times greater than that of the Sun. More observations are needed to understand the variety and nature of flare events. Previously, flare stars have been observed by fast two channel photoelectric photometry; typical timescales of flare phenomena are 1 ~ 100 seconds. We have started an observing program to monitor flare stars using a normal CCD camera via a method called "trailed mode photometry". For this program, we developed a fully automated trail photometry software. Our study will later be expanded into a general survey to search for unknown flare star population.
We describe here the details of our photometry software algorithm.

2. Outline of Photometry Procedure

The steps for our photometry can be outlined as follows. First, we obtain an image, as shown at the right of Figure 1 with a normal CCD camera; this is a "trail image" taken with the telescope tracking at a speed slower than siderial rate. After the basic pre-processing (bias, dark, flat field, bad pixel & cosmic ray removal), local peaks are detected using an algorithm similar to DAOPHOT. Then, by continuation of detected peaks, the locations of each trail in the image are defined. Finally, the light variations of each object along the trails are extracted.

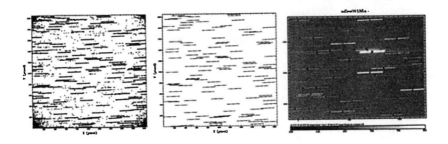

Figure 1.   Steps for detection trails. Left : peak detection. Middle :
Peak continuation. Right : Trail locations marked on the trail image.

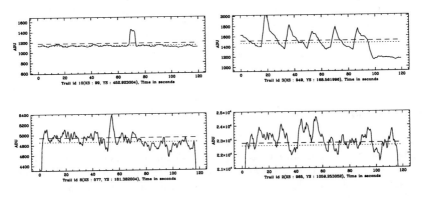

Figure 2.   Trail light curves of some fast varying objects. Flux variations (in ADU) versus time (in seconds) are shown. Dotted lines and
dashed lines show the average and the 3-σ threshold levels, respectively.

3.   Discussions

Our study can be summarized as follows. First, although CCDs are slow-readout
devices, it is possible to study fast-varying objects using trailed mode photometry. Second, with automation, it is possible to extract and examine light curves
for a large number of objects with a single trail image. Third, the application of
this kind of photometry includes, but is not limited to, the study of flare stars,
Gamma-ray Bursts, and other fast transient events. This method is however limited by fast variations caused by the earth's atmosphere, such as scintillation,
which needs further study.

Small–Telescope Astronomy on Global Scales
ASP Conference Series, Vol. 246, 2001
W.P. Chen, C. Lemme, B. Paczyński

Search for δ Scuti Type Pulsating Components in Eclipsing Binary Systems

S.-L. Kim, J. W. Lee, J.-H. Youn, H.-K. Moon, K. J. Choo

Korea Astronomy Observatory, Taejon, 305-348, Korea

Abstract. We present observational results of four eclipsing binary stars, Y Cam, RZ Cas, AS Eri and RU UMi, which were reported to have δ Scuti type pulsating components.

δ Scuti type pulsating components discovered in eclipsing binary systems are very attractive observing targets from an asteroseismological point of view, because they allow for the independent determination of masses and radii for each component. Nevertheless, few such stars have been reported so far and, furthermore, only two stars, AB Cas and Y Cam, have been studied in depth (Lampens & Boffin 2000 for a recent review). This is probably due to their too small amplitude oscillation features in comparison with large amplitude eclipsing phenomena.

We present preliminary CCD photometric results for a known eclipsing binary with a pulsating component, Y Cam, and three recently reported candidates, RZ Cas, AS Eri and RU UMi. The observations were done with a PM512 CCD camera attached to the 61cm telescope at Sobaeksan Optical Astronomy Observatory (SOAO) in Korea. Instrumental magnitudes were obtained using the ADPS (Automatized Differential Photometry System; Park 1993). In order to detect high frequencies resulting from pulsations, we subtracted low frequencies caused by orbital motions using the polynomial fitting technique. Light curves of two stars, Y Cam and RZ Cas, are shown in Figure 1, and our results are summarized in Table 1. We confirmed the previous results for three stars, Y Cam, RZ Cas and AS Eri. However, we could not find a definite oscillating frequency for RU UMi. Its pulsating features are thought to be uncertain.

We will start an observational survey program to search for δ Scuti type pulsating components in eclipsing binary systems after February 2001, using the SOAO 61cm telescope and the 2K SITe CCD camera (FOV of $20\rlap{.}'5 \times 20\rlap{.}'5$).

References

Broglia, P., & Conconi, P. 1984, A&A, 138, 443

Gamarova, A. Yu., Mkrtichian, D. E., & Kusakin, A. V. 2000, IBVS, no.4837

Lacorte, M. B., & Van Hamme, W. V. 1999, AAS, 195, 7615

Lampens, P., & Boffin, H. M. J. 2000, Delta Scuti and related stars, ASP conf. series, vol. 210, 309

Ohshima, O., Narusawa, S.-Y., Akazawa, H., Fujii, M., Kawabata, T., & Ohkura, N. 1998, IBVS, no.4581

Park, N.-K. 1993, Publications of the Korean Astronomical Society, 8, 185

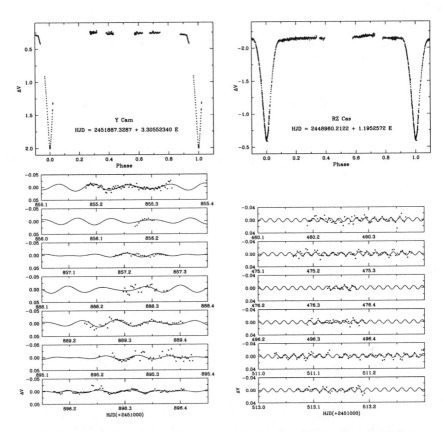

Figure 1.   (Top) Phase diagram of two eclipsing binary stars, Y Cam (left) and RZ Cas (right). (Bottom) Light variations of pulsating components in these two stars, using the out-of-eclipse data.

Table 1.   Basic parameters of four variable stars.

Name	V Sp. Type	$^\dagger P_{pulsating}$ $P_{eclipsing}$	$^\ddagger A_{pulsating}$ $A_{eclipsing}$	Previous results Reference
Y Cam	10^m50 A8 V	0^d0664 $3^d30552340$	$\sim0^m03$ 1^m74	0^d0664 (15.05 c/d) Broglia & Conconi 1984
RZ Cas	6^m18 A2.8 V	0^d0156 $1^d1952572$	$\sim0^m02$ 1^m54	0^d0156 (64.20 c/d) Ohshima et al. 1998
AS Eri	8^m29 A3 V	0^d0172 $2^d664152$	$\sim0^m01$ 0^m71	0^d0169 (59.03 c/d) Gamarova et al. 2000
RU UMi	10^m00 F5	0^d0269 (?) $0^d52492599$	$\leq0^m01$ 0^m66	0^d0194 (51.43 c/d) Lacorte & Van Hamme 1999

$\dagger$: period, $\ddagger$: amplitude

Small–Telescope Astronomy on Global Scales
ASP Conference Series, Vol. 246, 2001
W.P. Chen, C. Lemme, B. Paczyński

Long-period Red Variables in the Large Magellanic Cloud from the MOA Database

Mine Takeuti

Astron. Inst., Tohoku Univ., Sendai, 980-8578 Japan

Noda, S.

Solar-Terrestrial Environment Lab., Nagoya Univ., Nagoya, 464-8601 Japan

Abe, F.[1] Bond, I. A.[2], Dodd, R. J.[3], Hearnshaw, J. B.[4], Honda, M.[5], Jugaku, J.[6], Kabe, S.[7], Kilmartin, P. M.[2,4], Matsubara, Y.[1], Masuda, K.[1], Muraki, Y.[1], Nakamura, T.[8], Nankivell, G. R.[9], Noguchi, C.[1], Ohnishi, K.[10], Rattenbury, N. J.[2], Reid, M.[11], Saito, To.[12], Sato, H.[8], Sekiguchi, M.[7], Skuljan, J.[4], Sullivan, D. J.[11], Sumi, T.[1], Watase, Y.[7], Yanagisawa, T.[1], Yock, P. C. M.[2], and Yoshizawa, M.[13]

Abstract. We studied 147 long-period red variable stars in the Large Magellanic Cloud from the MOA database. Amongst them, seven red luminous stars are likely pulsating in a higher mode.

1. Observation and Reduction

The MOA is a massive photometry project designed to study the distribution of cosmic dark matter, based on observations of gravitational microlensing events.

[1]Solar-Terrestrial Environment Lab., Nagoya Univ., Nagoya, 464-8601 Japan

[2]Dept. of Physics, Univ. of Auckland, Auckland, New Zealand

[3]Carter National Obs., Wellington, New Zealand

[4]Dept. of Physics and Astron., Univ. of Canterbury, Christchurch, New Zealand

[5]Inst. Cosmic Ray Research, Univ. of Tokyo, Kashiwa, 277-8582 Japan

[6]Institute for Civilization, Tokai Univ., Tokyo, Japan

[7]KEK, Tsukuba, 305-0801 Japan

[8]Research Inst. Fundamental Physics, Kyoto Univ., Kyoto, 606-8502 Japan

[9]Lower Hutt, New Zealand

[10]Nagano Tech. College, Nagano, Japan

[11]Dept. of Physics, Victoria Univ., Wellington, New Zealand

[12]Tokyo Metropolitan College of Aeronautics, Tokyo, 140-0011 Japan

[13]National Astron. Obs., Mitaka, Tokyo, 181-8588 Japan

Millions of stars in the Large Magellanic Could are monitored by using a wide-field camera on a 0.61-m telescope at the Mount John Observatory of Canterbury University on the South Island of New Zealand.

Variable star research with the MOA project has been described in Hearn-shaw et al. (2000). In a recent study, we compare our results with those of the WFPC2 of the HST taking the effects of colour into account. Calibration is performed by a careful frame-to-frame check (Kato 2000). Approximately 300 observations, from January 1997 to December 1999, are used. The probable error of the intensity mean, $\langle V_m \rangle$ and $\langle R_m \rangle$ is less than $\pm$ 0.015 mag. We have analysed the V_m and R_m periodicity of stars that show a prominent variability by using a folding method, the phase difference minimization. Variables with a period of less than 30 d or with colour $\langle V_m \rangle - \langle R_m \rangle$ less than 0.4 are omitted. Because our time span is 1060 d, we also omitted stars with periods longer than 400 d. Finally, 147 stars with an amplitude ΔR larger than 1.3 mag are selected. Because the blue passband of the MOA system is very broad, the relation among the V_m, R_m, and K-magnitudes is different for M-type and C-type stars, due to the effect of the TiO band.

2. Results

The scattering of the period-colour diagram is smaller than in our preliminary report (Takeuti et al. 2000). Among the 147 stars, seven red and short-period variables are separated from the other variables. The K-magnitude-log P relation of M-Miras coincides with that presented by Feast et al. (1989). On this diagram, the seven stars form a group separated from the other stars. On the K-magnitude-colour diagram, the stars locate at the tip of the giant branch. This indicates that the radii of these stars are very large. These large, short-period stars are less regular in their periodicity, and their amplitudes are small. They may be a counterpart of the short period red (SP red) stars found in the study of Hipparcos parallaxes (Whitelock & Feast 2000), and also the stars suggested to be higher mode pulsators (Wood & Sebo 1996; Bedding & Zijlstra 1998).

References

Bedding, T. R., & Zijlstra, A. A. 1998, ApJ, 506, L47

Feast, M. W., Glass, I. S., Whitelock, P. A., & Catchpole, R. M. 1989, MNRAS, 241, 375

Hearnshaw, J. B., Bond, I. A., Rattenbury, N. J., et al. 2000, in ASP Conf. Ser. Vol. 203, The Impact of Large-scale Surveys on Pulsating Star Research, ed. L. Szabados, D. W. Kurtz (San Francisco: ASP), 203, 31

Kato, Y. 2000, Dissertation for Master Degree, Nagoya University (in Japanese)

Takeuti, M., Noda, S., Bond, I. A., et al. 2000, in ASP Conf. Ser. Vol. 203, The Impact of Large-scale Surveys on Pulsating Star Research, ed. L. Szabados, D. W. Kurtz (San Francisco: ASP), 120

Whitelock, P., & Feast, M. 2000, MNRAS, 319, 759

Wood, P. R., & Sebo, K. M. 1996, MNRAS, 282, 958

Small–Telescope Astronomy on Global Scales
ASP Conference Series, Vol. 246, 2001
W.P. Chen, C. Lemme, B. Paczyński

Spectroscopic Detection of an Extraordinary Flaring-Event on DF Tau

J. Z. Li[1,2], W. P. Chen[1] & W. H. Ip[1]

1. Institute of Astronomy, National Central University, Chung-Li 32054
(Email: ljz@astro.ncu.edu.tw)
2. Beijing Astronomical Observatory, National Astronomical
Observatory, Chinese Academy of Sciences, Beijing 100012, China

Abstract. An accretion intrigued flare-like brightening of DF Tau has been recorded spectroscopically, during which spectacular spectral changes took place. We inferred the possible formation of a transient envelope during the process of this episodic mass accretion of DF Tau.

1. Motivation of Study

With an initial motivation to study properties of hot spots on Classical T Tauri stars (CTTS), we have launched a spectroscopic monitoring campaign to study some CTTS known to display periodic or quasi-periodic light variations. DF Tau is one of the program stars.

2. Observations and Data Reduction

Intermediate resolution spectroscopic observations of DF Tau and several other CTTS (with a dispersion of 50 A/mm, 1.2 A/pixel and a 2.5"slit) were carried out by the 2.16m optical telescope of the Beijing Astronomical Observatory, in both the blue (3500A - 4700A) and red (5750A - 6950A) spectral ranges on a nightly basis. Each night, when all the program stars were exposed (once per CTTS) with the blue settings, the red settings were employed, or vise versa. An OMR (Optomechanics Research Inc.) spectrograph and a Tektronix 1Kx1K CCD detector were used during the observations from Jan. 7 to Jan. 14, 2000 (Jan. 11 was skipped for weather reasons).

The spectral data were reduced following standard procedures in the NOAO Image Reduction and Analysis Facility (IRAF, version 2.11) software packages.

3. Results and Discussion

1) Our 8-night campaign of these spectroscopic observations of DF Tau witnessed the entire event of an extraordinary flare-like brightening, with a remarkably sharp increase in brightness (about 4 magnitudes within one day in the B band), followed by a slow decay lasting for 5 days.

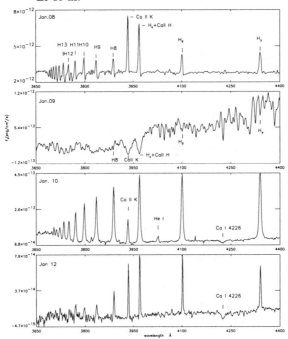

Figure 1.   Representative spectra of this brightening of DF Tau.

2) Spectacular changes took place between Jan. 8 and 9 in the continuum and line emissions, a major difference is that the latter seems to have lost its UV excess, while apparent excessive emissions still exist, which could be due to wavelength-dependent cooling efficiency as well as probable severe absorption encountered along the line of sight (Fig. 1). This could indicate the formation of a transient envelope, which serves just like an extra extension of the photosphere, where severe absorption takes place.

3) The amplitude of variation during this event reached up to 6 magnitudes in the B band and decreased as a function of wavelength to about 2 magnitudes in the V and R bands.

4) Apparent correlation between variations of line flux and veiling is displayed in this run of observations of DF Tau.

4.   Summary

1) An unprecedentedly large flare-like event was recorded spectroscopically on DF Tau. 2) Significant Y Y Orionis line profiles (Inverse P Cygni profiles), of supersonic origin, appeared on Jan. 8, suggesting inhomogeneous mass infall onto the contracting star. 3) We suggest the probable formation of a transient envelope in or over the shock-cooling region, which caused the spectacular spectral changes between Jan. 8 and 9, and its rapid, radial dissipation, which may have coherent relations to its intrinsic fierce turbulence (as indicated by the unusually extended CaII H & K absorption on Jan. 9 and broad Balmer emission lines on Jan. 10) and also the later emerging stellar wind.

Small–Telescope Astronomy on Global Scales
ASP Conference Series, Vol. 246, 2001
W.P. Chen, C. Lemme, B. Paczyński

Search for Variable Stars in the Open Cluster NGC 2539

K. J. Choo[1,2], S.-L. Kim[2], T. S. Yoon[1], M.-Y. Chun[2], H. Sung[2],
B.-G. Park[1,2], H. B. Ann[3], M. G. Lee[4], Y.-B. Jeon[2,4], I.-S. Yuk[2,4]

[1] Dept. of Astronomy and Atmospheric Sciences, Kyungpook National
Univ., Taegu, 702-701, Korea
[2] Korea Astronomy Observatory, Taejon, 305-348, Korea
[3] Dept. of Earth Sciences, Pusan National Univ., Pusan, 609-135, Korea
[4] Astronomy Program, SEES, Seoul National Univ., Seoul, 151-741,
Korea

Abstract. We report on the results of CCD photometric observations of
the open cluster NGC 2539. Eight new variable stars have been found in
the observed field of this cluster. However, no γ Doradus-type variability
was found among the member stars.

Since 1998 we have been performing a long term project for CCD photometric survey of open clusters, as well as time-series photometry, to search for
variable stars in the clusters at the Bohyunsan Optical Astronomy Observatory
(BOAO). As a part of this project we have carried out $UBVI$ CCD photometry
and V filter time-series photometry of the open cluster NGC 2539. The observations were done for ten nights between February and March of 2000, with
the 1.8m reflector and an SITe 2K CCD camera at BOAO. Figure 1 shows the
observed field of the cluster.

From the $UBVI$ photometry of NGC 2539, we have derived a reddening
value $E(B-V) = 0.06 \pm 0.03$, a true distance modulus $(V - M_V)_\circ = 10.2 \pm 0.1$,
and an age of $log\ t\ [yr] = 8.8$, using the empirical ZAMS (Sung & Bessell 1999)
and the theoretical isochrone with $Z = 0.019$ (Girardi et al. 2000).

We have examined the light curves of 583 stars from V filter time-series
images and have discovered eight new variable stars among them. Five of them
are classified to be eclipsing binary stars, and the others are: one δ Sct star,
one field γ Doradus star, and one field SPB star. Basic parameters of the eight
variable stars are summarized in Table 1. We could not find any γ Doradus-
type pulsating stars among the member stars in this cluster. This result is
consistent with Krisciunas & Patten (1999)'s suggestion that the γ Doradus-
type phenomenon occurs only in an open cluster younger than about 250 Myr.

References

Girardi, L., Bressan, A., Bertelli, G., & Chiosi, C. 2000, A&AS, 141, 371
Krisciunas, K. & Patten, B. M. 1999, IBVS, No. 4705
Sung, H. & Bessell, M. S. 1999, MNRAS, 306, 361

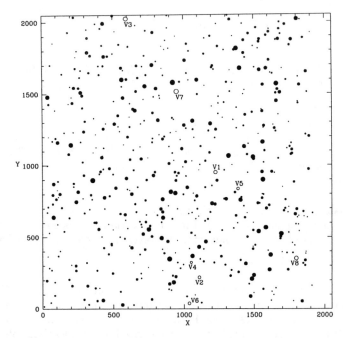

Figure 1.   Observed CCD field (11.6 × 11.6) of the open cluster NGC 2539. Eight new variable stars are represented by open circles with their IDs.

Table 1.   Basic parameters of eight new variable stars in NGC 2539

ID	V^1	$B - V^1$	Period	ΔV_{max}	Epoch[2]	Type
V1	13^m213	0^m255	0^d055	$\sim 0^m02$	2451597^d1575	δ Sct
V2	14^m290	0^m577	0^d352	$\sim 0^m04$	2451585^d13	field(?) γ Dor
V3	11^m847	0^m086	1^d092	$\sim 0^m06$	2451618^d05	field(?) SPB
V4	15^m652	0^m739	0^d292	$\sim 0^m25$	2451591^d025	eclipsing binary
V5	14^m584	0^m749	0^d340	$\sim 0^m13$	2451596^d06	eclipsing binary
V6	14^m301	0^m581	0^d945	$\sim 0^m13$	2451617^d99	eclipsing binary
V7	11^m050	1^m603	1^d964	$\sim 0^m07$	2451630^d07	field eclipsing binary
V8	12^m527	0^m250	0^d700	$\sim 0^m04$	2451629^d15	eclipsing binary

[1] average visual magnitude and color
[2] at the maximum brightness for pulsating stars and the minimum for eclipsing binary stars

Small–Telescope Astronomy on Global Scales
ASP Conference Series, Vol. 246, 2001
W.P. Chen, C. Lemme, B. Paczyński

The Ongoing Search for Variables in Young Clusters: Up-to-Date Results and Perspectives

A. Pigulski, G. Kopacki, Z. Kołaczkowski and M. Jerzykiewicz

Wrocław University Observatory, Kopernika 11, 51-622 Wrocław, Poland

Abstract. The goals, results and perspectives of the ongoing search for variables in open clusters and associations of the northern hemisphere, carried out at the Białków Observatory of the Wrocław University, are presented.

Since 1994 we have observed seven open clusters and one OB association and have discovered about 130 variables. The observations were carried out with a 60-cm Cassegrain telescope equipped with a $6' \times 4'$ field-of-view CCD camera, an autoguider, and two sets of filters: $BV(RI)_C$ and $H\alpha$. A typical run consisted of about 20 observing nights and, depending on the cluster, covered one or several fields.

1. The goals

The main goals of our search were the following: (1) To map the observational instability regions of the β Cephei, SPB and other pulsators of the upper part of the H-R diagram and to find the incidence of variability within them. (2) To find the characteristics of variable stars in young clusters at different evolutionary stages, including the pre-main sequence stage. (3) To select variables suitable for asteroseismology and/or distance determination. (4) To find the Be star content in clusters and OB associations. (5) To derive the cluster parameters such as distance, age, and reddening. (6) To derive reddening maps (in cases of substantial differential reddening).

2. Results

The most interesting results we obtained during the search can be summarized as follows:

- Observations of two young objects, the Cygnus OB2 association (Pigulski & Kołaczkowski 1998) and the open cluster NGC 6823 (Pigulski et al. 2000), show that, in comparison with the predictions of the linear pulsation theory, there is a deficiency of β Cephei-type pulsators among stars earlier than B0.

- Although ten β Cephei stars were observed in the clusters we monitored, it seems that they constitute only a small fraction of B-type stars falling into

213

the instability strip. The fraction is much larger in some southern clusters and could be the consequence of metallicity differences among clusters.

- Only 1–2% of all mid and late B-type stars show variability of the SPB type. All the stars we discovered of this type appear to be monoperiodic. It seems, therefore, that multiperiodicity is not a common feature of these stars as was initially thought.

- Two pre-main sequence δ Scuti stars were found in NGC 6823. Both are biperiodic and can be used to test the evolutionary period changes predicted by the theory.

- About 30 Be stars in six clusters and the Cygnus OB2 association were discovered by means of our Hα photometry. In this context, the most interesting is finding that NGC 7419 contains at least 31 Be stars, that is, about 36% of all early B-type stars in the cluster (Pigulski & Kopacki 2000). Apart from NGC 663, this is the only known cluster in the Galaxy with such a large fraction of Be stars. About 80% of all Be stars show some kind of photometric variability with ranges up to 0.5 mag in the I_C filter.

- We discovered about ten eclipsing binaries that have components of O or B spectral type. Some are bright enough to be studied spectroscopically with a large telescope. This would yield cluster distances.

3. Perspectives

There are at least 200 open clusters in the Northern hemisphere young enough to contain B-type pulsators. In addition, 29 OB associations are known in this part of the sky. All are potential targets for our investigation. The considered area ($30° < l < 210°$, $-10° < b < 10°$) covers about 3500 square degrees.

The most serious limitation of our search is the small field of view of our camera and the fact that only 15–20% of the nights are clear at our site. This enables us to acquire observations for at most two open clusters per year. The improvements we plan are the following:

1. To install a CCD camera with a larger field of view. At present, the camera's field-of-view covers only about 8% of the telescope's field-of-view. Attaching a new camera would increase the observed area several times. Still, some interesting objects (especially OB associations) will not be covered because of their large angular extent.

2. Since the regions around the galactic plane are heavily crowded and reddened, all-sky automated surveys similar to ASAS (Pojmański 1997) would reach only relatively near early B-type stars. In order to also observe these types of stars at larger distances, we plan to set up an inexpensive robotic telescope (or telescopes) at a good site, equipped with a large-format CCD camera and the optics with a spatial resolution on the order of $1''$ per pixel. Observations would be then restricted to selected areas containing some interesting clusters, OB associations and dense stellar fields in the galactic plane.

Acknowledgments. The work was partially supported by the KBN grant 2 P03D 006 19.

References

Pigulski A. & Kołaczkowski Z. 1998, MNRAS 298, 753
Pigulski A., Kołaczkowski Z. & Kopacki G. 2000, Acta Astron. 50, 113
Pigulski A. & Kopacki G. 2000, A&AS 146, 465
Pojmański G. 1997, Acta Astron. 47, 467

Small–Telescope Astronomy on Global Scales
ASP Conference Series, Vol. 246, 2001
W.P. Chen, C. Lemme, B. Paczyński

Variable Stars in the Globular Cluster M92

Grzegorz Kopacki

Wrocław University Observatory, Kopernika 11, Poland

Abstract. Results of a search for variable stars in the central region of the Oosterhoff II type globular cluster M92 are presented.

Out of the 28 variable and suspected variable stars listed in the Catalogue of Variable Stars in Globular Clusters (Clement 1997), only two were not observed. Surprisingly, almost half of the observed suspected variables did not show any evidence of variability. Only one out of the 11 candidate RR Lyrae variables of Kadla et al. (1983) appeared to be variable. Moreover, variable v7, until now classified as an RR Lyrae star with a period of about 0.515 d, turned out to be of the BL Herculis type, with the period approximately twice that long. In addition, six new variables were found in the very core of the cluster: four of the RR Lyrae type and the remaining two of the SX Phoenicis type. Altogether, light curves have been obtained for 20 variable stars. The total number of known RR Lyrae stars in M92 now equals 17; 11 of type RRab and the remaining 6, RRc.

RRc variable v11 shows changes in the light curve shape. Due to the aliasing problems, however, we can state only that this star is biperiodic. There is a possibility that v11 belongs to the new group of recently discovered double-period RRc variables pulsating in non-radial mode (Olech et al. 1999).

Period–shift analysis for M92 and M2 has resulted in finding that, despite a significant difference in metal abundance between these two clusters ([Fe/H] $=$ -2.24 for M92 and -1.62 for M2), there is no firm evidence for a shift in period.

Acknowledgments. This work was supported by Wrocław University grant 2041/W/IA/2000. GK gratefully acknowledges financial support through the IAU grant.

References

Clement, C. 1997, Preliminary Fourth Edition of the Catalogue on Variable Stars in Globular Clusters, electronic version.

Kadla Z.I., Yablokova N.V., Gerashchenko A.N., Spasova N. 1983, Peremennye Zvezdy, 21, 827

Olech A., Kałużny J., Thompson I.B., et al. 1999, AJ, 118, 442

Small–Telescope Astronomy on Global Scales
ASP Conference Series, Vol. 246, 2001
W.P. Chen, C. Lemme, B. Paczyński

Blazhko Cycles of ω Centauri RRab Stars

Johanna Jurcsik

Konkoly Observatory of the Hungarian Academy of Sciences

Abstract. The present observational strategy of globular cluster variables does not favour long observational runs, which are needed to study the modulation properties of Blazhko type RR Lyrae stars. The only globular cluster for which we have enough data (both photographic and CCD) to determine modulation cycles of RRab stars is ω Centauri. Any connection between the modulation periods and other stellar or cluster parameters may serve as a new clue to the explanation of the Blazhko phenomenon.

1. Introduction

RR Lyrae stars are amongst the most studied and best known pulsating variables thanks to the systematic photographic surveys of globular cluster variables prior to the quick spread of CCD technique in the 80s. Although, we have a clear picture of their global evolutionary and pulsational status, there are still unsolved problems in the details. There is also no fully consistent description of the Blazhko phenomenon. The recent finding of nonradial modes in RR Lyrae stars, both from the observational (Olech et al. 1999; Alcoock et al. 2000) and theoretical (Dziembowski & Cassisi 1999) points of view, does not help us indeed to get closer to an understanding of the Blazhko behaviour.

To study the long term (order of magnitude of 10-100 days) phase and/or amplitude modulation of these stars, extended photometric data are required which are available only in a few cases. Among field stars there are altogether 5 variables with known frequency pattern of the modulation: AH Cam (Smith et al. 1994); RV UMa (Kovács 1995); RS Boo (Nagy 1998); AR Her (Smith et al. 1999); RR Lyr (Szeidl & Kolláth 2000). All these cases indicate that analytically the Blazhko phenomenon can be described with the appearance in the Fourier spectrum of equidistant side-peak frequencies at $\Delta f = f_m = 1/P_{Bl}$ distant from the primary frequency and its harmonics in both directions, but not necessarily with the same height.

Regarding cluster variables, only the existence of the modulation of different percentages of variables in different clusters has been clearly shown up to now. As variables in a globular cluster represent a much more homogeneous group (both in terms of their chemical compositions and their ages) than the field stars, then if any connection between the Blazhko properties and other physical parameters exists, it will be the easiest to detect it in individual clusters.

The present observational strategy of globular cluster photometry however, does not favour to follow any type of multi-mode behaviour. The only globular cluster with extended enough CCD photometry to study Blazhko variables as yet is ω Cen (Kaluzny et al. 1997). These data together with earlier photographic measurements have enabled us to determine the modulation cycles of the ω Centauri Blazhko type RRab stars.

2. Data and Method

- All the available long term photometries (compiled in Jurcsik et al. 2001) of ω Centauri fundamental mode Blazhko variables (13 RRab stars).

- Fourier analysis (MUFRAN package: Kolláth 1990).

  A least square minimization of the residuals (Θ transforms) using different number of harmonics (4-8) of the dominant mode frequency, and 2-11 additional frequencies from a 'Blazhko type' frequency pattern.

- Δf, which appears uniformly in the Θ transforms of the different datasets of a given variable, is accepted as the true modulation frequency.

3. Results

Blazhko cycle lengths are determined for all the 13 stars and are compared with other known parameters of the stars. There are hints that P_{Bl} depends on $\langle V \rangle$ and/or P, or through the metallicity dependence of these parameters, possibly on [Fe/H].

Acknowledgments.   This work has been supported by the Hungarian OTKA grants No 30954 and No 24022.

References

Alcoock, C. et al. 2000, ApJ, 542, 257

Dziembowski, W. & Cassisi, S. 1999, Acta Astronomica Vol. 49, 371

Jurcsik, J., Clement, C., Geyer, H., & Domsa, I. 2001, AJ, 121, 951

Kaluzny, J., Kubiak, M., Szymański, A., Udalski, W., Krzemiński, W., & Mateo, M. 1997, A&AS, 125, 343

Kolláth, Z. 1990, The program package MUFRAN, Occ. Techn. Notes of the Konkoly Obs., No.1, http://www.konkoly.hu/staff/kollath/mufran.html

Kovács, G. 1995, A&A, 295, 693

Nagy, A. 1998, A&A, 339, 440

Olech, A., Kaluzny, J., Thompson, I.B., Pych, W., Krzeminski, W., & Schwarzenberg-Czerny A. 1999, AJ, 118, 442

Smith, H., Matthews, J., Lee, K., Williams, J., Silbermann, A. & Bolte, M. 1994, AJ, 107, 670

Smith, H., Barnett, M., Silbermann, A., & Gay, P. 1999, AJ, 118, 572

Szeidl, B., & Kolláth, Z. 2000, ASPCS Vol. 203, 281

Small–Telescope Astronomy on Global Scales
ASP Conference Series, Vol. 246, 2001
W.P. Chen, C. Lemme, B. Paczyński

A Near Infrared Camera Refrigerated by Two Stirling Machines – an Alternative to Robotic Telescopes

José K. Ishitsuka I.[1], Takehiko Wada[2], Fumihiko Ieda[3],
Noritaka Tokimasa[4], Takehiko Kuroda[4], Masaki Morimoto[4],
Takeshi Miyaji[5], Toshihiro Omodaka[3], Munetaka Ueno[1],
Wataru Hasegawa[3], Shin-ya Narusawa[4], and Yoshifumi Waki[6]

[1] *Department of Earth Science and Astronomy, University of Tokyo,
Komaba, Tokyo 153-8902*
[2] *Institute of Space and Astronautical Science, Sagamihara, Kanagawa
229-8510*
[3] *Faculty of Science, Kagoshima University, Kagoshima 890-0065*
[4] *Nishi-Harima Astronomical Observatory, Sayo, Hyogo 679-5313*
[5] *Nobeyama Radio Observatory, National Astronomical Observatory of
Japan, Minami-saku, Nagano 384-1305*
[6] *Association of Amateur Astronomers, Nishi-Harima Astronomical
Observatory.*

Abstract. We have developed and tested a new near infrared camera
equipped with a 512 × 512 PtSi CCD and cooled by two independent
Stirling Cycle refrigerators. The camera, installed on the 60 cm reflector
telescope of the Nishi-Harima Astronomical Observatory (NHAO) since
April 2000, has begun regular observations toward infrared objects. Since
the reasonable cost and lower maintenance needs of the camera make it
more attractive, we introduce it as an alternative to robotic telescopes.

1. Introduction

We took four years to develop the near infrared camera NIHCOS (Ishitsuka et
al. in preparation) that uses a 512 × 512 PtSi CCD proved for astronomical
applications. The project is supported by the Hyogo Prefecture Nishi-Harima
Astronomical Observatory, the University of Kagoshima, the National Astronomical Observatory of Japan, the Institute of Space and Astronautical Science
and the University of Tokyo. NHAO is a public observatory managed by a local
government, the Hyogo Prefecture. The main objective of the observatory is
education and popularization of Astronomy within the general public. But it
also performs scientific oriented observations.

2. Observations and Preliminary Results

Regular observations were begun from April 24th of 2000, we began infrared
photometry of the semi-regular variable star R Crateris. Simultaneously, the
University of Kagoshima with the 6 m radio telescope performed the monitoring

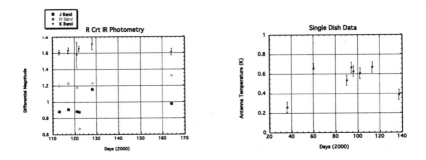

Figure 1.   Simultaneous IR photometry and microwave measurements of R Crt

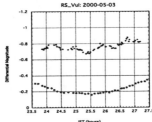

Figure 2.   RS Vul, V band and H band simultaneous photometry

of water vapor maser at 22 GHz (see Figure 1). The aim of the observations is to find the correlation between intensities of maser and dust emission and to determine the water maser radiation mechanism. On the other hand, the NHAO supports a group of amateur astronomers, and one of the members, Y. Waki, performed V band photometry of the binary star RS Vul, while infrared photometry was in progress. The equipment used for V band observations is a 7.5 cm refractor telescope with a ST-7 CCD. The results of the simultaneous photometry are presented in Figure 2.

3.   Conclusion

We successfully tested our infrared camera system and made photometric measurements, also imaged planets, galaxies, globular clusters, planetary nebulae, etc. The camera system is a prototype and could be an alternative to fully automated or robotic telescopes.

Acknowledgments.   The author gratefully acknowledges the financial assistance from the Japanese Foundation for Promotion of Astronomy (Travel grant) and the IAU grants that made it possible to attend the meeting.

Small–Telescope Astronomy on Global Scales
ASP Conference Series, Vol. 246, 2001
W.P. Chen, C. Lemme, B. Paczyński

Observations of Variable Stars by the 76-cm SuperLight Telescope of NCU

J. Z. Li[1,2],C.H. Wu[1],Z.W. Zhang[1],C.P. Chang[1], C.Y. Lin[1],H.H. Li[1],W.H. Tsay[1],W.H. Ip[1] & W.P. Chen[1]

1. Institute of Astronomy, National Central University, Chung-Li 32054
2. Beijing Astronomical Observatory, National Astronomical Observatory, Chinese Academy of Sciences, Beijing 100012, China
(Email: ljz@astro.ncu.edu.tw)

Abstract. The self-constructed Super-Light telescope of the National Central University is now ready for open use. Systematic studies of RR Lyraes stars and other variables are outlined and some preliminary results are introduced.

1. Introduction

During the last several months, the functionality of the newly constructed 76cm Super-Light Telescope of the Institute of Astronomy has been considerably improved. This small telescope has basically passed its testing phase and is now ready for scientific data acquisition. Light-variation of some CVs and RR Lyrae variables, obtained during the test mode observations of the telescope, is presented in this paper.

2. SLT Program of Variable Stars

- Study of some newly identified CVs by differential photometry

- Photometric study of a sample of unusual RR Lyrae stars from the Hipparcos Catalogue.

- Identification and study of a sample of RR Lyrae candidates from SDSS

- Monitoring of X-ray binaries as a part of our routine observations

3. Preliminary Results from the Test Mode Observations

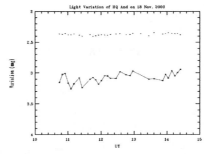

Figure 1. Light variations of a cataclysmic variable HQ And 15.3mag. (The time resolution of this run of observations of the CV is not high enough to resolve the rapid variations of the CV).

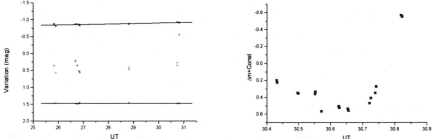

Figure 2. Composite light curve of an unusual RR Lyrae star , AA Cmi, from the fit of several days' discontinuous observations.

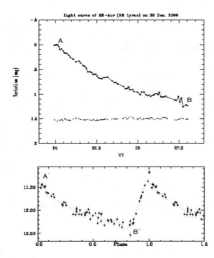

Figure 3. The light curve of another unusual RR Lyrae star, BH Aur, based on SLT observations (upper panel), as compared to data from the Hipparcos catalogue (lower panel).

Small–Telescope Astronomy on Global Scales
ASP Conference Series, Vol. 246, 2001
W.P. Chen, C. Lemme, B. Paczyński

Observations of Variable Stars With a Small Telescope at Tabriz University

D. M. Z. Jassur, F. Adabi, & N. A. Cham

Tabriz University, Iran

Abstract. For more than a decade a small group of Astronomers has been monitoring variable stars of different types. In the present contribution I will present the observations of chromospheric active binaries and near contact binaries and the results obtained from their analysis.

1. Introduction

The short-period group of the RS CVn stars, as defined by Hall (1976) and reviewed by Milano (1981), form a particular subgroup of binary systems that exhibit observational signs of magnetic activity. With orbital periods of less than a day, the stars in short-period group are relatively close to filling their Roche lobes. It has been proposed (e.g., Popper and Ulrich, 1977) that the evolution of RS CVn stars is characterized by at least one of the component stars evolving off the main sequence (but not to the point of Roche overflow). Hence, these stars would appear to be optimal candidates for examining the properties of cool stars prior to some dramatic change in their evolutionary status.

To help in determination of the physical properties of such lower main sequence stars, an analysis of their photometric light curves offers a well-known procedure. However, as has been long recognized (see Wood 1946), the "distortion waves" that are the photometric hallmark of the RS CVn systems complicate a reliable estimation of the system parameters that can be directly read from the optical light curves. These distortion waves do, however, convey information about the active regions of the photosphere. The basic problem in dealing with the photometry of these stars is to extract appropriately all the available information ideally to separate the effects of eclipses and normal proximity interaction from those associated with the "distortions". In this paper we attribute these latter effects to phenomena having methodologically desirable feature of formal representability, which we shall call "Starspots" maculation effects. Our main goals in this paper are to establish an effective procedure and baseline for analyzing the Starspot phenomena on the short-period RS CVn stars to reveal the nature of the activity cycles, and, as well, the physical properties of the stars themselves.

2. Observation

Five systems of the short-period RS CVn type eclipsing binaries, namely XY UMa, UV Psc, CG Cyg, RT And, and ER Vul were photoelectrically observed (e.g., Jassur et al 1993, 1994) during 1989 and 1995. The observations were made in yellow, blue and ultraviolet bands using the 40-cm Cassegrain telescope of Khadjeh Nassir-Addin Observatory of Tabriz University. The light was fed to a single channel photometer tube (RCA 1P21). Figure 1 shows UBV light curves of these stars.

3. Analysis

The analysis of light curves was carried out in the following three steps:
(i) The Wilson-Deviney light curve analysis method was used to find uncleaned parameters specifying the system. As much information as possible about each star was gathered from other published data. The values of limb darkening appropriate for the spectral type of respective components of each system were assumed and the available spectroscopic values for mass ratios were retained.
(ii) We deconvolved the theoretical light curve corresponding to our adapted model in the first step from the observed light curve. The object of this step was to obtain the intrinsic light variation of the systems. The quasi-sinusoidal light variation were interpreted in terms of cool star-spots covering a significant fraction of the stellar photosphere.
(iii) We cleaned the original observations by correcting for the presence of the distortion wave, adding the theoretical distortion wave effects of the data with opposite signs. Wilson-Deviney light-curve analysis method was then used to find cleaned parameters to provide us with information about the physical properties of the systems. To show the procedure the analysis of XY UMa is given below as an example.

4. XY UMa

XY UMa (BD+55° 1317) is a well known short-period binary system related to the RS CVn group. Geyer has observed it photoelectrically since 1995 and interpreted its unusual photometric behavior as being a result of a large-scale spot activity. The photoelectric observations of this star have been carried out at the Khajeh Nassir-Addin Observatory of Tabriz University and the light curves have been analyzed by the WD method. The results of this analysis are given in Tables Ia, Ib and II.

5. Near Contact Binaries

This is a group of binaries (Shaw, 1990) whose components are near enough to each other to have strong proximity effects like W UMA but are not in contact. They are divided into two subclasses according to their physical characteristics: V1010 Ophiuchi and F0 Virginis (named after the brightest member of each class). In former subclasses, the primary is at or near its Roche lobe while the secondary is inside its Roche lobe. But in the latter subclasses the primary is inside its Roche lobe and the secondary is at or near Roche lobe. Some members of V1010 Ophiuchis show the O'Connell effect that is max I being brighter than max II. If we accept that the contact binaries evolve from detached systems by means of AML via magnetic braking or via mass transfer/mass loss, then evidence for this might be the existence of near contact binaries. Our group has carried out observations of some members of near contact binaries (Figure 5a and 5b.) and some results have been obtained from light curve analysis.The observation and light curve analysis of TZ Lyrae a member of this group are discussed.

6. TZ Lyrae

The variable TZ Lyrae (BD+41° 3021), an eclipsing binary with a period of 0.5288 days is the brighter component of the visual double ADS 11219. In the NGC of double stars by Aitken (1932) the separation is given as a=2.87" in position angle 358.7° for the year 1917. D'Esterre (1914) discovered this star but Hoffmeisster determined the correct light elements. A photographic light curve of this Lyrae type system was published by Jordan (1929) from extrafocal images. Using this light curve orbital elements were published by Slonim (1934) and Krat (1935). The only photoelectric observations and light curve of this star were carried out by Binenndijk in 1970 through B and V filters who also obtained geometrical elements of the system from the analysis of his light curves (1972). In 1985, Kaluźny has used some normal points extracted from Binenndijk's observations and obtained a solution for this system.

The star was observed through BVR filters using essentially the same equipment employed for the observations of XY UMa. The faint component of the visual double was included in these observations. A total of 224 observations were obtained in each colour. According to our observations there is a significant shift in the epoch of zero phase with respect to the primary minimum (O-C=-0.005 day). The derived light curves of V, B and R colours are illustrated in Figure 5.

Analysis of the light curves was carried out by the differential corrections method. First we assumed that the system is detached and both components are inside their Roche lobes. So we employed mode II with starting values taken from Kaluźny (1985). After a few trials we noticed that a detached configuration is not a possible solution. Then we changed to mode IV, thus assuming that the system is a semidetached and the primary component fills its Roche lobe.After some trials the solution started to diverge. Finally mode V was chosen. The final results of this analysis are given in tables IIIa and IIIb. This solution indicates that the secondary component is in contact with its Roche lobe, while the primary is inside its Roche lobe (Figure 7.). The theoretical light curves are shown in Figure 6, overlayed with the observations.

References

Aitken, R. G. 1932, New General Catalogue of Double Stars (Carnegie Institution of Washington), 2, 988

Binenndijk, L. 1972, AJ, 77, 7

D'Esterre, C. R. 1914 Astron. Nacher.198, 163

Hall, D. S. 1976, in IAU colloquium 29, Multiple periodic variable stars, ed. W. S. Fitch. P. 287

Jassur, D. M. Z. and Kermani, M. H. 1993 IBVS, No. 3896

Jassur, D. M. Z. and Kermani, M. H. 1994, Astrophys. Space Sci., 219, 35-47

Jordan, F. C. 1929, Allegheny publ. 7, 146

Kaluzny, J. 1985, Acta Astron., Vol.35, 340-341

Krat, W. 1935, Russian Astron. J. 12, 25

Milano, L. 1981, in photometeric and spectroscopic binary systems, ed., E. B. Carling and Z. Kopal.331-360

Popper, D. M. and Ulrich, R.K. 1977, ApJ. (letters), 212, L131

Shaw, J. S. 1990, in Active Close Binary, ed. C. Ibanoglu. P.241

Slonim, E. 1934, Tashkent Bull. No. 4, 85

Wilson R. E., Devinney E. J. 1971, ApJ 166, 605

Wood, F. B. 1946, Contr. Princeton University Obs. No. 21

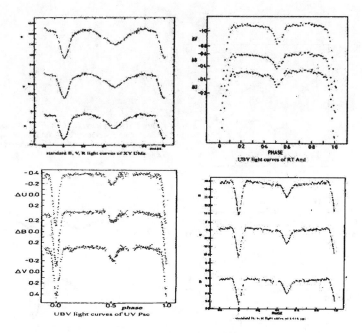

Figure 1.light curves of selected short-period RS CVn type eclipsing stars.

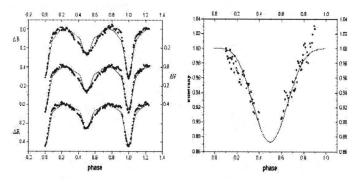

Figure 2. Initial light curve fit of XY UMa. The dots are individual observations. The solid lines are the theoretical light curves based on the parameters in Table Ia and Ib

Figure 3. Spot model fit (solid line) to the distortion wave calculation from the difference between the observational points and model in Fig. 2

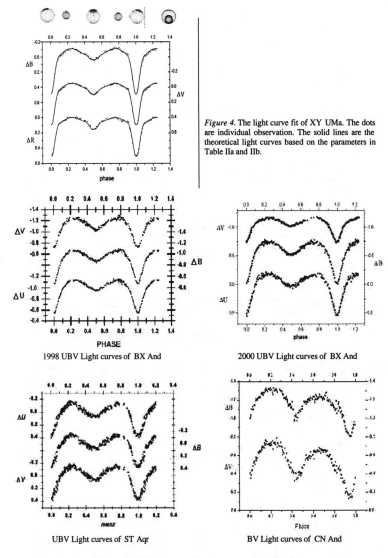

Figure 4. The light curve fit of XY UMa. The dots are individual observation. The solid lines are the theoretical light curves based on the parameters in Table IIa and IIb.

1998 UBV Light curves of BX And

2000 UBV Light curves of BX And

UBV Light curves of ST Aqr

BV Light curves of CN And

Figure 5a. Light curves of some near contact binaries

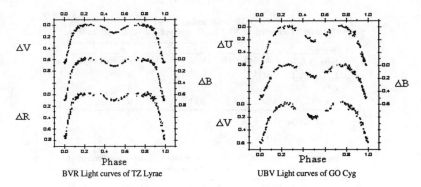

BVR Light curves of TZ Lyrae   UBV Light curves of GO Cyg

Figure 5b. Light curves of some near contact binaries

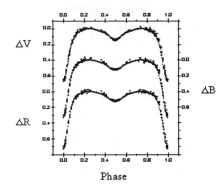

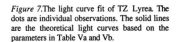

*Figure 6.*The light curve fit of TZ Lyrea. The dots are individual observations. The solid lines are the theoretical light curves based on the parameters in Table Va and Vb.

*Figure 7.*The light curve fit of TZ Lyrea. The dots are individual observations. The solid lines are the theoretical light curves based on the parameters in Table Va and Vb.

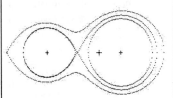

Table Ia (clean light curve)
Photometric Parameters of XY UMa (Wavelength - dependent parameters)

Parameter	B	V	R
$L_1/(L_1 + L_2)$	0.9506±0.0013	0.9219±0.0011	0.9196±0.0013
$L_2/(L_1 + L_2)$	0.0494	0.0781	0.0804
x_1 *	0.75	0.62	0.52
x_2 *	0.99	0.83	0.71

• Not adjusted

Table Ib (clean light curve)

Photometric Parameters of XY UMa (Non-Wavelength-dependent Parameters)

Parameter	B	V	R	Average
i	82.55°±0.20	82.65°±0.22	82.45°±0.25	82.55°
Ω_1	3.4250±0.0068	3.4450±0.0062	3.4580±0.0075	3.4427
Ω_2	4.1960±0.0155	4.2660±0.0126	4.3060±0.0133	4.2560
T_1	5872±35 K	5862±24 K	5822±23 K	5852
T_2	3965±13 K	3965±9 K	3985±10 K	3972
$q = \dfrac{m_2}{m_1}$	0.618±0.002	0.615±0.002	0.615±0.003	0.616
A_1 *	0.5	0.5	0.5	0.5
A_2 *	0.5	0.5	0.5	0.5
g_1 *	0.5	0.5	0.5	0.5
g_2 *	0.5	0.5	0.5	0.5
r_1 (pole)	0.3518±0.0008	0.3491±0.0007	0.3475±0.0009	0.3495
r_1 (side)	0.3654±0.0010	0.3622±0.0009	0.3604±0.0010	0.3627
r_1 (back)	0.3810±0.0011	0.3771±0.0010	0.3750±0.0012	0.3777
r_1 (point)	0.3969±0.0015	0.3920±0.0013	0.3892±0.0015	0.3927
r_2 (pole)	0.2043±0.0010	0.1989±0.0008	0.1963±0.0008	0.1998
r_2 (side)	0.2066±0.0010	0.2010±0.0008	0.1983±0.0008	0.2020
r_2 (back)	0.2107±0.0012	0.2047±0.0009	0.2018±0.0009	0.2057
r_2 (point)	0.2122±0.0012	0.2060±0.0009	0.2030±0.0009	0.2071

*Not adjusted

Table II parameters of the spot	
Colatitude of spot center λ	25°
Longitude of spot center β	180°
Angular radius of the spot γ	9%
Temperature of the spot	4682°

Table IIIa
Photometric parameters of TZ Lyrae (wavelength-dependent parameters)

Parameter	B	V	R
$L_1/(L_1 + L_2)$	0.9557±0.0132	0.9349±0.0151	0.9128±0.0125
$L_2/(L_1 + L_2)$	0.0443	0.0651	0.0872
x_1 *	0.60	0.5	0.35
x_2 *	0.81	0.66	0.57

• Not adjusted

Table IIIb
Photometric parameters of TZ Lyrae (non-wavelength-dependent parameters)

Parameter	B	V	R	Average
I	78.009±0.012	77.91±0.021	77.85±0.15	77.923
Ω_1	2.9524±0.00013	2.9566±0.0054	2.9517±0.0046	2.9537
Ω_2	2.7202	2.7251	2.7192	2.7215
T_1	8540±16k	8592±17k	8561±52k	8564.3
T_2	5152±5k	5152±14k	5153±18k	5152.3
$q = \dfrac{m_2}{m_1}$	0.4159±0.0029	0.4168±0.0041	0.4188±0.0035	0.4169
A_1 *	1.0	1.0	1.0	1.0
A_2 *	0.5	0.5	0.5	0.5
g_1 *	0.32	0.32	0.32	0.32
g_2 *	0.32	0.32	0.32	0.32
r_1 (pole)	0.3951±0.0014	0.3898±0.0015	0.3827±0.0020	0.3892
r_1 (side)	0.4061±0.0017	0.4083±0.0017	0.4102±0.0015	0.4082
r_1 (back)	0.4231±0.0021	0.4243±0.0021	0.4254±0.0055	0.4242
r_1 (point)	0.4435±0.0018	0.4445±0.0028	0.4458±0.0017	0.4446
r_2 (pole)	0.2854±0.0022	0.2871±0.0018	0.2818±0.0037	0.2848
r_2 (side)	0.2977±0.0018	0.2957±0.0019	0.2967±0.0011	0.2967
r_2 (back)	0.3302±0.0034	0.3161±0.0030	0.3152±0.0023	0.3205
r_2 (point)	0.3196±0.0027	0.3167±0.0039	0.3152±0.0026	0.3172

(Top; from left, facing) Woo, Choo, Chun, Kang, Park (standing), Kim
(Bottom; from left) H. K. Chang, (Guest of Chang), Peng, Chung, Hwang,
Kuan, H. L. Chiou, H. T. Huang

Small–Telescope Astronomy on Global Scales
ASP Conference Series, Vol. 246, 2001
W.P. Chen, C. Lemme, B. Paczyński

The Role of Small Telescopes in the Discovery and Follow-up of Near Earth Objects

Andrea Boattini

Osservatorio Astronomico di Roma, Sede di Monte Porzio Catone,
00040 Monte Porzio Catone (RM), Italy

Abstract. The discovered population of Near Earth Objects (asteroids and comets) consists of more than 1,300 bodies. Small telescopes provided with a large field of view have played a major role in the field, contributing to the discovery of about 95% of the total known population. We present a brief summary of the steps that have led to the present knowledge and suggestions to improve it. A proposal for accessing archival images to search for previous apparitions is presented.

1. Overview of Near-Earth Objects

One field of astronomical research where small telescopes have played a fundamental role is the discovery and follow-up of Near Earth Objects (NEOs).

NEOs are conventionally defined as those minor planets, asteroids and comets, whose perihelia $q \leq 1.3$ AU. NEOs have been divided into four categories, based on their osculating orbital elements (Shoemaker *et al.*, 1979; Tedesco *et al.* 2000):

Amors	$1.0 < q \leq 1.3 \ AU$
Apollos	$q < 1.0, \quad a \geq 1.0 \ AU$
Atens	$a < 1.0, \quad Q \geq 1.0 \ AU$
IEOs	$Q < 1.0 \ AU$

where a is the object's semi-major axis and Q its aphelion distance to the sun; **IEOs** stands for *Interior to the Earth Objects*. The cometary component, based on our current knowledge, is much less numerous and will not be addressed in this paper. Comets are conventionally divided into two groups, short and long-period. As of February 28, 2001, the asteroidal component of NEOs, Near Earth Asteroids (or NEAs), include almost 1,300 bodies, with sizes ranging from 30-km to less than 10 meters (The NEO Page).

The most recent studies estimate that there are about 1,000 NEAs of 1 km or bigger and a much higher number at smaller sizes. Only 40 % of the km-sized and a minor fraction of the smaller objects are known; furthermore good orbits are not available for all of them.

1.1. What is Needed for the Study of NEOs

Our knowledge of NEOs has been achieved through three major steps: discovery, follow-up and space studies. They can be briefly outlined according to the following scheme: a) **Discovery** - *photographic* (sporadic, systematic from 1973 to 1996), lim. mag. 15 - 18.5 V or *CCD* (from 1989, large field of view (FOV) from 1996), lim. mag. 18 - 22 V; b) **Follow-up** - *astrometric* - optical (phot., CCD), radar; *physical* - morphological, dynamical and miralogical characterization; *dynamical* - long and short-term evolution, collision analysis; c) **Space missions** - *physical studies* - radiometry, thermal infrared studies, near-IR spectroscopy, spectrophotometry; *threat mitigation* and *discovery*.

2. The Role of Small Telescopes in the Discovery of NEOs

Astrometric and photometric work is where small telescopes have played a leading role. NEO discovery is not an easy task to conduct successfully: because of their vicinity to the Earth during their visible apparitions, these objects can appear anywhere in the celestial sphere. They also tend to move pretty fast across the sky and remain visible for only a short period of time. In order to conduct a successful NEO search program the two most important requirements are: i) to sweep as much of the sky as possible in a limited amount of time and to inspect the data in near real time; ii) to reach relatively faint magnitudes with short exposures, since the limiting magnitude of NEOs is set by their apparent angular speed in the sky. For this reason Schmidt telescopes and astrographs have been among the most successful instruments when photography has been used, while other optical configurations, such as short focal length Cassegrain and Newton systems have proven their potential with CCD systems. NEOs have been discovered during the course of programs with different objectives (Helin and Dunbar 1990). We can determine three distinct historical periods:

1900-1970: incidental discoveries, made in the course of programs initiated for other objectives, such as mapping the sky in different colours, or studying the proper motions of stars. Most of these discoveries were achieved in the course of general asteroid surveys, not directly addressed to NEOs.

1970-1990: first NEO search programs. In the early 1970s the first dedicated NEO search program was started at Mount Palomar using the 0.46-m Schmidt. Two programs operated this telescope, PCAS and PACS and contributed to about 50 % of all the photographic discoveries. The Australian project AANEAS was also quite successful (Steel et al. 1997). In 1983 two NEAs were discovered by the infrared satellite IRAS. The most interesting result was to obtain a first inventory and characterization of the NEO population, its size distribution and a hazard assessment.

1990-2001: second period of NEO search programs. The transition between the first and the second period started about ten years ago, when the scientific community realized that in order to discover the great majority of km-sized NEOs, the ones that have the potential to cause a global catastrophe, it would have taken more than a century to accomplish this task with photographic methods. To shorten the time of such a survey, called the *Spaceguard Survey* (Morrison 1992), to only one or two decades, electronic devices such as large format CCDs or CCD mosaics have to be used because of their digital processing capacity, greater dynamic range and quantum efficiency. After the pioneering work of the Spacewatch program (Scotti 1993), the only CCD program between

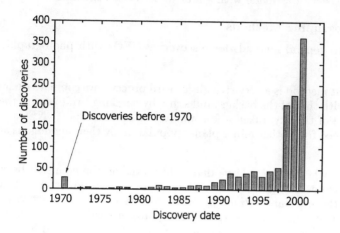

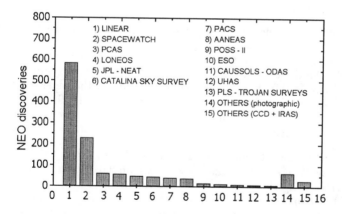

Figure 1.   The upper plot shows the number of NEA discoveries per year in the course of the last century. The first NEA, (433) Eros, was discovered in 1898. We can see a slow but constant rise in the discovery rate by the photographic programsin the 1970s and 1980s, before the great jump in 1997/1998, when the wide-field CCD programs started their operations. The lower plot shows the most successful programs as of February 28, 2001. LINEAR has been the most productive so far, and in the past three years has been credited with the discovery of two-thirds of all NEAs. CCD survey programs cover two-thirds of the visible sky to a magnitude of 18 - 19 V about twice a month.

1989 and 1996, further CCD projects were started and have provided very successful results (The NEO Page; JPL-NEOP). Only in 1997 there was the first NEO discovery with a telescope bigger than two meters and only in 2000 the first such discovery was made with a four meter class instrument.

2.1. Photographic Methods

There are two general methods for discovering NEOs with photographic techniques:

- The first method is a very straightforward process: we compare their movement with that of the background stars, by checking two plates of the same location of the sky, taken a few minutes or an hour apart. NEOs are distinguished from other minor planet population by their anomalous angular speed.

- In cases where objects leave detectable trails on the emulsion because of longer exposures and/or larger plate scales, a very powerful technique to use is the *exposure gating*: by breaking the exposure into two parts of unequal duration on a single search plate, separated by a short blank interval, minor planets can be unambiguously distinguished from defects, and their motions, both apparent speed and direction, can be determined (Helin and Dunbar, 1990).

2.2. CCD Methods

Similar to photography, there are two general search methods for CCD systems:

- **Stare-mode**: consists of comparing three to five frames of the same field taken 15-30 minutes apart. Detection is performed by using automatic and semi-automatic software. Its efficiency depends on the pixel scale, seeing conditions, number of frames used and the kind of algorithms set up for detection threshold. Special software has last been developed to detect long trailed images at very low sigma to noise ratios ($\sigma = 1$ or less).

- **Scan-mode**: the telescope does not track at all. When the CCD is operated in scan mode, it means that the accumulating electronic charge is transferred along the CCD rows in sync with the drift of the sky across the CCD. Scans are n pixel high in declination (where n is the size of the CCD in pixels) and an arbitrary number of rows long in right ascension. A survey region consists of three scans of the same length at the same location of the sky. The detection is made in the same way as described for the stare mode approach.

3. Pitfalls in the Current Knowledge of NEOs

There are three major pitfalls in the current astrometric activities on NEOs:

- The discovery activity is still not supported by equivalent follow-up work, both astrometric and physical. One of the reasons for this situations is a need of larger telescopes.

- The present knowledge of NEOs with small aphelia is very limited (*Atens*), or non existent (*IEOs*). The reason is that only in the last few years NEO searching has been extended to smaller solar elongations, but a lot more needs to be done: small telescopes can still contribute in this task.

- Discovery programs are not geographically well distributed and coordinated. One third of the visible sky is not surveyed since there are no NEO discovery programs in the southern hemisphere. The distribution in longitude is also critical because search efforts are essentially confined to the dry states of the USA.

4. Follow-up Work on NEOs

The reliable determination of the orbit of a newly discovered NEO is impossible without precise follow-up astrometry over a sufficient arc of observations (Ticha *et al.* 2000). When the technology developed that allowed a rapid diffusion of CCDs in the early 1990's, the quality of follow-up work improved significantly and the volume of data submitted to the Minor Planet Center (MPC) has increased exponentially. This was also possible thanks to the availability of much better astrometric catalogues. The result was that even amateur astronomers have been able to set up competitive equipment, at least for the follow-up process of NEOs. About one hundred well-equipped stations from all over the world take part in the process of follow-up astrometry of NEOs and a few of them also conduct photometric studies. The typical size of an *average* instrument used in this task is only 0.4-m or 0.5-m. This allows NEOs to be followed up to magnitude 20. Fainter magnitudes can be reached only when observing time at better equipped professional facilities is allocated.

The Spaceguard Central Node (SCN), and other centers (NEODyS; Koehn and Bowell 1999) help observers to devise their observing plans. The SCN, for example, provides a few prioritized lists, updated on a daily basis, where observers can select those objects in greatest need of observations. It also organizes direct observing campaigns when objects particularly dangerous risk to be lost prematurely.

5. The Importance of Astronomical Archives for Cataloging NEOs

The remarkable increase in the NEO discovery rate and significant improvements of the methods developed by the teams that study collision analysis and short-term evolution, has led to the discovery and the recognition of a few NEOs with small collision probabilities. In order to assess this danger, the orbits of these bodies must be known with great accuracy and this only can be achieved by obtaining astrometric positions over a long period of time.

The inspection of archival material where one or more images of NEOs, previously undetected, may be present, offers several advantages: i) very good orbits can be obtained soon after discovery; ii) two km-sized NEAs with small collision probabilities were *mitigated* using data from photographic precovery images (1997 XF11 and 1999 AN10); iii) a great deal of telescope time, as well as money is saved; iv) this can be a day-time activity, unaffected by the vagaries of weather;

Although a few dedicated teams routinely look for pre-discovery images of NEOs in a few high quality archives (DANEOPS, ANEOPP), the great bulk of the estimated two million photographic plates/films from professional collections is not easily accessible. In addition, CCD archives are quickly developing that promise to represent a very potential resource in future decades.

5.1. A Proposal for International Collaboration

The recent development of NEO scientific work imposes a higher degree of coordination among all the tasks involved to obtain knowledge of these bodies. For the reasons discussed above, the archival work, both collection and organization, needs to be improved. The SCN is trying to organize a proper interface for NEO work. There is a need for an international center, defined by the IAU, eventually sponsored also by other Institutions and agencies that could organize all this archival material, both photographic and CCD. For NEO work it is essential to meet the following requirements:

- A complete list of CCD surveys and various sky patrols initiated for diverse objectives.

- For each program a list of catalog files reporting the epoch of the exposure, the exposure time, the position, the field of view (FOV), and the limiting magnitude

- Easy retrieval of desiderable data in digital form.

The goal of such an effort would be to complete the inventory of NEOs (NEOs with good orbits!) in much less time and to assess the danger posed by these objects using all the available resources.

6. Future Programs

Many new projects all around the world are on the horizon for the near future. Asia is starting several programs, with the BATTeRS project in Japan taking a leading role (Isobe, this meeting). I'm involved in an Italian project, CINEOS, that will survey regions at small solar elongations to look for Atens and IEOs. The UK and other European countries are working on additional projects. Australia should see the first NEO survey in the southern hemisphere later this year, as part of the Catalina Sky Survey. In the USA, significant upgrades are expected on the short term at Mount Palomar and in the course of the Spacewatch program. Small telescopes are going to play a leading role, even from space: in order to address the danger of Atens and IEOs more efficiently and to better characterize the physical properties of all NEOs, many space agencies are undertaking a few dedicated missions that will start operations within a decade.

References

ANEOPP, http://www.arcetri.astro.it/science/aneopp/

DANEOPS, http://earn.dlr.de/daneops/

Helin E.F. and Dunbar S.R. 1990, Vistas in Astronomy, 33, 21-37

Koehn B., and Bowell E. 1999, Asteroid Observing Services,
     http://asteroid.lowell.edu

Minor Planet Center, http://cfa-www.harvard.edu/iau/mpc.html

Morrison D. (editor) 1992, The Spaceguard Survey, NASA Report

NASA-JPL NEO Program, 2001, http://neo.jpl.nasa.gov

NEODyS, http://newton.dm.unipi.it/neodys

The NEO Page, Minor Planet Center,
     http://cfa-www.harvard.edu/iau/NEO/TheNEOPage.html

Scotti J.V., Asteroids, Comet, Meteors, 1993, 17-30

Shoemaker E.M., Williams J.G., Helin E.F., Wolfe R.F. 1979, Asteroids, 253-282

Spaceguard Central Node, http://spaceguard.ias.rm.cnr.it

Steel D.I., McNaught R.H., Garradd, G.J., Asher D.J., Russell K.S. 1997, Austr.
     J. Astron., 7, 67-77

Tedesco E.F., Muinonen K., Price S.D. 2000, P&SS, 48, 801-816

Ticha J., Tichy M., Moravec Z. 2000, P&SS, 48, 787-792

Small–Telescope Astronomy on Global Scales
ASP Conference Series, Vol. 246, 2001
W.P. Chen, C. Lemme, B. Paczyński

NEOPAT: Near-Earth Object PATrol program

Hong-Kyu Moon[1], Moo-Young Chun[1], Yong-Ik Byun[2], Wonyong Han[1], Seung-Lee Kim[1], Young-Beom Jeon[1], Yong-Woo Kang[2]

[1] *Korea Astronomy Observatory, Taejon, 305-348, Korea*
[2] *Dept. of Astronomy, Yonsei University, Seoul, 120-749, Korea*

Abstract. In 2000, Korea Astronomy Observatory launched the Near-Earth Object Patrol (NEOPAT) program. NEOPAT has conducted follow-up observations of NEOCP (NEO Confirmation Page) objects and discovered 52 new main-belt asteroids during the observation runs. We initiated collaboration with the Yonsei Survey Telescopes for Astronomical Research (YSTAR) team for NEO search. Wide-field of view, fast readout time, and fully autonomous data pipeline will enable us to detect and track NEOs with a high efficiency. Scheduled to begin active operations in mid-2001, our survey system is going to be the first network of robotic telescopes for NEO search with automatic access to both hemispheres.

1. Introduction

Among the enormous numbers of bodies orbiting the sun, only a tiny fraction of them follow paths which bring them to the near-Earth space. About 90 percent of the Near-Earth Objects (NEOs) are near-Earth asteroids (NEAs) or short-period comets, and the other 10 percent are long-period comets. The objective of an NEO survey is to detect these objects during their periodic approaches to the Earth, to calculate their long-term orbits, and to verify the potential threat over the next few centuries. According to Chamberlin (2001), the statistics of NEAs discovered by predominant search programs in 2000 is as follows; LINEAR: 160, NEAT: 14, Spacewatch: 27, LONEOS: 38, Catalina: 13, others: 12. The total number of NEA is a steeply increasing function of time where the cumulative total in 2000 was 1244, while it was only 350 in 1995.

Being aware of past impacts and their destructive power, the general public as well as the astronomical community began to take anxious interest. The collision of the comet Shoemaker-Levy 9 with Jupiter, and such Hollywood films as "Armageddon" and "Deep Impact" added to the concern. Moreover, the first orbital calculations of two NEAs, 1997 XF11 and 1999 AN10 indicated that they could impact Earth in the next 50 years. However, further study has dismissed this probability. Recently, 2000 SG344 was covered on the evening news in Korea, with an animation which depicted the extinction of the dinosaurs in an exaggerated manner. The issue of an asteroid impact has widely attracted public interest, and encouraged the National Assembly and the Korean government to support astronomers to conduct NEO search. In 2000, Korea Astronomy Observatory launched the Near-Earth Object Patrol (NEOPAT) program designated as National Research Lab by Ministry of Science and Technology. NEOPAT is now teamed up with YSTAR (Byun et al. 2001) for collaboration in NEO research. On the other hand, NEOPAT is conducting follow-up astrometry of newly discovered NEOs with existing facilities.

2. Follow-up Observations

Since early 2000, NEOPAT has been reporting on the results of follow-up observations of NEOCP (http://cfa-www.havard.edu/iau/ NEO/ToConfirm.html) objects to Minor Planet Center. The observations are carried out with the Sobaeksan Optical Astronomy Observatory (SOAO) 61cm telescope and the Bohyunsan Optical Astronomy Observatory (BOAO) 1.8m reflector with CCDs. The NEOCP gives access to ephemerides for newly-discovered moving objects in need of confirmation. SOAO allocates 4-5 days every month for NEO follow-up observation. With a PM512 CCD camera mounted on the 61cm reflector, we monitor the orbit of updated NEO candidates with $17 \leq m_R \leq 19$. The observations are made more than three times over an arc of several hours during the course of at least one night, with a typical exposure time of 200-300 sec. Although we have yet applied for BOAO 1.8m telescope time, follow-up astrometry is being performed at BOAO before astronomical twilight, or in bad seeing conditions in the middle of the regular observation run. Employing the 1.8m telescope, follow-up observations are made for 18-20 mag NEO candidates in V or R band with 30~300 sec exposures in binning mode.

In each image, one can determine accurate position and magnitude of a suspected NEO with ASTROMETRICA using field stars with known coordinates. The coordinates of the comparison stars are taken from the HST Guide Star Catalogue or USNO 2.0. In Table 1, we list the results of follow-up observations performed at SOAO and BOAO from January to December, 2000. The total number of observations reported to MPEC is 1060 during the season.

Table 1.   Observation summary of NEOCP objects

Month (2000)	SOAO (345)			BOAO (344)			Reference
	No. of data	No. of objects	MPEC	No. of data	No. of objects	MPEC	
2	0	0	0	42	6	1	MPC38264
3	35	9	4	73	13	4	MPC38972
4	25	5	3	0	0	0	MPC39716
5	33	5	4	16	5	4	MPC40595
6	0	0	0	66	6	4	MPC40751
9	13	3	2	23	8	0	MPC41267
10	0	0	0	21	6	6	MPC41456
11	8	2	2	482	44	1	MPC41642
12	0	0	0	337	60	2	MPC41805

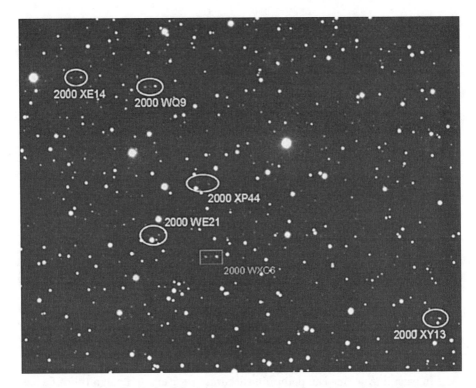

Figure 1.   A CCD image obtained at BOAO on Dec. 7, 2000. Five new asteroids and one previously known asteroid are displayed in the same field. This is a color-coded image made from three consecutive V band CCD frames in which the asteroids were moving relative to the background stars.

3. Discovery of New Asteroids

Since May 2000, we have discovered 52 new asteroids during the follow-up observation of NEOCP objects, and also from color-coded images of the Crab nebula (M1) taken for public release. The first asteroid, 2000 KJ4 was found on May 28, 2000, and turned out to be a new main-belt asteroid with an absolute magnitude of (H)∼15.6, after confirmation observations on consecutive nights. The name, date of designations, and observers of these serendipitous discoveries are summarized in Table 2.

Figure 1 shows a discovery image of Dec 7, 2000 taken from the BOAO 1.8-m reflector equipped with a SITe 2K CCD camera. It reveals five newly discovered asteroids, 2000 XE14, 2000 WQ9, 2000 XP44, 2000WE21, and 2000 XY13 marked with an ellipse, together with a known asteroid at the location shown by a box in the same field. This is a color-coded photograph made from three consecutive V band CCD frames where the asteroids were moving relative to the background stars, so they look slightly trailed with different colors. We conducted confirmation observations of the neighboring fields, and within

Table 2.　New main-belt asteriods discovered at BOAO

Designation				
2000 KJ4	2000 WQ9	2000 WE21	2000 WD21	2000 WR21
2000 WV28	2000 WV50	2000 WU50	2000 XY13	2000 XZ13
2000 XA14	2000 XB14	2000 XC14	2000 XD14	2000 XE14
2000 XJ15	2000 XK15	2000 XL15	2000 XM15	2000 XA44
2000 XB44	2000 XC44	2000 XD44	2000 WZ26	2000 XJ2
2000 XK2	2000 XO44	2000 XP44	2000 XQ44	2000 XR44
2000 XL53	2000 XM53	2000 XN53	2000 XO53	2000 XP53
2000 XQ53	2000 XT53	2000 YE04	2000 YB16	2000 YC16
2000 YD16	2000 YS30	2000 YDA1	2001 AN19	2001 AO19
2001 AP19	2001 AQ19	2001 AR19	2001 AS19	2001 BB03
2001 BC03	2001 BZ10			

a month, found forty new moving objects. The newly discovered objects are also known to be main-belt asteroids with semi-major axes, $2.2 \leq a \leq 3.4$AU, and ellipticities of the orbits ranging from 0.0 to 0.25. The reasons we discovered dozens of new objects in a relatively short period of time are that: (1) December is the most favorable season for astronomical observation in Korea, and period when the ecliptic rises high above the horizon, (2) We can reach as faint as 20th magnitude, provided that we make use of the 1.8-m telescope with a suitable exposure time. For the time being, there is no dedicated 2m class telescope in regular service for asteroid search.

4.　NEOPAT-YSTAR Collaboration

A robotic telescope is suitable for repetitive and routine work such as discovery and tracking of NEOs with maximum efficiency. Undoubtedly, a survey project can be successfully accomplished with a global network of telescopes of this kind. NEOPAT team is developing such system in collaboration with YSTAR.

We employ wide field optics and a fast, fully automated computing pipeline. The 0.5m f/2.0 primary optics identical to the one prepared for the TAOS project (King 2001), provides 3.5 square degrees FOV onto an AP10 14μm CCD chip. The telescopes are being subjected to remote observation and safety tests. Our first telescope has just begun remote autonomous observations at Yonsei University Observatory, and after its launch in mid 2001, we plan to install the second telescope in Sutherland, at the South African Astronomical Observatory (SAAO). Before the SAAO installation, the NEOPAT-YSTAR team will complete the fine-tuning and development of the 3TB data storage system, the data pipeline, the observation planner/scheduler, and the off-site image archiving system. We intend to archive the data at the Korea Astronomy Observatory and to build a common database for ease of future analysis.

In a couple of years, we will expand the number of survey telescopes and place them home and abroad to increase the detection and tracking efficiency by multiplying the coverage. As such a network of survey telescopes is expected to provide a huge search volume, a large number of new NEOs are expected to be found far from the ecliptic plane. Moreover, orbits of long-period comets are so elongated and inclined to the ecliptic plane that the warning time for objects of this kind would be as short as a year compared to decades or centuries for asteroids. In spite of their small apertures, our survey telescopes offer more frequent coverage of the entire sky for effective search of long-period comets.

5. Conclusions

The NEOPAT program has carried out confirmation observation of NEOCP objects and discovered 52 new asteroids, among them forty were found within a month. In 2000, the joint program between NEOPAT and YSTAR was launched. We have just started routine autonomous observation with our first telescope and the second one is scheduled to be installed at SAAO in summer 2001. In the coming years, a network of survey telescopes will be constructed around the globe for more efficient monitoring of NEOs.

References

Byun, Y.-I. et al. 2001, in these proceedings
Chamberlin, A. 2001, http://neo.jpl.nasa.gov/missions/stats.html
King, S. K. 2001, in these proceedings

Small–Telescope Astronomy on Global Scales
ASP Conference Series, Vol. 246, 2001
W.P. Chen, C. Lemme, B. Paczyński

An Education Program Using Tera-Byte NEA Observation Data

A. Asami, D.J. Asher, T. Hashimoto, S. Isobe, S. Nishiyama,
Y. Ohshima, J. Terazono, T. Urata, and M. Yoshikawa

Japan Spaceguard Association

Abstract. There are three wide-field telescopes at the Bisei Spaceguard Center operated by the Japan Spaceguard Association. These telescopes are dedicated to detect near-earth asteroids and produce several tera-byte data per month. Since these data contain many main-belt asteroids, we will use them for an education program that will allow school pupils and the general public to find new main-belt asteroids. We are now developing a new software for its purpose.

1. Introduction

There are several teams developing automatic remote operation telescopes called robotic telescopes. These telescopes make day-time observing possible because the telescopes are far apart in longitudes, well suited for day-time astronomy education. However, within this system, one telescope is occupied by only one team and many telescopes should be prepared to make many teams satisfy. This situation brings a limitation on usage of the telescope system.

An effective way in astronomical education is to offer pupils and others good astronomical exercises. If these exercises contain only ready-made data, their interest in astronomy can not continue to flourish. It is certainly better to use original data than ready-made data. Our telescopes are producing so many original data which can be distributed to more than a few thousand teams per month. Therefore, we are developing a new asteroid detection software by which even non-professionals can find a number of moving objects from several given images obtained by the BATTeRS (Bisei Asteroid Tracking Telescope for Rapid Survey) team under the Japan Spaceguard Association.

Here, we will describe our telescope system and the newly developed educational software.

2. The Near-Earth Asteroid (NEA) Project at the Bisei Spaceguard Center

During the last decade, a possibility of NEA collisions with the Earth and their hazardous consequences to human-beings has been studied and naturally we first have to detect these NEAs and determine their orbits in order to avoid the hazard. At this stage there are 5 US teams which work mainly for this detection observations. However, these are not enough to detect all the NEAs with a

Figure 1.   An over-view of the Bisei Spaceguard Center. A 1-m telescope will be set in the dome and the 25-cm and 50-cm telescopes are
operated in the sliding roof building.

diameter large than 0.5 km within decades, and therefore it is highly desirable
within the NEA community to develop more NEA detection telescopes.

    In Japan, we have set up a non-profit organization, the Japan Spaceguard
Association, and lobbied for new NEA telescopes.   These activities brought
successful results and now new NEA telescopes are under construction at the
Bisei Spaceguard Center (Figure 1) (Isobe 2000; Isobe and Williams 2000; Isobe
and Japan Spaceguard Association 2000). There are three telescopes with wide
fields of view: 25 cm and 50 cm telescopes, with a two degree field (Figure 2)
and 100 cm telescope with a three degree field. The first and second telescopes
are being tested and have so far detected over one hundred new asteroids. The
third telescope will begin operation in summer, 2001.

    Table 1 shows some parameters of the 100 cm telescope which has a size
for the focal plane of 162 mm and 3 degree. To cover this large field, 10 2 k
x 4 k CCD chips are set as shown in figure 3. Our target magnitude is 20.5,
with an exposure time of about 30 seconds. Considering the read-out time, the
data-transfer time, and the telescope pointing time, one should get 160 Mbyte
of data per minute, which gives about 2 Tera-byte per month. This number is
very big and therefore we are developing automatic reduction software.

    We are interested in detecting NEAs in our observational data. However,
our data contains much more information which is astronomically useful. In

Figure 2.   A view of the 50-cm telescope having a short focal length.

Table 1.   Some specific parameters of the 1-m telescope.

Primary mirror	100 cm
Focal ratio	F/3
Focal plane size	162 mm
Field size	3 °
Spectral range	470 nm - 800 nm
Point source energy within 15 μm x 15 μm	80 %
Typical image size at the site	2 arc seconds
CCD pixel size	15 μm
Focal length	3 m
Diameter of a 1 arc second image	15 μm
CCD pixel number	2k x 4k
Number of CCD chips covering 3 ° field	10

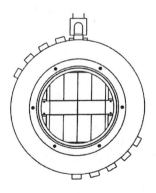

Figure 3.   A distribution of 2 k x 4 k back-illuminated SITe CCD chips covering 3 degree field.

Table 2.   Schedule of data distribution.

Data Distribution	
0 - 1 day	Detection of NEAs and Space Debris
1 - 14 days	Educational Program
15 - 28 days	Open the Data to Public
28 days	Delete the Data

Figure 4.   The front page of the software "Asteroid Catcher B-612".

recent years, there has been much development in high speed data transfer systems through Internet and fiber optic lines. Therefore, we decided to distribute the data as outlined in Table 2. We will release all the data two weeks after the observations to registered scientists. During these two weeks, the data will be used for educational programs.

3.   An Educational Program Using the NEA Observational Data

The idea of our software for educational program is simple. For the NEA observations we usually take 30 second exposures several (3 to 5) times with an interval of 15 minutes to 1 hour for a celestial field. One downloads these image data into the memory of a computer and blinks these images after adjusting stellar positions top on top. Moving or variable objects are easily identified within the star field.

To do this properly, one has to develop a system with 1) a fiber optics line, 2) a user's registration home page, 3) a software to identify which team gets which sets of data, 4) a software manual, and 5) a parameter summary of detected objects.

1) The Japan Global Network (JGN) project started a few years ago. The Okayama prefecture where our telescopes are located is expanding its own Okayama Giga-bit Network (OGN) connected to the JGN. Fortunately, this OGN passes close to our site and we can connect our system to it with a small

amount of cost. It has a capacity of 600 Mbps which is enough for our data transfer.

2) The goal of our homepage is not only to publicize our program but also to identify each applicant and his category of data access right. Certainly, this process needs some minimal manual actions.

3) After a team obtained an identification number and password, they can request our data. However, since there is a criteria for data distribution as shown in Table 2, we have to identify what class of data access right they have.

4) A team can see a data list of exposure images including observing dates, exposure times, filters, central right ascensions and declinations, etc. and then requests a preview image since the whole image would be too large. After choosing the necessary image(s), they can down-load the image data through the Internet. For educational program, we are preparing a software called "Asteroid Catcher B-612" (Figure 4), which is used to detect moving objects and which also contains a manual describing how to use the software.

5) Detected positions of the moving objects should be relayed to the Minor Planet Center (MPC) of the International Astronomical Union (IAU) through the JSGA. All the information should match the format of the MPC well. Our software shows school pupils and the general public how to write up the information and to send it to us.

All this is shown on our homepage : http://www.spaceguard.or.jp.

4. Methods to Publicize our Program

The JSGA has its own homepage and a program "Spaceguard Detective Agency - Tracking down the asteroids" and tries to publicize it through public lectures and international conferences. To make our program more efficient, we collaborate with the Liverpool John Moore University (JMU) with an interface of the British Council. The JMU is developing a 2 m robotic telescope in Canarie Island and an educational software. Both teams share their own software and programs. The British Council organized two international meetings, an Asian one in Kuala Lumpur and a European one in Liverpool, in 2000, where representatives from 14 countries got detailed information about the joint project called the International Schools' Observatory (ISO) (Figure 5).

The program will start in the summer of 2001 and some test programs are already underway. In Japan, we will invite 500 groups of school pupils and public people to join the program with collaboration of the Yomiuri Newspaper Co. Ltd, and as a test program, they will get all the software and image data on CD-ROMs.

After this is successful, many people will know about our program and hopefully will decide to join. At the full operation phase, we have the possibility of distributing to nearly 500 teams per full clear night.

5. Conclusion

All the teams which join this program can handle real astronomical data and have a high possibility to detect new asteroids. Then, they can get names of

The
International Schools'
Observatory

is operated by
British Council Japan

in co-operation with
Japan Spaceguard Association
Liverpool John Moores University

Figure 5.   A front page of the program "International Schools' observatory", which is now under development.

discoverers for those objects. It is an exciting experience and can be a real scientific education tool. We are intending to proceed with this program on an international basis.

References

Isobe, S. 2000, The position of the Japan Spaceguard Association with regard to NEO problems, Planetary and Space Science, 48, 793-795

Isobe, S. and Williams, R. 2000, Education and research opportunities for mining the data of the Bisei Spaceguard Center, in Proceedings of SPIE, Vol. 4010, 168-171

Isobe, S. and Japan Spaceguard Association 2000, Japanese 0.5 m and 1.0 m telescopes for near-earth asteroid observations, Ap&SS, 273, 121-126

Small–Telescope Astronomy on Global Scales
ASP Conference Series, Vol. 246, 2001
W.P. Chen, C. Lemme, B. Paczyński

CCD Photometry of Two Asteroids (895) Helio and (165) Loreley

H.-S. Woo[1], S.-L. Kim[2], M.-Y. Chun[2], M.-G. Park[1]

[1]Dept. of Astronomy and Atmospheric Sciences, Kyungpook National University, Taegu, 702-701, Korea
[2]Korea Astronomy Observatory, Taejon, 305-348, Korea

Abstract. We have presented photometric results for two asteroids (895) Helio and (165) Loreley. The observations were performed from 2000 Oct. to 2001 Jan. using the 61cm telescope installed at Sobaeksan Optical Astronomy Observatory in Korea.

The asteroid (895) Helio was classified as FCB type and (165) Loreley as CD type (Tholen 1989). Their diameters are known to be 147km for Helio and 160km for Loreley (Clifford 1988).

Our observations were carried out in the V and R bands on Oct. 30, Nov. 2, Nov. 30, Dec. 1 and Dec. 3 for the asteroid (895) Helio. The asteroid (165) Loreley was observed intermittently for the latter three days, and on Jan. 16. The observing conditions for two asteroids are given in Table 1.

For the first two days, we chose GSC 02853-00901 as the comparison star for the asteroid (895) Helio. Another comparison star GSC 02322-00635 was used for the last three days. In order to correct the magnitude difference of the two comparison stars, we observed the first one during the last three nights. The comparison star for the asteroid (165) Loreley is in the USNO-A2.0 Catalogue, $\alpha=08^h53^m14.5$, $\delta=+19°15'07.''04$. Differential magnitudes of the asteroids were derived using the classical differential photometric method.

We estimated the rotational period of the asteroid (895) Helio to be about 27.792hr with a maximum amplitude of 0.16±0.01mag which is quite different from Danforth's result, 9.67hr and 0.21mag. We also obtained the rotational period of the asteroid (165) Loreley to be about 7.224hr(=0.301day), with a maximum amplitude of 0.17±0.01mag. That was the same period as derived by Schober et al. (1988), and our lightcurve was roughly similar to theirs.

References

Clifford, J. C. 1988, in Introduction to Asteroids, William-Bell Inc. Press, p.143

Danforth, C. W. 1994, Minor Planet Bull. 21, 4

Schober, H. J., Di Martino, M., and Cellino, A. 1988, A&A, 197, 327

Tholen, D. J. 1989, in Asteroids II, eds. R. P. Binzel, T. Gehrels, M. S. , Matthews (University of Arizona Press) p.1139

Table 1.   Observing conditions for the asteroid (895) Helio and (165) Loreley.

Object	Date(UT)	†r(AU)	‡Δ(AU)	α(2000)	δ(2000)
(895) Helio	Oct. 30	2.816	1.908	$02^h43^m53^s6$	$+42°34'17''$
(895) Helio	Nov. 02	2.813	1.896	$02^h42^m07^s4$	$+42°09'58''$
(895) Helio	Nov. 30	2.789	1.897	$02^h19^m01^s2$	$+36°38'01''$
(895) Helio	Dec. 01	2.788	1.902	$02^h18^m30^s9$	$+36°24'04''$
(895) Helio	Dec. 03	2.786	1.910	$02^h17^m35^s3$	$+35°56'06''$
(165) Loreley	Nov. 30	3.380	2.807	$08^h53^m13^s6$	$+19°19'01''$
(165) Loreley	Dec. 01	3.380	2.794	$08^h53^m12^s1$	$+19°17'24''$
(165) Loreley	Dec. 03	3.380	2.768	$08^h53^m05^s3$	$+19°14'22''$
(165) Loreley	*Jan. 16	3.377	2.403	$08^h29^m42^s0$	$+19°04'39''$

† : distance between the Sun and the Asteroid, ‡ : distance between the Earth and the Asteroid, * : data in 2001.

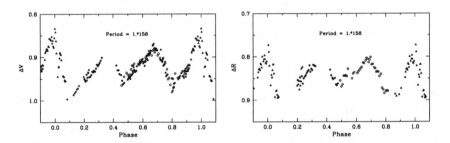

Figure 1.   Phase diagram for the asteroid (895) Helio (Epoch : H.J.D=2451848.100). Left is V band; right is R band. The symbol is different for each day.

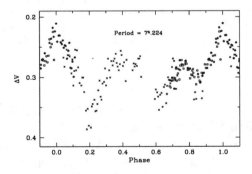

Figure 2.   Phase diagram for the asteroid (165) Loreley (Epoch : H.J.D=2451880.3087).

Small–Telescope Astronomy on Global Scales
ASP Conference Series, Vol. 246, 2001
W.P. Chen, C. Lemme, B. Paczyński

The Taiwan-America Occultation Survey for Kuiper Belt Objects

Sun-Kun King[1]

Institute of Astronomy and Astrophysics, Academia Sinica, P.O. Box 1-87, Nankang, Taipei, Taiwan 115, R.O.C.

Abstract. The purpose of the TAOS project is to directly measure the number of Kuiper Belt Objects (KBOs) down to the typical size of cometary nuclei (a few km). In contrast to the direct detection of reflected light from a KBO by a large telescope where its brightness falls off roughly as the fourth power of its distance to the sun, an occultation survey relies on the light from the background stars thus is much less sensitive to that distance. The probability of such occultation events is so low that we will need to conduct 100 billion measurements per year in order to detect the ten to four thousand occultation events expected. Three small (20 inch), fast (f/1.9), wide-field (3 square degrees) robotic telescopes, equipped with a 2,048 × 2,048 CCD camera, are being deployed in central Taiwan. They will automatically monitor 3,000 stars every clear night for several years and operate in a coincidence mode so that the sequence and timing of a possible occultation event can be distinguished from false alarms. More telescopes on a north-south baseline so as to measure the size of an occulting KBO may be later added into the telescope array. We also anticipate a lot of byproducts on stellar astronomy based on the large amount (10,000 giga-bytes/year) of photometry data to be generated by TAOS.

1. Introduction

More than three hundred small planetary bodies with radii larger than 100 km have been detected beyond Neptune using large telescopes, since the discovery of 1992 QB_1 (Jewitt & Luu 1993) eight years ago. Pluto and its satellite Charon are probably the largest members of this family, the so called "Kuiper Belt Objects" (Edgeworth 1949; Kuiper 1951), which is believed to be the source of most short period comets that return to the inner solar system every few years.

In general, a region close to Neptune is dominated by the gravitational perturbation of this giant planet. Hence, a depletion from a smooth extrapolation of

---

[1]Other TAOS team members: Alcock, C., Lehner, M. (U Penn), Axelrod, T. (ANU), Byun, Y.-I. (Yonsei), Chen, W.-P., Ip, W., Lemme, C., Tsay, W.-S. (NCU), Cook, K., Marshall, S., Porrata, R. (LLNL), Lee, T., Wang, A., Wang, S.-Y., Wen, C.-Y., Yuan, C. (ASIAA), Lissauer, J. (NASA/Ames), de Pater, I., Liang, C., Rice, J. (UC Berkeley)

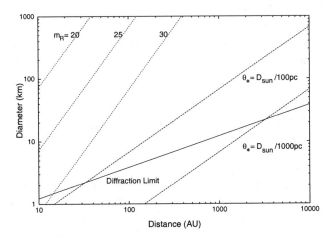

Figure 1.   The minimum size of an object at a given distance that can be detected by an occultation survey is determined by the combinational effect of diffraction (solid line) and the finite angular size of the target star (long dash lines). Roughly speaking, an object located above these lines can be detected. On the other hand, the detection limit of a direct detection, which depends on the light reflected from an object, is shown as short dashed lines with a red magnitude of 20, 25 and, 30 respectively. Here, an albedo of 0.04 is assumed.

the surface density of the planetary system is expected (Weidenschilling 1977). In a more distant region, the gravitational influence of Neptune and the giant planets should be insignificant. However, observations indicate a further depletion for large objects beyond 50 AU (Allen & Bernstein 2000). An alternative study of this region is demanded.

TAOS (Taiwan America Occultation Survey) is a joint project among Lawrence Livermore National Laboratory (through its Institute of Geophysics and Planetary Physics), National Central University (Institute of Astronomy), and Academia Sinica (Institute of Earth Sciences and Institute of Astronomy and Astrophysics). The purpose of this project is to directly measure the number of these KBOs down to the typical size of cometary nuclei (a few km). This knowledge will help us understand the formation and evolution of comets in the early solar system as well as to estimate the likelihood of their impacting our home planet.

2.   The Strategy

Moving in between the earth and a distant star, a KBO will block the starlight momentarily. A telescope monitoring the starlight will thus see it blinking. In contrast to direct detection of reflected light from a KBO by a large telescope where its brightness falls off as roughly the fourth power of its distance to the sun, an occultation survey (Bailey 1976; Dyson 1992; Cook et al. 1995) relies on the light from the background stars thus is much less sensitive to this distance.

The detection limit of an occultation survey is determined by the diffraction effect and the angular size of the target star. These are shown in Figure 1. That limit would be mostly set by diffraction for sun-like background stars located at 1,000 pc (m $\sim$ 15). It is obvious that occultation is a superior method for the detection of comet size objects beyond Neptune. As for the diffraction itself, though, the size of a KBO, s, is small compared to its distance, r, from us, $s^2/(\lambda r)$ could be larger than unity. Thus, Fresnel diffraction applies.

Our design goal is the capability to measure a rate of stellar occultations by comets of 1 per 1,000 stars per year of observing time. The typical relative velocity between a KBO and an observer is around 20 km/s. For a 3-km comet an expected duration of 0.2 seconds follows. Hence, there are approximately (3,000 stars)$\times$(10% $\times 3 \times 10^7$ sec)/(0.2 sec) $\sim 10^{11}$ photometric measurements per year, from which we want to derive less than one spurious occultation. The "false alarm" probability per observation must therefore be exceedingly small if the results of the experiment are to be interpreted with confidence. False alarms can arise from a variety of causes, a major one being the Poisson statistical fluctuation in the number of photons detected from a light source of constant intensity. False alarms due to this can be controlled only by detecting a sufficiently large average number of photons from a star. This clearly sets the brightness limit of the target stars, and hence the telescope aperture required. Extreme atmospheric fluctuations may also pose a threat of false alarms. In addition, false alarms can occur when a star is occulted by terrestrial objects, such as insects, birds, bats, aircraft, and even orbital debris. In practice, birds and swarms of insects appear to be the most damaging sources due to their potentially great numbers, whereas orbital debris can be neglected due to the short time scale (less than 10^{-3} s) of any occultation event. We require from our system a false alarm probability per measurement of $\sim 10^{-12}$, resulting in a false alarm probability per year of approximately 0.1. A combination of at least three telescopes monitoring the same fields at the same time was proposed.

Table 1.   This is an estimate of TAOS event rate. r is KBO's distance to the sun. $\Sigma(r)$ is the surface density of the solar nebula. s is the diameter of a KBO.

r (AU)	30 $\sim$ 50	50 $\sim$ 100	Note
condensible mass	24 M$_\oplus$	44 M$_\oplus$	minimum solar nebula $\Sigma(r) \propto r^{-1.5}$
event/year	1200	420	size distribution: $dN \propto s^{-3.5}$
$\times$ 1/100	12	(not depleted ?)	depletion due to Neptune

Using an initial solar nebula model (Hayashi 1981, 1985) considering its condensible parts (Weidenschilling 1977; Pollack 1996; Jewitt 1999) and an assumption of the size distribution of KBOs, we may obtain an estimate of the occultation events that could happen at a specific site with a few observational constraints. An example is shown in Table 1. A depletion of 100 times within 30 to 50 AU is expected. However, a higher rate from the outer region is possible if there is no depletion due to the gravitational perturbation of the Neptune, for example. Other assumptions are: the number of target stars per field (3,000), duty cycle in a year of observation (10%), disk thickness (±20°), phase angle of KBO (at opposition), mass density of a KBO (1 g/cm^3), and wavelength of observation (500 nm). One should note that the event rate listed in Table 1 is somewhat different from the expected detection rate. The latter depends on the site conditions, various noises from hardware, and our photometric algorithm. Nevertheless as serious as the issue is, it is beyond the scope of this brief report.

The probability of such occultation events is so low that we will need to conduct 100 billion measurements per year in order to detect the ten to a few thousand occultation events expected. The large range in this estimate reflects our ignorance in how to extrapolate from large KBOs to small ones. Evidently, there is an urgent need to conduct a census. An occultation survey seems to be the only way to tell if an outer edge of the solar system is observed. Thus, instead of doing a "pencil-beam survey" with large telescopes for several days a year, we use small telescopes to constantly monitor with wide fields a sufficient number of target stars. This fully robotic system including its real-time control, analyzing software, and large data archive handling capacity only has recently become technically feasible.

Though, a larger object has a better chance of occulting a distant star, an analysis of the occultation duration shows that most of the events are due to small comet-size objects. Chiang & Brown (1999) observations yielded $dN \propto s^{-3.6}$. There could be thousand times more events from objects of one tenth of their size if a size distribution of $dN \propto s^{-4}$ (Kenyon & Luu 1999) is assumed. Thus, a long occultation will be a very rare event which is worth follow-up observations with large telescopes. Grazing events will reduce the detection rate further. For spherical objects with a size cut-off and a size distribution given above, up to 40% of the events might be lost as being too short to be detectable.

3. The Project

Three small (20 inch) fast (f/1.9) wide-field (3 square degrees) robotic telescopes, each equipped with a 2,048 × 2,048 CCD camera, will be deployed along a 7 km east-west baseline. They will operate in a coincidence mode so that the sequence and timing of the three separate blinkings can be used to distinguish real events from false alarms. The three partners will each contribute one telescope and work together to set up the automatic observatory on peaks at 3,000 m elevation in or near the Yu-Shan (Jade Mountain) National Park in Taiwan. The three robotic telescopes will automatically monitor 3,000 stars every clear night for several years, consult among themselves to reject false events, and notify us via telecommunication when a possible event is found. For each telescope, a weather station is integrated with the robotic enclosure which can be shut down during a power failure.

All telescopes shall monitor the same field at the same time. The selection of star fields for an occultation survey is subjected to the following factors. First of all, photometry is a concern. The performance of the optical system, electronics, and the site set a limit on the magnitude of the target stars. Simulation and on-site testing are needed in order to determine a realistic value. A star too bright should be avoided as well as the moon light. The second is to have the density of target stars high enough so that one could have a reasonable event

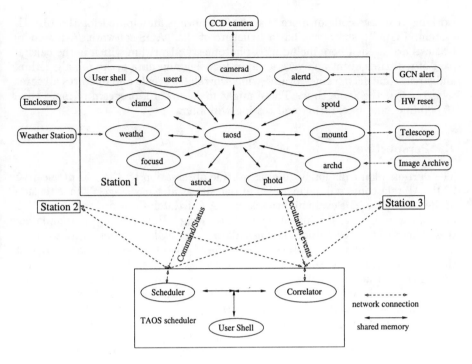

Figure 2.   TAOS software scheme

rate. A number density around 800 to 1,300 stars per square degree might serve this purpose. To have a higher probability of occultation, one looks for an event near the ecliptic plane and near the opposition position. Other criteria such as a request to search for a possible Neptune Trojan, to conduct a survey away from the plane to understand the inclination distribution, or, to distinguish a distant KBO from a nearby asteroid might also affect the selection of the observing field.

To spot an occultation event, one straightforward solution is to operate in "drift mode". This means that after pointing a telescope towards a selected field, the tracking would be turned off so that stars drift across the image frame from east to west at the sidereal rate. Some issues are of concern. For example, re-pointing the telescope on a specific field would be needed. The image motion and a shutter operating on and off at a rate of around $10^5 \sim 10^6$ per year might pose a problem. Aiming at the detection of a 5 Hz signal, a special mode of operation could be helpful. The so called "zipper mode" is now under consideration (Axelrod 1998; Liang 1999). Several versions are possible. The basic idea is to keep tracking at the selected field, but to shift the whole frame electronically a few pixels every 0.2 second. A detailed analysis on the overlapping of a crowded field and other issues are being undertaken.

The control and analysis software is technically the most challenging part of the TAOS project. We tap the experience of LLNL's MACHO project. State of art hardware and software are both certainly necessary. Based on the program structure of ROTSE project, on which one of our collaborators (LLNL) is working, a source code of more than 30,000 lines is anticipated for the LINUX system. Figure 2 shows the basic structure of the TAOS software. A few more features are needed here including a photometry algorithm which is the central part of real-time event analysis, an auto-focusing and data archiving algorithm, and a real-time networking "brain" which correlates events and operates all three telescopes at the same time. These might make up of a program of more than 10,000 lines and integrating and testing will follow.

4. Perspective

The current plan calls for beginning routine observation by the end of the year 2001. All telescopes will be installed at one site at an early stage. The extension of the baseline will be evaluated thereafter. A fourth telescope on a north-south 2 km spur to refine the size information of occulting KBOs is being contemplated. It will help to further reduce the false alarm rate. Follow-up observations using large telescopes at major observatories around the globe attempting to detect the reflected sun light from the KBO, hence its orbit and distance, are being organized. In addition, the capability of reacting to GCN alerts will be implemented as shown in Figure 2. We anticipate a lot of byproducts on stellar astronomy based on the large (10,000 giga-bytes/year) photometry data bank to be generated by TAOS. The monitoring of a few selected fields which contain some other interesting objects is considered. No doubt, with these small telescopes, a project like this could be very fruitful and exciting in the next few years.

References

Axelrod, T. 1998, personal communication

Allen, R. L. & Bernstein, G. M. 2000, astro-ph/0011037

Bailey, M. E. 1976, Nature, 259, 290

Chiang, E. I. & Brown, M. E. 1999, AJ, 118, 1411

Cook, K. H., Alcock, C., Axelrod, & Lissauer, J. 1995, BAAS, 27, 1124

Dyson, F. J. 1992, QJRAS, 33, 45

Edgeworth, K. E. 1949, MNRAS, 109, 600

Hayashi, C. 1981, Prog. Theor. Phys. Suppl., 70, 35

Hayashi, C. 1985, in Protostars and Planets II, 1100

Jewitt, D. C. & Luu, J. X. 1993, Nature, 362, 730

Jewitt, D. C. 1999, Annu. Rev. Earth Planet. Sci., 27, 287

Kenyon, S. J. & Luu, J. X. 1999, AJ, 118, 1101

Kuiper, G. P. 1951, in Astrophysics: A Topical Symposium, ed. J. A. Hynek, (McGraw-Hill, New York), 357

Liang, C. 1999, thesis (in preparation), UC Berkeley, USA

Pollack, J. B., Hubickyj, O., Bodenheimer, P., Lissauer, J., Podolak, M., & Greenzweig, Y. 1996, Icarus, 124, 62

Weidenschilling, S. 1977, Ap&SS, 51, 153

Small–Telescope Astronomy on Global Scales
ASP Conference Series, Vol. 246, 2001
W.P. Chen, C. Lemme, B. Paczyński

Distinguishing KBO from Asteroid Occultations in TAOS

Claudia Lemme

Institute of Astronomy, National Central University, 32054 Chung-Li, Taiwan

Chyng-Lan Liang

Department of Statistics, University of California, Berkeley, USA

Abstract. The goal of the Taiwan-America Occultation Survey (TAOS) project is to estimate directly the number of Kuiper Belt Objects (KBOs) by measuring the rate of stellar occultations. Occultations of distant stars may also be caused by asteroids and this article describes how both types of occultations can be distinguished on the basis of their shadow ground speed on Earth and the observing angle.

1. Introduction

A flattened annulus beyond the orbit of Neptune, named the "Kuiper belt", is thought to be the source of short-period comets. Containing the most pristine material, this region provides valuable diagnostics for the early stages of Solar System formation. As the point of origin of the short-period comets, it is also relevant for calculations on the number of Earth-crossing bodies and thus for the frequency of biological extinction events. Since observational success was first achieved in 1992, more than 300 Kuiper belt objects (KBOs) have been found. So far however, direct imaging has been sensitive only to the larger members of the belt, those with diameters greater than around 50 km. The extent, size distribution and total mass of the Kuiper belt are all still unknown, and are of especial interest considering the opposing theories that abound (e.g. Jiang, I.G., this volume). The Taiwan-America Occultation Survey (TAOS, see King, S.K., this volume) will be able to measure the number of these objects down to $\sim$3 km, by their chance occultations of background stars. Several telescopes will operate in coincidence mode to distinguish real occultations from false events, that result from noise or occultations by terrestrial objects. However, a fraction of the real occultations is expected to be produced by asteroids, and not by KBOs. To distinguish these, we consider the speed of the shadow ground track on Earth.

2. Results

The speed of the shadow ground track is basically given by the relative velocity with respect to Earth, perpendicular to the line of sight of an object at distance

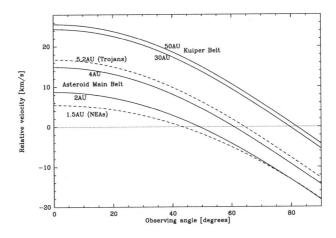

Figure 1.   Relative velocity (corresponds to the speed of the shadow ground track on Earth) as function of observing angle for the Kuiper belt between 30 and 50 AU and the asteroid main belt between 2 and 4 AU. In addition, the relations for the Trojan asteroids and a typical Near-Earth asteroid are marked by dashed lines.

r and observing angle ϕ. For circular, plane orbits this is given by:

$$RV(r, \phi) = v_e \cos \phi - v_e \sqrt{(\frac{r_e}{r})(1 - (\frac{r_e}{r})^2 \sin^2\phi)}$$

where r_e is the heliocentric distance of Earth and v_e the orbital velocity of Earth. The observing angle is measured clockwise from opposition (when the Earth is located between the object (KBO/asteroid) and the Sun). Figure 1 reveals:
1. For observing angles $\phi < 42^o$, both KBOs and asteroids move retrograde (RV>0, this corresponds to an East→West movement of the shadow on Earth).
2. Each asteroid has an "angle of quadrature" (RV = 0) between $\phi \sim 42^o$ and 65°. The asteroid's speed is 0 (or close to 0) at this angle.
3. For $\phi > 65^o$, all the asteroids move prograde.
4. For $80^o < \phi < 82^o$ KBOs reach their angle of quadrature.
5. For $\phi > 82^o$ both KBOs and asteroids move prograde.
But most importantly for distinguishing the two groups: for any observing angle, there is always a difference in the shadow ground speed. Around opposition this is ~13 km/s (and at least 7 km/s between the outermost Trojans and the innermost KBOs). For a ϕ between 60° and 80° the absolute speed for both can be similar, but the asteroids move prograde and the KBOs retrograde. The difference in speed affects occultation duration and arrival time at the different telescopes, and the arrival time allows us to determine whether the occultation was caused by an asteroid or a KBO. Note that the distinction can not be made by a single telescope alone and we need telescopes separated far enough apart from each other in East-West direction. We expect to detect about 1 asteroid occultation every 10 or 100 KBO occultations; the range reflects the uncertainty in the total number of asteroids and KBOs.

Small–Telescope Astronomy on Global Scales
ASP Conference Series, Vol. 246, 2001
W.P. Chen, C. Lemme, B. Paczyński

The Humps of KBO's Size Distribution

Cheng-Pin Chen and Ing-Guey Jiang

Institute of Astronomy and Astrophysics, Academia Sinica, Taiwan

Abstract. We study the possible humps or deviation from the single-power law for the size distribution of the Kuiper Belt Objects (KBOs). Both the current observational data and theoretical simulations show evidence of such humps. We conclude that this is an imprint of the depletion of the outer Solar System in the Kuiper Belt region.

1. Introduction

It is known that the outer Solar System beyond Neptune is populated by small bodies after the first KBO was discovered (Jewitt & Luu 1993). One of the most important properties of KBOs is their size distribution because this gives us some hint of the evolutionary history of the Solar Nebula. Kenyon & Luu (1999) did planetesimal accretion calculations for a mass-conserved single annulus at 35 AU and found that all models produce power-law size distributions.

2. Current Observations and Theoretical Simulations

We obtained the data of currently known KBOs from the daily updated list of Transneptunian Objects at the web site: http://cfa-www.harvard.edu/iau/lists/ TNOs.html. The size distribution was determined directly from both the data of 27th July 2000, when there were 282 discovered KBOs and the data of 13th Dec. 2000, when there were 346 discovered KBOs. Figures 1(a)-(b) show that there are humps for the size distribution and the least-square fitting curve of the single-power law fails to be located within the error bars for most of the bins. Is this just a bias effect or is it an important hint for KBOs' history?

Of course, the current observational data may not be statistically significant enough to determine the true size distribution because only a small fraction of the total KBO population has been discovered. In particular, the current data has no constraint or information about the size distribution of smaller KBOs, i.e. objects a few km to 150 km in size. This is the region that the TAOS project (Chen 2000, King 2001) should be able to contribute to.

Therefore, we investigate the humps of the size distribution by the theoretical simulations. We randomly place KBOs in the region between 30 to 50 AU and assume that they follow a single-power-law size distribution. Our calculations show that the size distribution will keep changing as some KBOs escape from this region. This is because the growth of the KBOs causes local gravitational instability, which depends on the size of the objects. This result implies that the

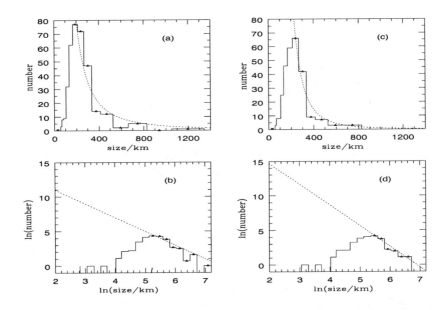

Figure 1.   The observational size distributions. Each triangle point is the mean of each bin and these data points are used to determine the power-law least-square fitting curve (dotted line). Panel (a) shows the data from 13th Dec. 2000 and Panel (b) is the same thing but in ln-space. Panel (c) shows the data from 17th July 2000 and Panel (d) is the same thing but in ln-space.

size distribution should have changed during the depletion of the Kuiper Belt region. Moreover, these humps or deviation from the single-power law, as can be seen in both the simulations and the observational data, could be an imprint of the depletion of the Kuiper Belt.

Acknowledgments

We are grateful to the questions from Hsiang-Kuang Chang, Wen-Ping Chen, Charles Alcock, and also the comment by K.Y. Lo during the oral presentation of this paper. We acknowledge the financial support of the Academia Sinica.

References

Chen, W.-P. 2000, Proceedings of The Transneptunian Population, 24th IAU General Assembly, Manchester, U.K., in press

Jewitt, D., Luu, J. X. 1993, Nature, 362, 730

Kenyon, S. J. & Luu, J. X. 1999, AJ, 118, 1101

King, S.-K. 2001, this volume

Small–Telescope Astronomy on Global Scales
ASP Conference Series, Vol. 246, 2001
W.P. Chen, C. Lemme, B. Paczyński

The True Colors of KBOs

Hui-Chun Hsu and Wing-Huen Ip

Institute of Astronomy, National Central University, 32541 Chung-Li, Taiwan

Abstract. The existence of a population of large planetoids outside the orbit of Neptune predicted by GP Kuiper, KE Edgeworth and JA Fernandez has been confirmed by ground-based observations. The physical properties of these Kuiper Belt Objects (KBOs) remain elusive. Photometric measurements have indicated that they have diverse color variations. A theoretical model is formulated to simulate the evolution of the surface materials of the KBOs under the influence of cosmic ray irradiation and meteoroid impacts. The long-term goal is to couple this theoretical model to observations and laboratory experiments such as LARA (Laboratory Astrochemistry and Astrophysics).

1. KBOs in the Trans-Neptunian Region

Since the first discovery of QB1 by David Jewitt and Jane Luu in 1992, ground-based observations have now discovered nearly four hundred KBOs with sizes ranging between 100 km and a few hundred km. Pluto is the largest object in this trans-neptunian population. It has been estimated that there should be about 70,000 objects with diameters larger than 100 km. The KBOs are believed to be remnants from the accretional process of Uranus and Neptune. They are also the main supplier of short-period comets. Investigation of the KBO population is therefore essential to our understanding of the solar system origin.

2. Surface Irradiation History

The chemical composition of the KBOs is most likely a mixture of water ice, organic matter of hydrocarbon nature and rocky material. The exposure of the surface material to cosmic rays and solar radiation could lead to the build up of a crust of dark matter with very low albedo. It is known from laboratory experiments that organic material made up of complex polymers could change its color from being reddish red to darkest dark by energetic irradiation. On the other hand, occasional impact collision with other small bodies could lead to the formation of craters or gardening (re-surfacing) of the subsurface material. This stochastic bombardment effect would hence produce layers of icy material of different cosmic-ray exposure ages on top of the old surface (Luu and Jewitt, 1996). It is perhaps for this reason that photometric and spectral measurements have indicated a wide range of surface colors and chemical compositions of the KBOs (Luu and Jewitt, 1996; Tegler and Romanishin,1998).

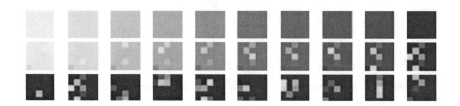

Figure 1. Schematic illustration of the surface color variations of KBOs: (a)upper row: a KBO without impact; (b)middle row: a KBO with impacts; (c)lower row: ten different KBOs. Figure courtesy of Z.W.Zhang.

3. Stochastic Impact Model

In order to trace the surface "weathering" process of the KBOs so that their evolutionary history might be eventually deciphered, a theoretical model is being developed to simulate the cumulative effect of cosmic ray exposure and random meteoroid impacts. The basic principle of our model can be demonstrated by considering the addition of patches of resurfaced material at different times of the solar system history. Figure 1 illustrates the time evolution of the surface color of a KBO in 10 time steps. It can be seen that the general area not subjected to impact cratering will have its color turning from bright to dark. On the other hand, localized regions rejuvenated by meteoroid collisions will first regain their bright color while turning to a darker color at a later time. It is in this manner that the KBOs could have gained such a diverse range of surface reflectance spectra and colors.

4. Future Work

(1) The size distribution of the interplanetary meteoroids
(2) The size distribution of the impact craters and possible overlapping
(3) The orbital evolution of KBOs (i.e., classical KBOs vs. scattered KBOs)
(4) A comparison with spectroscopic and photometric observations
(5) The inputs from laboratory experiments of ice irradiation

References

Luu, J.X. and Jewitt, D.C. 1996. Astron. J., 112, 2310
Tegler, S.C., and Romanishin, W. 1998. Nature, 392, 49

Small–Telescope Astronomy on Global Scales
ASP Conference Series, Vol. 246, 2001
W.P. Chen, C. Lemme, B. Paczyński

Revealing Variety of Comets by Long-Term Monitoring Observation with a 50-cm Telescope

Jun-ichi Watanabe, Hideo Fukushima

National Astronomical Observatory of Japan, Mitaka, Tokyo 181-8588, Japan

Abstract. We have been monitoring comets using the 50-cm telescope at the Mitaka campus of the National Astronomical Observatory, Japan since 1996. Over 20 comets, including bright ones such as comet C/1995 O1(Hale-Bopp), have been observed over long-term periods, namely over large heliocentric distance scales. Our samples show variety of comets in terms of both morphological and temporal variations. Several typical examples of various morphology, probably due to dust, are shown to emphasize the importance of long-term monitoring of comets.

1. Introduction

Comets are the representative solar system objects which often show drastic time variations in both brightness and morphology. As most of the readers of these proceedings recognize, small telescopes are quite appropriate to monitor such time-variable objects, mainly because we can get access to plenty of time without heavy discussion with the time allocation committee. Moreover, recent developments of high efficient detectors such as the CCD made it possible to see fainter objects than before. Since 1995, we have started a long-term monitoring program of comets, simultaneously targeting as many comets as possible, which indicate the existence of a variety of cometary activities at a wide range of heliocentric distances. Some typical examples are introduced in this paper.

2. Observations

Our image monitoring has been carried out using the 50-cm telescope at the Mitaka campus of the National Astronomical Observatory of Japan, located near Tokyo, Japan (35.66°N, 139.55°E, H=59m). A liquid-nitrogen cooled CCD camera (Astromed Type 3200) is attached to the cassegrain focus(f/12) of the telescope, giving a field of view of 14.8 × 9.9 arcmin^2. We use another portable telescope in order to complement the observations in case of severe observational conditions. The installed CCD is EEV type P88231, which has 770 × 1152 pixels. The pixel size is 22.5 × 22.5 μm. In order to reduce the background sky level from the severe light pollution, and to study dust activity of the coma, R-band or I-band filters are usually applied. The telescope is tracked by following the apparent motion of the comet. Table 1 lists the comets observed to the end of 2000. Most of the observed images have been released on our web page at http://www.nao.ac.jp/pio/Comets/.

Table 1. Comets monitored up to the end of 2000

C/1995O1(Hale-Bopp)	C/1996B2(Hyakutake)	C/1998M5 (LINEAR)
21P/Giacobini-Zinner	C/1998K5 (LINEAR)	52P/Harrington-Abell
C/1998T1 (LINEAR)	C/1999S4 (LINEAR)	C/1999L3 (LINEAR)
C/1999E1 (Li)	P/2000B3 (LINEAR)	C/1999J2 (Skiff)
C/1999H3 (LINEAR)	C/1999T2 (LINEAR)	29P/Schwassmann-Wachmann 1
C/1999Y1 (LINEAR)	P/2000S1 (Skiff)	C/2000W1(Utsunomiya-Jones)

3.   Comet C/1995 O1(Hale-Bopp) – Activity at Large Heliocentric Distances

The first example is comet C/1995 O1(Hale-Bopp), which was one of the intrinsically brightest comets in the last millennium. The apparent magnitude of this comet reached -1 when the geocentric distance was about 1 A.U. The orbit of this comet was close to the Earth's orbit, so there was a chance of a close encounter. If this comet had come four months earlier, its apparent magnitude would have been about -6. It is interesting to monitor such a bright comet at large heliocentric distances. The results of our image monitoring before the perihelion passage are shown in Figure 1; one can see the gradual and steady evolution of the cometary activity with its approach. Many straight jet-like structures are interpreted as the edges of cones produced by dust jets from the rapidly rotating nucleus.

After the perihelion passage, we detected several eruptive ejections of huge dust clouds from the nucleus as shown in Figure 2 (Yamamoto & Watanabe, 1997).

4.   Comet C/1996 Q1(Tabur) and C/1999S4(LINEAR) – Complete Disruption of Nuclei?

The next example is comet C/1996 Q1(Tabur). This comet appeared to be normal until the end of October 1996, when the central part of the coma started to elongate, and began to fade-out. In November we could see only a faint, diffuse cloud of dust tail as shown in Figure 3.

We do not know what happened in this comet. Fulle et al (1998) analyzed this fade-out and suggested a sudden cessation of evaporation from the nucleus, but nobody knows the truth. Four years later, another comet C/1999 S4(LINEAR) played an important role in solving the nature of such behaviour. This comet also seemed to be normal until mid-July 2000, then showed quite a similar morphological variation together with a brightness change at the end of July. One thing that differed from comet C/1996Q1(Tabur) was that Comet C/1999S4(LINEAR) was bright enough, that many large telescopes, including HST and VLT, could observe it, and witnessed the complete disruption of the nucleus. This lends strong support to a complete disruption of the nucleus of

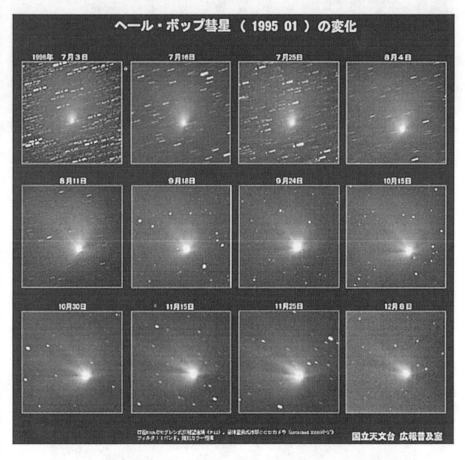

Figure 1.   The I-band images of Comet C/1995 O1(Hale-Bopp) from July 3 through December 8, 1996.

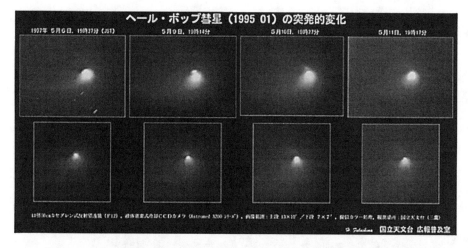

Figure 2. Time variation of the huge dust cloud of Comet C/1995(Hale-Bopp) on May 9-11 1997. An eruptive ejection of the dust cloud can be recognized as a knot just above the central condensation in the image of May 9 (second panel from the left). The field of view of is 13'x10' for the top panels and 7'x7' for the bottom panels.

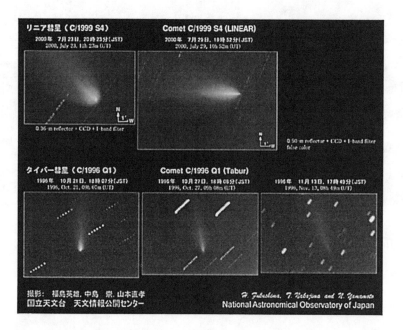

Figure 3.   Morphological variation of Comet C/1996 Q1(Tabur) and C/1999S4(LINEAR)

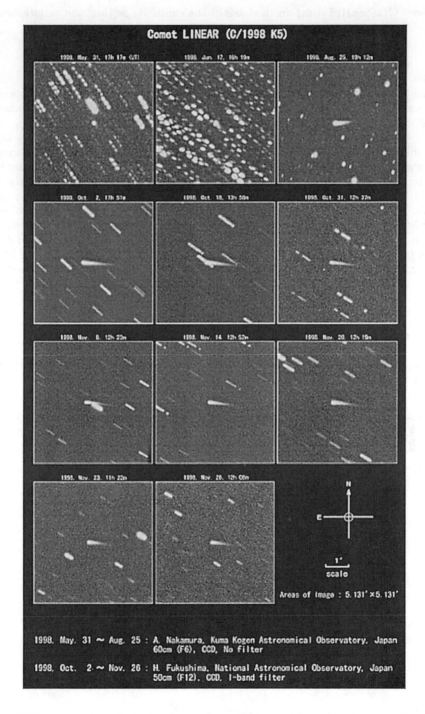

Figure 4.   Morphological variation of Comet C/1998 K5(LINEAR) from May 31 through November 26 1998.

comet C/1996Q1(Tabur) on the basis of the similarity in the time variations of both comets.

5. Comet C/1998 K5(LINEAR) – a Strange Tail?

The last example of an unusual feature revealed by our monitoring observations is comet C/1998 K5(LINEAR). Figure 4 shows the time variation of this comet; the spindle-like structure should be noted. Usually the dust tail of a comet of this brightness class should be faint and diffuse. The width of the dust tail should also be roughly proportional to the distance from the nucleus. However, the wedge-shaped tail of this comet was narrower with the distance. In order to determine the origin of this structure, information on time variation is important. Our long-term monitoring has indicated that this tail began to extend from July 1998, and interestingly, had been quite stable until the end of November. Although there is no space to describe in detail the possible origin here, from such time variation we can promptly eliminate the several possibilities including ion tail. The only solution to explain this structure is the dust tail formed by very limited large-size particles ejected at a certain epoch. We do not know why this happened. Could it be dust mantle peeling off the nucleus? Impacts of meteoroids? No clear explanation can be offered yet.

6. Conclusion

It is clear that long-term monitoring for many comets, which can be realized only by small telescopes, is important to delineate the nature of comets.

References

Fulle, M., Mikuz, H., Nonino, M., & Bosio, S. 1998, Icarus, 134, 235

Yamamoto, N., & Watanabe, J. 1997, Earth, Moon and Planets, 78, 229

Small–Telescope Astronomy on Global Scales
ASP Conference Series, Vol. 246, 2001
W.P. Chen, C. Lemme, B. Paczyński

A Simulation of Shell Structures of Comet Hale-Bopp in February 1997

J. Tao & B.C. Qian

Shanghai Astronomical Observatory, Chinese Academy of Sciences, Shanghai, 200030

Abstract. We present a simulation based on the morphology of the inner coma of comet Hale-Bopp(C/1995 O1) in February 1997. The synthetic images can fit the observations well, suggesting a simple spin mode. The rotation parameters, obliquity of orbital plane to equator and argument of subsolar meridian at perihelion, are 71 and 81 degrees respectively.

1. Introduction

Comet Hale-Bopp(C/1995 O1) has been one of the active comets in recent years. The comet shows various coma morphology, such as fans and radial jets. Coma structures may be an indicator of the spin axis (Sekanina & Larson 1984, 1986; Sekanina 1996, 1998a, 1998b; Vasundhara et al. 1999; Samarasinha et al. 1997-1999; Samarasinha 2000, Jorda et al. 1997-1999; Wang 1999). Most of these models have similar physical concepts: sources on the rotating nucleus are ejecting continuously from local sunrise to sunset. The basic concept of our model comes from Sekanina's work.

2. Observations

Optical observations of comet Hale-Bopp were carried out with the 1.56m Telescope and a CCD camera (1024*1024pixels) at She-shan Station, Shanghai Astronomical Observatory. The camera offers a field of view of 13'*13'. The cousins filter I is used for direct imaging because this waveband is related to cometary dust structure. In Feb. 1997, we observed for 8 days and obtained CCD images of this comet. Each CCD image is corrected for bias, dark, and flat field. In order to depress the coma background and to clearly show the near nuclear structure, we processed the images with the wavelet method (Tao et al. 1997; Tao et al. 2000; Marchis et al. 1999). The coma structures can be clearly seen in Figure 1 (white patterns).

3. Results and Discussion

The obliquity of the orbit plane to the equator, is less than 90 degree for the prograde sense of rotation and ranges from 90 degree to 180 degree when rotation is retrograde relative to the comet's orbital motion. The phip is the argument of the subsolar meridian at perihelion. The coma morphology suggests the presence of two active regions. Three point sources for each region were used to simulate the coma patterns. The distance between each point source is 5 degree and these point sources are lined in the longitude direction.

By variation of the parameters listed in Table 1, we modeled the shell structure of the dust coma until the simulated result (black dots in Figure 1) provided optimum agreement with the observed shell features. Table 1 lists the results of spin orientation and position of the active sources. Our results are consistent with others (e.g. Sekanina 1998a,1998b; Jorda et al. 1997-1999; Vasundhara et al. 1999; Licandro et al. 1997-1999).

Table 1.   Simulation result

α	δ	Obliquity	phip	1st latitude	2nd latitude
250	-60	71	81	30	0

References

Jorda, L., Rembor, K., Lecacheux, J., et al. 1997-1999, Earth, Moon, and Planets, 77, 167

Licandro,J., Bellot Rubio, L.R., Casas, R., et al. 1997-1999, Earth, Moon, and Planets, 77,199

Marchis,F. et al. 1999, A&A, 349, 985

Samarasinha, N. H. 2000, ApJ, 529, L107

Samarasinha, N.H., Mueller, B. E. A., Belton, M. J. S. 1997-1999, Earth, Moon, and Planets, 77, 189

Sekanina, Z. 1996, A&A, 314, 957

Sekanina, Z. 1998a, ApJ, 494, L121

Sekanina, Z. 1998b, ApJ, 509, L133

Sekanina, Z & Larson, S.M. 1984, AJ, 89, 1408

Sekanina, Z & Larson, S.M. 1986, AJ, 92, 462

Tao, J., Qian, B., Tang.Y., et al. 1997, Progress in Astronomy, 15, 68

Tao, J., Qian, B., Gu, M., et al. 2000, Planetary and Space Science, 48, 153

Vasundhara, R.& Chakraborty, P. 1999, Icarus, 140, 221

Wang, Y. 1999, Master dissertation, Nanjing University

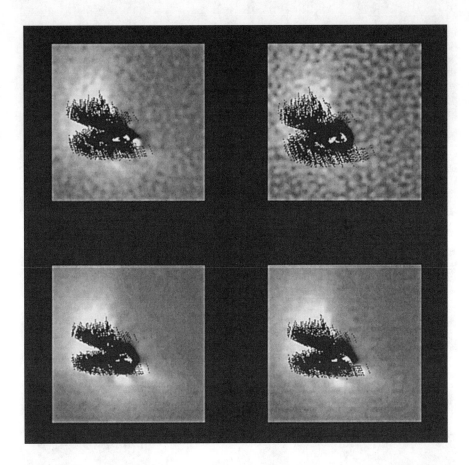

Figure 1.   Observation images of Comet Hale-Bopp with synthetic features superposed. The observations were taken on Feb. 12 (upper left), Feb. 16 (upper right), Feb. 8 (lower left) and Feb. 10 (lower right) 1997. West is up and south to the left. The FOV is 2'.3 square size. An I filter is used.

(Top, from left) Samus, C. Scarfe and A. Scarfe;
(bottom, from left) Chen and Paczyński

Small–Telescope Astronomy on Global Scales
ASP Conference Series, Vol. 246, 2001
W.P. Chen, C. Lemme, B. Paczyński

Cometary Polarimetry

Asoke K Sen

Dept. of Physics, Assam University, Silchar 788011, India.
aksen@dte.vsnl.net.in

Abstract. Comets are known to have a high (2-10 percent) polarization, caused mainly due to dust scattering and resonance fluorescence emission. Since near earth comets are generally bright (integrated visual magnitudes of 10 or even brighter) and as they show high polarization near earth-sun location, one can try to perform imaging polarimetry of such objects with small (40 cm or even smaller) telescopes. By using a dichroic polaroid sheet either in front of the telescope tube or before the imaging detector at the Cassegrain plane, one can record cometary images with a good signal to noise ratio. By rotating the dichroic sheet in three discrete steps and then reducing the corresponding three comet images to a single image, one can determine the linear polarization value at each pixel location on the image. The error in polarization will typically be the inverse of the 'signal to noise ratio'. Such polarization images of a comet help us to determine its dust properties and also to look for possible dust jet activities. This type of work is possible with small telescopes and minor instruments.

1. Introduction

There are several mechanisms which produce linear and circular polarizations in astrophysical environments, such as synchrotron emission, dust scattering, electron scattering etc. Polarimetry consists of mainly two classes: photographic imaging polarimetry with a low precision (1 %) and modulator polarimetry with a high precision (0.01 %). In Modulator Polarimeters two orthogonal polarization beams are measured with the same detector, within a very short time interval. Photographic low precision imaging polarimetry is suitable for high polarization objects (viz. comets, nebulae) and relatively bright objects ($\delta p \sim \frac{\delta I}{I}$).

2. Measurement of Cometary Polarization

Cometary polarization is mainly caused by the scattering of solar radiation by cometary dust and resonance fluorescence emissions in the molecular band. (Leborgne et al. 1987; Sen et al 1989). Cometary polarization in the continuum is a function of the size and composition of cometary dust and the scattering angle (θ). The study of cometary polarization places useful constraints on the size and composition of the dust particles (Mukai et al. 1987; Sen et al. 1991).

However, since dust properties vary within the comet, one should aim at imaging polarimetry rather than aperture integrated polarimetry.

3. Some Cometary Polarimetry Results

Aperture polarimetry: The heterogeneity of a cometary coma was first found in comet Halley, by Bastein (1986) and Dolfus & Suchail (1987), by using aperture scan polarimetry. By using all the available polarization data, Chernova et al. (1993) and Levasseur-Regourd et al. (1996) established synthetic curves. Using their data on comet Halley, Mukai et al. (1987) and Sen et al. (1991) commented that its grains are either dirty ice/silicate/CHON type.

Imaging polarimetry: Work on comet Halley by Eaton et al. (1988) and Sen et al. (1990), comet Okazaki-Levy-Rudenko by Eaton et al. (1991), comet Ashbrook-Jackson by Renard et al.(1996) etc. indicate that high polarization dust blobs characterised by fine grains were detected. Eaton et al. (1988) used modulator type two-channel polarimeters, whereas Renard et al. (1996) used a non-modulator type instrument (polarizer+ CCD) and Sen et al. (1990) used only a polarizer and a photographic plate.

Sen et al (1990) placed a polaroid sheet in front of the tube of a Celestron-14 telescope. Polarization (p) was then calculated from three images (I1,I2,I3) recorded at three positions on the polaroid sheet, with p expressed as :

$$p = \frac{2\sqrt{I1(I1-I2)+I2(I2-I3)+I3(I3-I1)}}{I1+I2+I3}$$

This work has shown that a small (submeter class) telescope with only minor instruments can be a good facility for performing 'cometary polarimetry', resulting in useful science.

References

Bastien, P., Menard, F., Nadeau, R. 1986, MNRAS,223,827

Chernova, G P, Kiselev,N.N., Jockers, K.,1993, Icarus, 103,144

Dollfus A., & Suchail, J.L.1987, A & A, 187, 669

Le Borgne, J.F., Leroy, J.L., Arnaud, J.1987, A &A, 173, 180

Levavasseur-Regourd, A.C., Hadamick, E., & Renard, J.B., 1996, A & A, 313, 327

Eaton, N., Scarrott, S.M., Warren-Smith, R.F. 1988, Icarus, 76,270

Eaton, N., Scarrott, S.M., & Wolstencroft, R.D. 1991, MNRAS, 250, 654

Mukai, T., Mukai, S., Kikuchi, S., 1987, A & A, 187, 650

Renard, J.B., Hadamcik, E., & Levasseur-Regourd, A.C. 1996, A & A, 316, 263

Sen, A.K., Joshi,U.C., & Deshpande, M.R.1989, A &A,217,307

Sen,A.K., Joshi,U.C.,Deshpande,M.R., & Debi P. C.1990, Icarus,86, 248

Sen, A.K., Deshpande, M.R., Joshi, U.C., Rao, N.K., & Raveendran, A.V. 1991, A & A, 242, 296

Small–Telescope Astronomy on Global Scales
ASP Conference Series, Vol. 246, 2001
W.P. Chen, C. Lemme, B. Paczyński

Observation of Comet C/1999 S4 (LINEAR)

C.Y. Lin and W.H. Ip

Graduate Institute of Astronomy, National Central University, 32054 Chung-Li, Taiwan

Abstract. Comet C/1999 S4 (LINEAR) with a maximum magnitude of 6 mag was the brightest comet in 2000. The 24" telescope on NCU campus was used to observe its coma behavior in June and July of 2000 as it approached perihelion. The observations were carried out using the comet filters designed by the European Space Agency for the ground-based observational program of Comet Wirtanen (target comet of the Rosetta mission). Comet C/1999 S4 (LINEAR) ended up disrupting into many small fragments near perihelion. We are in the process of analyzing the time variations of the cometary dust coma and dust tail to examine the physical condition prior to the destruction of the nucleus of this intriguing comet. An image processing method will be developed to trace small-scale structures in the dust tail.

1. Observation in Taiwan

Comet C/1999 S4 (LINEAR) was observed in June and July 2000 using the 24" Cassegrain reflector on NCU campus (Fig 1.a - 1.d). Besides the I-band filter, the comet filter set provided by ESA was used to carry out photometric imaging observations. The unexpected breakup of this bright comet near perihelion between July 6 and July 24 has attracted wide-ranging interest. This is because the dynamical behavior of the dust tail and the gas coma carries a wealth of important information on the composition and internal structure of the cometary nucleus. The present work serves as a pilot project for our participation in a world-wide network of cometary research.

2. Data Analysis

Spectroscopic observations indicated that the CN emission was deficient in this comet (Schleicher 2000). For this reason, we have not been able to obtain a good-quality coma image in CN emission. We have, however, made a series of narrow-band images of Comet C/1999 S4 (LINEAR) in C_2 emission. We are in the process of making careful subtraction of the continuum emission representing the dust component. A ring-mask image processing technique has also been developed to enhance the features of the dust and gas comas (see Fig 2.a and 2.b — kindly provided by Z. W. Zhang).

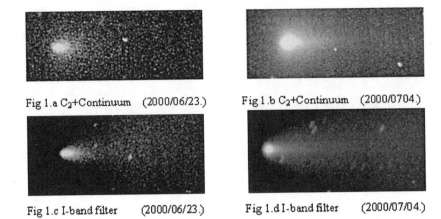

Fig 1.a C$_2$+Continuum   (2000/06/23.)   Fig 1.b C$_2$+Continuum   (2000/07.04.)

Fig 1.c I-band filter   (2000/06/23.)   Fig 1.d I-band filter   (2000/07/04.)

Figure 1.   Photometric images of Comet C/1999 S4 (LINEAR) taken at NCU observatory on June 23 and July 4, 2000. (a) and (b) C$_2$ + continuum filter images; (c) and (d) I-band filter images.

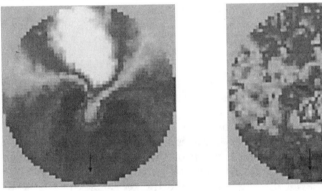

Fig 2.a Ring-masked image I-band   Fig 2.b Ring-masked image C$_2$+cont.

Figure 2.   Ring masked images of (a) Dust continuum and (b) C$_2$+continuum. The solar direction is indicated by the arrow.

References

Schleicher, D. G. 2000, IAU Circ, 7342.

Small–Telescope Astronomy on Global Scales
ASP Conference Series, Vol. 246, 2001
W.P. Chen, C. Lemme, B. Paczyński

High Spatial Resolution Observations of the 1998 and 1999 Leonid Meteors

X.J. Jiang[1,2] and J.Y. Hu[1,2]

1. Beijing Astronomical Observatory, Chinese Academy of Sciences, Beijing, 100012, P.R. China

2. National Astronomical Observatories, Chinese Academy of Sciences, Beijing 100012, China

Abstract. We observed the Leonid meteor storm in 1998 and 1999. The light curves extracted from the 2-dimensional images of the bright Leonid meteors clearly show the sub-structures, which are consistent with the generally accepted dustball meteor model (Hawkes & Jones, 1975).

1. Introduction

During November 17-19 of each year, the Earth passes through the debris cloud from comet Temple-Tuttle. This event generally creates a meteor shower as the particles enter the Earth's atmosphere and burn up. Since the debris appears to be coming from the direction of the constellation Leo, it is called the "Leonid shower".

Every 33 years, the Earth passes very close to Comet Temple-Tuttle's orbital path not long from when the comet has recently passed by. As a result, the Earth is likely to experience a meteoroid "storm" as it passes through the cloud of debris following closely behind the comet. The most recent Leonid meteor storm arises between 1998-2001, which provides a good opportunity to study the cometary originated meteoroids.

2. Observations and Discussions

Figure 1 shows the meteor light curves. Panels a,b,and c were taken on November 18, 1998 at XingLong Station, Beijing Astronomical Observatory with a 16-bit TEK 1024 CCD and a Carl Zeiss Jena Schmidt-Cassegrain camera (f.l. 150mm, f/1.5). The center of each field is about 10° away from the Leonid radiant, the field of view of the raw CCD image is 9.4° × 9.4°, and the resolution is 33″/pixel. The exposure time was fixed to 300 seconds. Panel c was taken on November 18, 1999 with SBIG ST-8 CCD and a 50mm f/1.8 lens, The field of view of the raw image is 15.8° × 10.6°, and the resolution is 37″/pixel.

The raw data were reduced by using the IRAF program package. The CCD reductions included bias subtraction and flat field correction, then the light curves were extracted from the two-dimensional images.

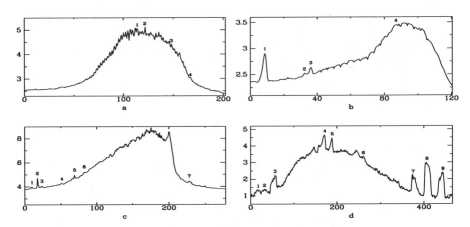

Figure 1.   Light curves of the four bright Leonid meteors. The relative intensity is plotted against the trajectory in arc minutes. The numbered peaks were caused by the background stars

Since all the images were taken by fixing the CCD camera on a camera tripod, some background star trails overlaid on the meteor's light curves. We found that it is very difficult to remove the background stars from the light curves without leaving artificial marks, so we decided to leave all the star trails and mark them on the light curves.

Besides the features caused by the background stars, there are still many sub-structures on the light curves. We consider that the rises and falls are due to the intrinsic properties of the meteoroids, i.e., the silicate and metallic grains described in the dustball model.

To confirm the idea, we first measured the intensities of these features, finding that they are much higher than the noise level of the light curves. Then we extracted random sky backgrounds from the original images, finding that the rises and falls in the sky backgrounds are much lower in intensity and less in density compared with the features that appeared on the meteor's light curves, which means these features are not from the background sky.

The typical flux of the split particles is about 0.1% of the total flux of the host meteoroid, implying that the mass of these particles is typically 0.1% of the whole meteoroid. We also inspected all the four images. No evidence of transverse spread or jet-like features as observed by LeBlanc et al (2000) was found, though our spatial resolution and detector's dynamic range are higher.

References

Hawkes, R.L., Jones, J. 1975, MNRAS, 173, 339

LeBlanc, A.G., Murray, I.S., Hawkes, R.L., Worden, P., Campbell, M.D., et al. 2000, MNRAS, 313, L9

Monet, D. et al. 1999, USNO-A2.0 (Washington: U.S. Naval Observatory)

Small–Telescope Astronomy on Global Scales
ASP Conference Series, Vol. 246, 2001
W.P. Chen, C. Lemme, B. Paczyński

Project MONICA for the Study of Time-Variable Phenomena of the Jovian Sodium Cloud and the Io Plasma Torus

Chien-Pang Chang and Wing Ip

Institute of Astronomy, National Central University, Chung-Li 32054, Taiwan

Abstract. Because of active volcanism, large amounts of gas and dust particles are being injected from the Galilean satellite, Io, into the Jovian system. The neutral cloud of sodium atoms and the plasma torus of sulfur ions provide very useful information on Io's interaction with the Jovian magnetosphere. A program called MONICA (Monitoring of Neutral and Ionized Atoms Clouds) was established at NCU with a view to participate in an international campaign during the flyby of the Jovian system by the Cassini spacecraft in December, 2000. Spectrographic observations were carried out using the 2.16m spectrograph of the Beijing Astronomical Observatory in Xing-Long. A progress report is presented here.

1. The Atomic Sodium Cloud and Jets

As a result of surface sputtering by the Jovian energetic ions, Io emits a large quantity of atmospheric gas (SO_2 and Na) into the circumplanetary region. From ground-based observations, it is known that there is a banana-shaped sodium cloud composed of slow-moving atoms with a relative velocity of 2~3 km/s with respect to Io (Brown, 1974; Goldberg et al., 1984). High spatial resolution imaging observations have also shown the presence of a narrow sodium jet moving at a speed of a few tens km/s (Pilcher et al., 1984; Schneider et al., 1991). This fast atomic beam might be produced by direct atmospheric interaction of the Na^+ ions in the Jovian magnetosphere. These fast atoms in turn create a giant sodium nebula surrounding the Jovian system (Mendillo et al.,1990); see Fig. 1.

2. Observational Method and Preliminary Results

In order to provide scientific input to the international Cassini Jupiter Flyby campaign in December 2000, we have acquired several nights (December 16, 18 and 25) of observations on the 2.16 m spectrographic telescope for Io observations. Some of the preliminary results are shown below in Figure 2.

With proper subtraction of the continuum background produced by the scattered light of Io's disk, the spectrographic images taken at different slit positions can be used to construct a 2D image of the atomic sodium cloud. The spatial distribution so derived would be useful in estimating the production rate of sodium atoms in the fast jet during the observational time interval.

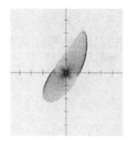 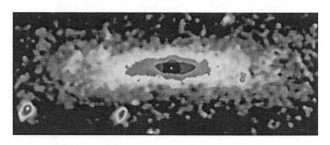

Figure 1.   The left-hand-side picture depicts the banana-shaped sodium cloud moving with a velocity of about 2~3 km/s. This figure is a 2D N-body simulation result. If the ejection velocity of particles is greater than about 7 km/s, the escaping atoms will form a giant sodium nebula surrounding Jupiter. The right-hand-side picture was taken form Mendillo et al. (1990)

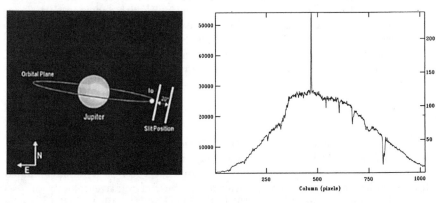

Figure 2.   The left-hand-side figure shows the position of the slit of the BAO observations. A strong sodium D-line emission can be seen superimposed on the continuum. There spectra can be used to build a 2D image of the sodium cloud.

References

Brown, R.A. 1974, IAU Symposium No. 65, Exploration of the Planetary System, p.527

Goldberg, B. A., Garneau, G. W., Lavoie, S. K. 1984, Science **226**, 512

Mendillo, M., Baumgardner, J., Flynn, B., and Hughes, W.J. 1990, Nature **348**, 312

Pilcher, C.B., Smyth, W.H., Combi, M.R., and Fertel, J.H. 1984, ApJ **287**, 427

Schneider, N.M., Hunten, D.M., Wells, W.K., Schultz, A.B., and Fink, U. 1991, ApJ **368**, 298

Small–Telescope Astronomy on Global Scales
ASP Conference Series, Vol. 246, 2001
W.P. Chen, C. Lemme, B. Paczyński

Long-Term Coudé Radial-Velocity Studies With a 1.2-m Telescope

C.D. Scarfe

Department of Physics and Astronomy, University of Victoria, Victoria, B.C., V8W 3P6, Canada

Abstract. I have used the 1.2-m telescope and coudé spectrograph of the Dominion Astrophysical Observatory for more than 30 years in a program of radial-velocity observations of binary stars. The program was begun with photographic plates as detectors, but for 20 years the primary detector has been the radial-velocity scanner, which cross-correlates stellar spectra with an artificial mask.

Since some of the binaries under observation have periods of several years, the instrument's stability is an important consideration. I have therefore been obliged to observe standard stars and asteroids to check its performance. These observations are of relevance to efforts to improve the IAU standard star system.

I will describe the telescope, spectrograph and scanner, and will briefly discuss some of the results obtained for a selection of binary and multiple stars.

1. Introduction

I have the good fortune to live only about 15 km from the Dominion Astrophysical Observatory, which forms the largest component of the National Research Council of Canada's Herzberg Institute of Astrophysics. I have thus had the opportunity to use that observatory's 1.2-m (48-inch) reflecting telescope and associated coudé spectrograph for most of those instruments' 38-year history, to pursue an observing program on binary and multiple stars. I have chosen to study systems many of which have periods of several years, which make it difficult to study them by infrequent visits to distant facilities.

In this paper I would like to outline the past history and present status of the telescope and the instruments I use with it, and to discuss a selection of the observations I have made.

2. The 1.2-m Telescope

The telescope was constructed by Grubb-Parsons and saw first light in 1962. It is a standard off-axis equatorial, with a primary focal ratio f/4. The single pier and conveniently located counterweight make it easy to reach the pole, or to switch from side to side of the polar axis. The original primary mirror was replaced in 1985 by a low-thermal-expansion borosilicate glass one. The telescope is normally operated at the coudé focus, using four additional mirrors to feed a horizontal spectrograph one floor below. In the late 1960's the original four mirrors were replaced by turrets of three mirrors each, with each of the three mirrors in each turret having a coating giving high reflectance in a specific wavelength region (Richardson 1968). They were necessarily small, and the beam between them is nearly parallel at f/145. It is restored to f/30 to match the spectrograph by a lens following the last mirror.

3. The Coudé Spectrograph

The spectrograph fills a heavily insulated room mounted on a concrete pad. Its collimator focal ratio is f/30, and it includes cameras of focal length 2.44 m and 0.81 m and gratings with 831, 600 and 1200 rulings per mm. Four identical replica gratings with the first ruling are mounted together as a mosaic, used with the longer camera to give a reciprocal dispersion of 0.24 nm mm^{-1} in the second order, for which they are blazed at 400 nm. The camera mirror is 0.91 m in diameter to avoid vignetting over a wavelength range of 70 nm.

Light enters the spectrograph by way of a superpositioning image slicer (Richardson 1968). A cylindrical lens makes the beam slightly astigmatic so that it will pass first through the horizontal entrance slot of the image slicer and then through its vertical exit slit. The latter is narrow enough to provide good resolution, and significant amounts of light fall on its jaws rather than passing through. But that light may not be wasted; it is reflected back by small mirrors to another set of mirrors mounted inside the image slicer near the entrance aperture, from which it is reflected toward the slit again. The slicer is designed to distribute the light on the collimator so as to avoid losing light on the back of a plate holder or pick-off mirror. Thus not only does the slicer transmit more light for the same seeing conditions than does a conventional slit, but it also distributes that light somewhat better. The overall gain in speed over a slit is of the order of one magnitude.

The spectrograph was originally designed for use with photographic plates, but has been used with a Reticon detector, and more recently with a variety of CCD's. Experience with plates (Scarfe, Batten & Fletcher 1990) showed that it has excellent stability for radial-velocity measurements, giving standard errors near 150 m s^{-1} for non-variable stars over many years. The velocities obtained from a carefully chosen set of reliable spectral lines have been found to agree well for bright asteroids with those predicted from their orbits (Scarfe 1985).

4. The Radial-Velocity Scanner

The scanner has been described by Fletcher et al. (1982) and McClure et al. (1985). It is based on a prototype built by Stilborn, Fletcher & Hartwick (1972) following the concept of Griffin (1967), but it incorporates several unique features. Light passes through a mask modelled on the spectrum of a star, idealized so as to give full transmission in line regions and zero transmission in the continuum. The transmitted light is collected by a Fabry mirror system which forms an image of the mosaic grating on the cathode of an EMI 9835/350 photomultiplier tube. The cathode is of such small area (9 mm by 9 mm) that it does not require cooling to yield an acceptably low dark current, but to bring all the light to such a small area the Fabry mirror system is necessarily complex, as described by McClure et al. (1985).

The masks in regular use are based on the spectra of Arcturus (K2 III) and Procyon (F5 IV-V). They cover the spectral range 400-460 nm. The Arcturus mask has about 750 slots each of which is 0.080 mm wide; those in the Procyon mask are slightly fewer and wider. The masks are scanned in dispersion by a stepping motor and the light transmitted is measured at each step and recorded automatically by a computer. For repeated scans the measurements at each step are binned in the computer so as to build up signal and average out the effects of seeing noise. The accumulated signal is displayed on an oscilloscope; the display is brought up to date after each scan, so that the observer can watch the progress of the observation and can terminate it when a satisfactory signal is built up. Drifts during the night are monitored by observations of an iron-argon hollow cathode lamp; the mask incorporates slots matching the lamp's spectrum.

The slots are arranged to make the spectral dispersion vary with height on the mask, and the mask is actually scanned at 45° to the dispersion direction, so that the mask dispersion varies, as it must, as a function of the velocity being measured. In this way the mismatch discussed by Hearnshaw (1977) and Griffin (1977) is avoided, and velocities of several hundred km s^{-1} can be measured without systematic error. It is of course necessary to adjust the zero point of the mask's height at the start of each night; this is readily done with the hollow cathode lamp, by covering alternately the red and blue halves of the mask, and ensuring that they give the same velocity.

Early experiments by McClure showed that the scanner's limiting magnitude could be increased by using an image slicer with an exceptionally wide exit slit, but at considerable cost in guiding error introduced to the radial velocities. This error was traced to systematically uneven illumination of the slit and collimator, and to avoid it a Guiding Error Minimizer (GEM) was introduced. This is a rotatable roofless Abbé prism placed in the beam just after the image slicer, which rotates the beam 180° after every five scans. The computer control of the scanner was arranged to permit scanning to stop only after multiples of ten full scans whenever the GEM was in use. Guiding error is much less of a problem for the narrower image slicer that I normally use, but the GEM still gives a modest increase in the reliability of the data.

5. Uncertainties of Measurement

Experience has shown that the uncertainty of a measurement with the scanner is slightly less than 0.50 km s^{-1} for bright stars with sharp spectral lines, independent of spectral type, between F5 and M2. It is larger, however, for fainter objects and for those whose lines are weak, such as halo stars. Rapid rotation broadens the signal and increases the uncertainty. Indeed the minimum becomes almost impossible to detect for $v \sin i \geq 30$ km s^{-1}.

Part of the uncertainty may be ascribable to night-to-night variations in the instrument. But I have never been able to convince myself of the reality of any such effect; to do so would require an observer to devote a disproportionate amount of time to observations of standard stars. I have instead made a modest number of such observations - typically five per night, accumulating to about 100 or 150 per year - and have used them to determine the long-term observational uncertainties reliably instead. They have also provided reliable estimates of the zero-point differences between masks. These amount to as much as 1.25 km s^{-1} in extreme cases. I have also used some 150 observations of asteroids to determine the overall zero-point of the system (Scarfe 2000).

It should be noted that Skuljan, Hearnshaw & Cottrell (1999) found significantly smaller uncertainties than those above from their observations with the same instrument. But their results are based on long runs of consecutive nights when the instrument was undisturbed, whereas mine are from many individual nights between which instrumental changes were made within the spectrograph. They also found a substantial year-to-year difference between runs; such differences may well be contributing to my larger uncertainties.

The instrumental profile of the scanner has maxima on either side of the central minimum, and shows additional, but smaller, fluctuations farther away in either direction. The maxima are due to the structure of the masks, which are opaque for a short distance on either side of each transparent region. This results in a complete mismatch on either side of the minimum, but random coincidences farther afield result in lower mean transmission there. The maxima and the unevenness, particularly the former, can cause systematic errors in the velocities measured for double-lined binaries, particularly for faint secondary stars. It is necessary to reduce traces of such objects more carefully than the usual method of fitting a parabola to the lower portion of the traces, which works well for single-lined objects. A program for such a reduction has been written by R.E.M. Griffin (Griffin 1991), who has kindly provided it for my use, for example, for the reduction of tracings of HR 6469 (Scarfe et al. 1994) as well as other similar systems.

6.   Some Illustrative Examples

Several of the more interesting systems I have studied are triple. Triple systems usually contain a subsystem of short period and large velocity amplitude, which may permit us to avoid some of the blending of spectra that bedevils long-period binaries. But they often present complications of their own. We discuss three very different triple systems here, as examples.

6.1.   HR 6469

This triple system, in which the distant component is an apparently spotted giant, was the subject of three consecutive papers in AJ (Wasson et al. 1994, Van Hamme et al. 1994, Scarfe et al. 1994). The close, eclipsing, main-sequence pair, whose combined mass is greater than that of the giant, was analysed photometrically in the second paper. The inclination of that pair, which is quite different from that found for the wide pair from speckle interferometry, allowed us to find all three masses, even though we could not detect the spectrum of the faintest component. We were able to find the distance to the system with an accuracy about 1.3%, and reliable radii for the eclipsing components, but the masses are accurate only to about 5% for each star. Moreover, the largest contributor to these uncertainties is the amplitude of the motion, in the long-period orbit, of the centre of mass of the eclipsing pair, although the inclination of the long-period orbit also contributes substantially to the parallax uncertainty and to that of the combined mass of the eclipsing system and hence to the individual masses of its components. Nevertheless, this system did yield the mass of a giant star, reliable determinations of which remain scarce. And it provided luminosities and effective temperatures sufficiently reliable to make comparison with published evolutionary tracks worthwhile. That comparison in the end led us to a rough estimate of the system's age.

6.2. HD 202908

This system contains three similar main-sequence F-stars, and has been the subject of papers by Fekel (1981) and Fekel et al. (1997). A close, but apparently non-eclipsing pair in a four-day orbit has a distant companion moving around it in 78.5 years. The long-period system passed periastron in 1986, and was followed by no less than four independent radial-velocity observers and the CHARA speckle group, all of whose observations were pooled for the joint 1997 paper. In the end we managed to obtain a distance for the system accurate to better than 1.5%, but the masses of the stars only to about 2.7%. The principal source of the uncertainty in the distance is that of the angular major semiaxis, but that in the masses is derived predominantly from the velocity amplitudes in the long-period orbit. Once again, better interferometry and better radial velocities are both needed to reduce our uncertainty.

6.3. 64 Orionis (HR 2130)

This system, consisting of three B-stars, was discussed by Fekel & Scarfe (1986) and in more detail by Scarfe, Barlow & Fekel (2000). Two sharp-lined objects form a pair with a period of two weeks, and a third body shares with them an orbit of period 13 years. This object is somewhat enigmatic in producing virtually no measurable spectral lines, despite contributing so much continuum light as to weaken the other components' lines considerably. Its luminosity is thus apparently comparable to that of either component of the short-period pair. The system has been resolved by a lunar occultation but is a difficult object for speckle interferometry. Our inability to measure the distant component's radial velocity means that we cannot determine the stars' masses without some assumption. However, the assumption that all three stars lie within the populated band of the mass-luminosity relationship defined by Andersen (1991) and Popper (1980) yields a strong constraint on the masses, and allows us to obtain them accurate to about 7%. But new observations with very high signal-to-noise ratio will be needed, as well as additional interferometric observations of the wide pair, if this system is to provide us with the accurate B-star masses it seems capable of doing.

7. Conclusions

Many other objects are being followed in this program. Several new variables have been discovered among naked-eye stars. Patience is required, since among them are unresolved systems with periods in excess of ten years. Indeed a few have shown monotonic variations of radial velocity over intervals approaching two decades, and their periods are thus still unknown. Many of these systems should be targets for long-baseline interferometry, although some may prove in the end to have too large a magnitude difference for either interferometric resolution or spectroscopic detection of secondaries. But combination of these techniques should continue to yield accurate stellar masses, distances and luminosities. Such data are well worth pursuing.

Acknowledgments. I am most grateful to Murray Fletcher for correcting some factual errors in an earlier version of this paper. Any that remain are solely my own responsibility.

References

Andersen, J.A. 1991, A&ARev 3, 91

Fekel, F.C. 1981, ApJ, 248, 670

Fekel, F.C., & Scarfe, C.D. 1986, AJ, 92, 1162

Fekel, F.C., Scarfe, C.D., Barlow, D.J., Duquennoy, A., McAlister, H.A., Hartkopf, W.I., Mason, B.D., & Tokovinin, A.A., 1997, AJ 113, 1095

Fletcher, J.M., Harris, H.C., McClure, R.D., & Scarfe, C.D. 1982, PASP 562, 1017

Hearnshaw, J.B. 1977, Observatory 97, 5

Griffin, R.F. 1967, ApJ 148, 465

Griffin, R.F. 1977, Observatory 97, 9

Griffin, R.F. 1991, Observatory 111, 67

McClure, R.D., Fletcher, J.M., Grundmann, W.A., & Richardson, E.H. 1985, in Stellar Radial Velocities (Proc. IAU Colloquium 88), ed. A.G.D. Philip & D.W. Latham, L. Davis Pres, Schenectady, 49

Popper, D.M. 1980, ARA&A 18, 115

Richardson, E.H. 1968, JRASC 62, 313

Scarfe, C.D. 1985, in Calibration of Fundamental Stellar Quantities (Proc. IAU Symposium 111), ed. D.S. Hayes, L.E. Pasinetti & A.G.D. Philip, Dordrecht: Reidel, 583

Scarfe, C.D. 2000, Standard Star Newsletter 29, 10

Scarfe, C.D., Barlow, D.J., & Fekel, F.C. 2000, AJ 119, 2415

Scarfe, C.D., Barlow, D.J., Fekel, F.C., Rees, R.F., McAlister, H.A., Hartkopf, W.I., Lyons, R.W., & Bolton, C.T. 1994, AJ 107, 1529

Scarfe, C.D., Batten, A.H., & Fletcher, J.M. 1990, Pub. DAO 18, 21

Skuljan, J., Hearnshaw, J.B., & Cottrell, P.L. 1999, in Precise Stellar Radial Velocities (Proc. IAU Colloquium 170), ed. J.B. Hearnshaw & C.D. Scarfe, ASP Conf. Ser. 185, 98

Stilborn, J.R., Fletcher, J.M., & Hartwick, F.D.A. 1972, JRASC 66, 49

Van Hamme, W. et al. 1994, AJ, 107, 1521

Wasson, R. et al. 1994, AJ, 107, 1514

Small–Telescope Astronomy on Global Scales
ASP Conference Series, Vol. 246, 2001
W.P. Chen, C. Lemme, B. Paczyński

INTEGRAL and Small Telescopes

Nami Mowlavi, Peter Kretschmar, Marc Türler, Nicolas Produit

INTEGRAL Science Data Center
16 ch. d'Ecogia, CH-1290 Versoix, Switzerland

Abstract. The future INTEGRAL satellite, to be launched in 2002, will observe the universe in gamma-rays. Parallel observations in other wavelengths are desirable for many gamma-ray objects, the most famous being certainly the gamma-ray bursts.

In this article, we present the INTEGRAL mission and its scientific objectives. We then give three examples of astrophysics fields for which combined gamma ray and optical observations are recommended. Small telescopes would be ideally suited for many of the studies.

1. Introduction

If human eyes were sensitive to gamma-ray photons ($E \geq 100$ keV), an astronaut in space would see a much more variable sky than what we see in the optical. The high energy universe is populated with objects whose luminosities vary on a wide time scale from milliseconds to years. The most famous of these objects are certainly the gamma ray bursts (GRBs), due to the mystery which veiled their very nature since their discovery in the fifties up to recent years. A better understanding of these objects was only made possible when the optical counterpart of a GRB was first observed simultaneously with the GRB event (van Paradijs et al. 1997). This stresses the importance of simultaneous multi-wavelength observations, particularly for high energy astrophysics where the duration of an event can be less than a day. The role of small (< 1 m) class telescopes is essential in this respect.

The next big gamma-ray mission is the INTErnational Gamma-Ray Astrophysics Laboratory (INTEGRAL). It is planned to be launched in 2002 by the European Space Agency. Its nominal life mission is two years, extendible to five years. The wide interest of the astrophysical community in this mission is attested by the large number of proposals submitted during the first announcement of opportunity (due on February 2001). The total requested observing time for the first year of the mission exceeded the available open time by a factor of 19.

In this contribution, we emphasize the role that small telescopes can play during INTEGRAL's mission. Section 2 describes the instruments on board of INTEGRAL and Sect. 3 the scientific objectives of the mission. Three examples of high energy astrophysics fields of study of potential interest for small telescopes are presented in Sect. 4. Conclusions are drawn in Sect. 5.

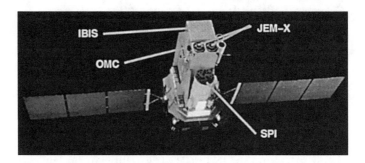

Figure 1.   INTEGRAL with its instruments.

2.   INTEGRAL and its Instruments

The INTEGRAL satellite is shown in Fig. 1. It comprises two main γ-ray instruments, a spectrometer (SPI) and an imager (IBIS), and two secondary instruments, an X-ray telescope (JEM-X, which is actually duplicated) and an optical monitoring camera (OMC).

The SPI spectrometer is sensitive from 20 keV to 8 MeV. It is composed of 19 cooled, hexagonally shaped, high purity Ge detectors, designed to achieve an optimal spectral resolution of 3 keV. Its field of view covers 16 degrees with an angular resolution of 2 degrees.

The IBIS imager is designed to achieve an optimal spatial resolution of 12 arcmin. It is made of thousands of small, fully independent pixels, and operates in the energy range 20 keV to 10 MeV.

The two identical X-ray imagers are built to extend the γ-ray observations down to energies of 3 keV. The angular resolution is 1 arcmin.

Finally, the optical monitoring camera enables to observe stars up to a magnitude of 19.7. It has a field of view of 5 degrees and a resolution of 17.6 arcsec.

All three high energy telescopes (SPI, IBIS and JEM-X) are coded mask instruments to localize the direction of a source. This technique is required because of the difficulty to focalize γ-ray photons. The coded mask apertures of SPI is clearly visible in Fig. 1 (mosaic on top of the instrument).

During the mission, the data from the satellite is received in near real time at the INTEGRAL Science Data Center (ISDC) located in Geneva, Switzerland, where it is analyzed for fast detection of GRBs and transients. Consolidated data is later processed and archived, and private data distributed to the observers in the scientific community. Further information can be found at *http://isdc.unige.ch.*

3.   Science with INTEGRAL

INTEGRAL deals with energies from about 1 keV to 10 MeV. This corresponds to temperatures T from 10^7 to 10^{11} K. The physical conditions which are able to produce such energetic photons can be identified from a rough estimation of the kinetic, gravitational and nuclear energies. These are considered in turn in the following paragraphs. The actual physical processes which convert these energies into γ-ray photons are mentioned at the end of this section.

The kinetic energy of a particle of rest mass m_0 can be estimated by equating the thermal energy $\frac{3}{2}kT$ (k is the Boltzmann constant) with the relativistic kinetic energy $(\gamma - 1)m_0c^2$ (γ being the Lorentz factor $1/\sqrt{1 - v^2/c^2}$, with c the speed of light). This shows that $v/c \simeq 0.1$ for protons at 10 MeV, while electrons are relativistic at energies above 1 MeV. INTEGRAL will thus deal with relativistic electrons.

Let us now consider the energy acquired by a particle in a potential well around an object of mass M and radius R. Equating the potential energy $G\, m_0\, M/R$ (G being the gravitational constant) to $3/2\ kT$ gives the relation $M/R = 3.1 \times 10^{-9} \times T/m_0$, masses, radius and T being expressed in g, cm and K, respectively. To reach an energy of 10 MeV ($\sim 10^{11}$ K), a proton must thus be in the potential well of an object with $M/R \simeq 1.8 \times 10^{26}$ g/cm ($\simeq 3.4 \times 10^{29}$ g/cm for electrons). White dwarfs have $M/R \simeq 4 \times 10^{23}$ g/cm, and are thus not dense enough to meet the requirement. Neutron stars, with $M/R \simeq 2 \times 10^{26}$ g/cm, are good candidates to accelerate protons to the required energies. The γ-ray astrophysical fields of interest to INTEGRAL will thus include accretion phenomenon around neutron stars or black holes. We note that clusters of galaxies also provide a deep potential well able to accelerate particles to γ-ray energies.

Transitions between two nuclear states of a given nucleus also provide a source for γ-ray emission. An example is the 1.8 MeV line from ^{26}Al disintegration.

For highly magnetic neutron stars ($B \gtrsim 10^{11}$ Gauss), the spectra are also influenced by the quantization of energies at these field strength (Landau levels) leading to observable line features.

Emission processes at the origin of γ-ray photons include:

- black body radiation (surface of neutron stars);
- bremsstrahlung (electrons accelerated in the Coulomb field of atomic nuclei);
- synchrotron radiation (relativistic electrons in a magnetic field);
- Compton processes (electron-photon diffusion) with either thermal or relativistic electrons;
- relativistic jets that boost either process;
- nuclear line emission (also from annihilation processes).

4. Examples of High Energy Astrophysical Objects

4.1. Active Galactic Nuclei

Active galactic nuclei (AGN) are very compact and luminous sources located at the very center of galaxies. The most distant and luminous AGNs are called quasars (contraction of quasi-star because they appear in the visible as faint stars of $m_V > 12.5$). They emit about 10^{48} erg/s and are thus the most powerful persistent sources in the universe.

Some quasars have the peculiarity to emit over the whole electromagnetic spectrum, from the radio to the gamma-ray domain (Fig. 2). This very broad spectrum can only be explained by many different emission processes arising at

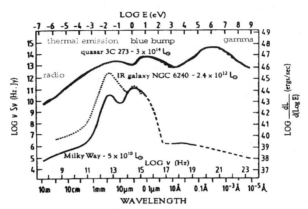

Figure 2.   Comparison of the spectral energy distribution of a typical quasar (3C 273), an infrared galaxy (NGC 6240) and a normal galaxy (the Milky Way). Adapted from Camenzind (1997).

different places around a central super-massive black hole (e.g. Robson 1996). In current models, the radiation from the near-infrared to the ultra-violet would be due to thermal emission from an accretion disk and a dusty torus surrounding the back hole. The radio component is most probably synchrotron emission arising in the jets observed to stream out from the central source. Hard X-ray and gamma-ray radiations are thought to be inverse-Compton emission due to thermal or synchrotron photons scattered by the relativistic electrons in the jet.

The luminosity of the so-called blazars, a subtype of quasars, varies by a factor of two or more on a time scale between a few days to a few months. Although variability is observed at all wavelengths, the amplitude and the variability pattern can be very different from one spectral band to another (Türler et al. 1999). This differential variability helps to distinguish the various emission components of the source. The study of multi-wavelength variability thus has the potential to put strong constraints on the various emission processes at work in AGN. For the bright quasar 3C 273, for instance, this approach has already proven to be successful for the study of the blue-bump (Paltani et al. 1998), the millimeter-to-radio emission from the jet (Türler et al. 2000) and the relationship between the infrared and the X-ray emission (McHardy et al. 1999).

The continuous monitoring of AGN at all possible wavelengths is the key towards a better understanding of the physics at work in these still poorly known objects. In this respect, small telescopes are ideally suited to monitor AGNs in the optical domain simultaneously to the gamma-ray observations performed by INTEGRAL.

4.2.   X-Ray Binaries

X-ray binaries are composed of a compact object — usually a black hole or a neutron star — which accretes mass from its binary companion. The binary companion is in most cases still on or close to the main sequence. There are three basic accretion mechanisms. The first one is the Roche-lobe overflow from the companion when the latter has evolved enough to increase its radius beyond the Roche lobe. It is the only mechanism which allows mass transfer from low-mass companions. The second accretion mechanism is wind accretion from the strong stellar wind of a massive star, and the last one is accretion from the equatorial outflow disk of a Be star.

The enormous gravitational potential energy released by mass accretion on these compact objects ($> 0.1\,mc^2$ compared to $0.007\,mc^2$ for nuclear burning of hydrogen) powers the production of high energy radiation. For black holes, the observed radiation must be produced in the accreted matter while it is drawn in, which is usually performed through an accretion disk. The dramatic spectral changes that are observed for many black hole candidates can be explained by changes in the properties of the accretion flow (Tanaka & Lewin 1995).

For neutron stars, the behavior is strongly influenced by the magnetic field. In sources with strong fields ($B \sim 10^{12}$ G) the accreted matter will be swept up by the magnetic field lines several hundred radii from the neutron star and led to the magnetic poles. There, all the energy is released in a very compact region — about $1\,\mathrm{km}^2$ — either on the surface of the neutron star or in an extended column above the poles delimited by a shock front in the plasma flow. The radiation spectrum is nonthermal and may include cyclotron line features as direct signature of the magnetic field. If the magnetic poles are not aligned with the rotation axis, pulsations can be observed (White, Nagase & Parmar 1995).

In sources with weak fields the accretion disk extends close to the neutron star and can even reach its surface. The spectrum from those objects can often be modeled by adding the spectral contribution from the accretion disk — the sum of blackbody spectra at various temperatures — and a contribution from a boundary layer between the disk and the star. Bursts are observed in some of these sources, which can be explained by the sudden ignition of the accreted mass at the surface of the star (Lewin, van Paradijs, Taam, 1995).

At optical wavelengths, emission from the companion star and sometimes from the accretion disk can be observed. Regular variations in the optical lightcurve are often caused by the orbital period of the system. However, in some cases, other phenomena can also be at the origin of the optical variations, such as a precessing warped accretion disk as observed in Her X-1 (Shakura et al 1998). Long-term optical monitoring of Be X-ray binaries has also shown feedback cycles between the neutron star and the mass donor (Negueruela et al. 2001). For black hole systems, a systematic comparison of the erratic variations in the lightcurves at different wavelengths is used to distinguish between different models of the accretion process.

4.3. Gamma Ray Bursts

Good spatial position accuracy of GRBs and a fast reaction time after their detection are crucial factors for the multiwavelength analysis of these objects and the determination of their distance.

INTEGRAL aims at a spatial accuracy of 1 arc minute and a reaction time of a few seconds. In order to achieve these goals, the data sent by the satellite is directly forwarded to the ISDC and analyzed in near real time (see Sect. 2). A dedicated software, the INTEGRAL Burst Alert System, automatically searches for the occurrence of a GRB, and issues an alert on a positive detection. Internet packets are simultaneously (within a few seconds after the occurrence of the GRB) sent to a pre-registered list of machines around the world giving the position of the GRB. A scientist on duty at the ISDC then analyzes the data to confirm (or possibly invalidate) the alert.

Any interested person can register to the GRB alert system. There are several alert trigger levels, according to the probability at which the alert system estimates the event to be a GRB. Small telescopes around the earth are particularly well suited to respond to the alert and search for an optical counterpart, because they are much more numerous and well-dispersed around the globe than big ones (this increases the probability of having several telescopes which can observe the event under clear night sky) and because they could react quickly if prepared to do so.

5. Conclusions

INTEGRAL will continue to nourish a fascinating branch of modern high-energy astrophysics which can greatly benefit from simultaneous multiwavelength observations. World-wide small telescopes may be efficiently involved in those efforts.

Several studies are being conducted by scientists at the ISDC. We have briefly mentioned three of them in Sect. 4. Interested readers involved in small telescopes are welcome to contact us in order to perform optical observations of some of the objects observed by us with INTEGRAL.

References

Camenzind M. 1997, in Les noyaux actifs de Galaxies, Springer-Verlag.

Lewin W.H.G., van Paradijs J. & Taam R.E., in Lewin W.H.G., van Paradijs J. & van den Heuvel E.P.J., eds, X-Ray Binaries, Cambridge Astrophysics Series 26, 1995, pp 175–228

McHardy I., Lawson A., Newsam A., et al. 1999, MNRAS 310, 571

Negueruela I., Okazaki A.T., Fabregat J. et al. 2001, A&A,369, 117

Paltani S., Courvoisier T.J.-L., Walter R. 1998, A&A 340, 47

Robson I. 1996, Active Galactic Nuclei, John Wiley & Sons Ltd.

Shakura N.I., Ketsaris N.A, Prokhorov M. & Postnov K.A. 1998, MNRAS, 300, 992

Tanaka Y. & Lewin W.H.G., in Lewin W.H.G., van Paradijs J. & van den Heuvel E.P.J., eds, X-Ray Binaries, Cambridge Astrophysics Series 26, 1995, pp 126–168

Türler M., Courvoisier T.J.-L., Paltani S. 2000, A&A 361, 850

Türler M., Paltani S., Courvoisier T.J.-L., et al. 1999, A&AS 134, 89

van Paradijs J., et al. 1997, Nature 386, 686

White N.E., Nagase F. & Parmar A.N., in Lewin W.H.G., van Paradijs J. & van den Heuvel E.P.J., eds, X-Ray Binaries, Cambridge Astrophysics Series 26, 1995, pp 1–49

Small–Telescope Astronomy on Global Scales
ASP Conference Series, Vol. 246, 2001
W.P. Chen, C. Lemme, B. Paczyński

The Carl Sagan Observatory: A Telescope for Everyone

J. Saucedo-Morales, A. Sánchez-Ibarra

Area de Astronomía, DIFUS, Universidad de Sonora, Hermosillo,
Sonora, México

D. Lunt

Coronado Instruments Group

Abstract. The Carl Sagan Observatory is a new project for a remote observatory that will be built at the summit of Cerro Azul (a 2480 m mountain located near Magdalena. Sonora, México). It will include one 55 cm and four 14 cm telescopes. The 55 cm telescope will be dedicated to supernovae research. One of the 14 cm Maksutov telescopes will be used as an autoguider for the stellar observations. The other 14 cm telescopes will feature different narrow band filters that will be used for solar research. The observatory will be controlled from the campus of the Universidad de Sonora in Hermosillo, Sonora, México ($\sim$ 200 km from the site). A prototype of the observatory building has been built on campus and first light is expected by the end of May of 2001. We expect to have an operating mountain observatory by the end of 2002. Some of the unique technical aspects of this observatory, which we believe can be a model for future small telescope observatories are discussed in this work.

1. Introduction

The Astronomy Area of the Center for Research in Physics, was established in 1990. It operates the only solar observatory in México, and a 41 cm stellar telescope, which is mostly used for educational purposes. The main research topics of the Area are: the activity of the solar corona and extragalactic astronomy. In 1996 a project for a remotely controlled solar-stellar telescope to be built at "Cerro Azul" was presented by the Astronomy Area to the Universidad de Sonora. After considering different options, it was decided in 1998 to commend the design and construction of the telescopes and optical instruments for the Carl Sagan Observatory Project (CSOP) to David Lunt (Coronado Instruments Group). The main features of the CSOP can be summarized as follows:

- A very good astronomical site
- Full use of technological innovations
- Low cost of construction and operation
- Outstanding optics in telescopes and instruments
- Maximum use of observing time (day and night operation)

• Maximum use of the data (The observations will immediately be put on the Internet). This means that people from all over the world will have the same priority for the data use as the people on the staff from the CSOP.

• Aside from the scientific aspects of the project, a strong impact on educational and outreach programs is expected. It is for this reason that the new observatory will be named in memory of the great American astronomer and science educator Carl Sagan.

This paper concentrates on a discussion of the technical aspects of the telescopes, instruments and logistics of the CSOP. The scientific and educational objectives of the project, as well as additional information about the CSOP, can be found in Sánchez-Ibarra & Saucedo-Morales (2001), as well as on the WEB page of the Astronomy Area (http://www.cosmos.astro.uson.mx). In sections 2 and 3 we discuss the different telescopes and instruments that will be available at the Carl Sagan Observatory. The logistical problems associated with a remote observatory (communications, electricity, control and maintenance) are briefly discussed in section 4. The conclusions are given in section 5.

2. The Stellar Telescope

The aperture of this telescope is 0.55 m and the focal length, 4.4 m. The basis of the design is a folded, F/4 paraboloid with the image primary transferred by an ancillary optic lens group to a point behind the primary mirror which simultaneously doubles the focal length to the required 4.4 m. Both the primary mirror and the flat are made of Zerodur. The primary has a thickness of 0.55 mm and this, combined with an open frame support system will allow the mirror to rapidly come to thermal equilibrium. The ancillary, transfer lens group includes a field lens and aperture stop to effectively eliminate spurious light from reaching the image plane. In order to keep the moments of inertia of the telescope to a minimum, the main tube consists of an octagonal section completely situated between the tines of the fork mount. The secondary mirror is supported on eight small diameter carbon fiber rods from the main box section resulting in a total mass of the secondary assembly of only 7 kgs. The primary mirror is mounted centrally within the box section on an open design support cell. The box section has 'shutter' sections to prevent sunlight from reaching the primary mirror when the instrument is used in solar mode. The mount is an equatorial fork. The main right ascension bearing is a 'Kaydon' precision turntable bearing. This provides four point support for the balls in the bearing. The force and moment specifications of the bearing are approximately 10 times the loads to be encountered in this instrument. The azimuth drive is a 1.75 inch precision worm wheel. The drive is a 30 VDC stepping motor. The declination bearings are also 'Kaydon'. Both are 6-inch in diameter; one is of the same four point construction as the AZ bearing, the other is an axial support bearing.

The main instrument to be used in the stellar telescope will be a direct CCD camera, which will be used to image about 200 galaxies to search for supernovae every night. The CCD will be a high QE Apogee AP8 with a 1024×1024 format and a pixel size of 24μm. With these characteristics, the stellar telescope will have a total field view of $\sim 19' \times 19'$, and a scale of $1.1''\text{pix}^{-1}$ on the sky. The second instrument of the stellar telescope will be a spectrograph, whose main purpose will be the classification of supernovae. The rear end of the telescope will allow for rapid interchange from direct imaging to spectrograph mode by rotating a mirror.

3. The Solar Telescopes

The solar modules will rest on alternate sections of the octagonal box. This means that all the telescopes will share the same mount, which will greatly simplify the control of the telescopes. One of the most remarkable aspects of these modules is the surface smoothness of their optical components, which will be kept below 0.05 nm RMS to minimize the scatter from spurious sunlight. The first CCD detector to be installed in the solar modules is an Apogee Instruments AP10, which consists of a 2048×2048 Thompson CCD. One of the main reasons for choosing this camera for the solar telescope is its fast read out time which will make possible to study fast changes in the surface of the Sun. The plate scale for the solar modules will be $105''\mathrm{mm}^{-1}$, with a pixel size of 14 μm, with which the field of view will be large enough to cover the full disk of the Sun. Other instruments, such as a Kodak Megaplus camera are also expected to be installed at the telescope in the near future. This camera will make possible the transmission of the solar images in almost real time.

4. The Site and Logistical Aspects of the CSOP

Cerro Azul is located at West longitude 110 34', North latitude 30 44', with an elevation of 2480 m above sea level, in the Sonora-Arizona desert. It is 36 km from Magdalena, and close to 200 km from Hermosillo. It is a fairly isolated peak that it is expected to have a seeing of $\sim 1''$ (because the prevailing winds are expected to arrive at the observatory with close to laminar flows). The low level of light contamination at Cerro Azul was one of the reasons for choosing it as the site for the CSOP. There is however, a price one has to pay for this, since there are no electric power lines nor any other kind of basic services. Electric power will be obtained from solar energy. A network of solar cell panels and batteries will be installed outside the observatory building. The total power consumption is expected to be less than 2 KW. A back-up diesel generator, to be used in emergency situations, will also be available at the site.

Communications from Cerro Azul to the mountain called "Cerro de la Madera" (located ~ 20 km to the west of Cerro Azul) will be through radio-modem transmission. From "La Madera" to Hermosillo the communication will be through optical fibers (that are property of the Mexican telephone company TELMEX). The transmission bandwidth will be about 256 KBPS at the beginning of the life of the observatory, but this is expected to increase by the time all the instruments are in full operation. The CSOP will also have a weather station, which will permit the computer at the site to close the dome and turn off the equipment and any instruments that are likely to be affected by bad weather. The weather station will play a more important role whenever the communication between Hermosillo and the observatory fails.

The Carl Sagan Observatory will be completely robotic and will be controlled from the University of Sonora in Hermosillo. We are considering several options for the software that will be used to control the observatory. At the moment, we are more inclined to use the Observatory Control and Astronomical Analysis System (OCAAS), sold by Torus Technologies. There will be a maintenance and alternative control station in Magdalena de Kino, Sonora. Service visits to the observatory site are expected to occur about every other week. No technical personnel are expected to be continuously present at the mountain once the CSOP is fully operational. This will permit to significantly reduce the cost of operation of the observatory. Due to the complexity of the problems related to the control of a mountain remote observatory, the telescopes, instruments, as well as all the equipment and materials that are going to be on the mountain, are first being thoroughly tested at the Universidad de Sonora. The pier, the prototype building (a cylindrical 3-D structure), and the dome (bought from Astrohaven) were built on campus to the same specifications as the ones that will be built on the mountain.

5. Conclusions

The CSOP is a big challenge, undertaken by a small university from a third world country interested in contributing to the overall knowledge of Astronomy. Aside from the scientific research that can be done with small telescope projects like the CSOP, these types of telescopes can be very powerful in increasing the interest in science. We also think that sharing information can be very profitable to both observatories and individuals interested in Astronomy. The total cost of the project will be around $250,000 dollars, an extremely low cost compared with those of similar projects elsewhere. Taking into account the considerable return that can be obtained from a project like this, it should be expected that the number of remote observatories with small telescopes will grow in the near future.

Acknowledgments. The CSOP would not have been possible without the collaboration of several individuals. Here is a list of some of the people that have made the most substantial collaborations to the Project up to this time: José Martines Rocha, Carlos Méndez-Peón, Benito Noriega, José Fahra, Fernando Avila Castro, Fernando Diaz and Fernando Félix. We also like to acknowledge the continuous financial support from the Universidad de Sonora.

References

Sánchez-Ibarra, A. & Saucedo-Morales, J. 2001, in ASP Conf. Ser. Vol. 225 eds. Robert J. Brunner, S. George Djorgovski, and Alex S. Szalay, in Press

Small–Telescope Astronomy on Global Scales
ASP Conference Series, Vol. 246, 2001
W.P. Chen, C. Lemme, B. Paczyński

The NCU Lu-Lin Observatory

Wean-Shun Tsay, Alfred Bing-Chih Chen[1], Kuang-Hsiang Chang and Huan-Hsin Li

Institute of Astronomy, National Central University, Chung-Li 32054, Taiwan

Abstract. The NCU (National Central University) Lu-Lin observatory is located at Mt. Front Lu-Lin, 120°52'25"E and 23°28'07" N, a 2862-m peak in the Yu-Shan National Park. The construction of Lu-Lin observatory was finished on January 14, 1999. The initial assessment of Lu-Lin site started in 1989, after which a three-year project was founded by the National Science Council (NSC) to support a modern seeing monitoring program. The average seeing at Lu-Lin is about 1.39 arc-second with an average of 200 clear nights annually. The sky background is 20.72 mag/arcsec^2 in V band and 21.22 mag/arcsec^2 in B band.

The Lu-Lin observatory is for both research and education. A home-made 76-cm Super Light Telescope (SLT) and four TAOS 50-cm robotic telescopes for a survey on Kuiper Belt Objects will be the two major research facilities. The pilot program for SLT consists of observations of time-varying astrophysical phenomena. The TAOS #1 telescope was installed at Lu-Lin in March 2000. A 90 KW/240 VAC power line and a water pipe system have been pulled to the site in early 2001. A wireless Network system through A-Li Shan has been operating at Lu-Lin observatory while a faster wireless Network system with 11.5 Mbit/sec bandwidth is under consideration and may be available in the near future for remote observing.

1. Introduction

We started the first site survey for a modern observatory in Taiwan in 1989. The original plan was to move the NCU 24" telescope from the campus to a suitable site in the central high mountain range of Taiwan. Weather station data and infrared satellite cloud cover data for Taiwan were studied to identify potential sites (Table 1) for further investigation.

Mt. Front Lulin (Figure 1), due to its flat summit and relatively easy access, was eventually selected and a three-year seeing study followed. Lu-Lin, located within the vast region of the Yu-Shan (Mt. Jade) National Park and Forest Reserves, at an elevation of 2862 m, is a five-hour drive from the NCU campus.

---

[1]Current Address: Department of Physics, National Cheng-Kung University, Tainan 70101, Taiwan

Mt. Front Lu-Lin	2862 m	Mt. Lu-Lin	2860 m
Mt. Stone-Water	2770 m	Mt. Ali	2406 m
Mt. Small Snow site #1	2550 m	Mt. Small Snow site #2	2990 m
Mt. Pear	2600 m	Mt. Ho-Huan	3400 m

Table 1.   Potential Sites for Taiwan optical telescopes

Figure 1.   **Left**: A bird's-eye view of the Lu-Lin summit. A trail is visible to the lower left.
Right: A close-up view of the 6m dome that houses the SLT. The small enclosure to the central right houses one of the TAOS telescopes.

2.   Seeing Study

The visible seeing α is defined as

$$\alpha = 2 \times 10^5 \cdot (\lambda/r_0)$$

where α is in arc-seconds, λ in μm , and r_0 in units of cm is called the Fried parameter (Fried 1965, 1966), defined as the coherence radius of the wavefront distorted by the turbulent atmosphere. The Fried parameter, r_0, is thus a function of the turbulence, the refractive index, the wavelength of the transmitted light, and the air-mass above the telescope. Typically, at a reasonably good site, α is about one arc-second or less.

Seeing is considered as the utmost important factor for most observations (Woolf 1982). For direct imaging, spectroscopy, photometry, and interferometric observations, seeing determines the spatial resolution, limiting magnitude and instrument speed. There are many methods to estimate the seeing such as direct visual estimates, star trailing, high-speed photoelectric image scanning, monitoring of differential image motion and image profile width, speckle interferometry plus pupil imaging, and shearing interferometry (Ardeberg 1987).

For the Lu-Lin site survey, we designed a portable optical seeing monitor with a Celestron C-14 mounted on a Takashi NJP equatorial mount and a high-sensitivity CCD TV-camera (Philips Amprex NXA 1031/01) as detector (Figure 2, right). This seeing monitor recorded real-time images of stars whose profiles were then analyzed by an IBM-AT/compatible clone equipped with a video digitizer board at a rate of $\sim 1/30$ sec. After a three-year seeing study at Lu-Lin, we found the average seeing to be about 1.39 $\pm$ 0.34 arc-second, based on 757 CCD observations of 682 single star and 75 binary-star measurements

Figure 2.   Parts were carried up by manpower to the summit (left) to set up the seeing monitoring station (right).

(Figure 3). The annual number of clear nights at Lu-Lin is about 200. The sky background is 20.72 mag/arcsec^2 in V band and 21.22 mag/arcsec^2 in B band.

3.   Site Development

Negotiations with the Yu-Shan National Park, the Ministry of the Interiors and the Taiwan Forest Bureau took about one year for the permission to construct a site survey station within 100 square meters on the summit. Since there was no road access to the site, all the construction material for the site survey station and the instruments had to be carried up by manpower from the base of the mountain to the summit (Figure 2, left). The observing station was built in 1991. Power supply for all equipments relied solely on generators.

After the site survey was completed in 1993, we proposed that the Lu-Lin site be developed to host a medium size 2-m telescope. More land was contracted for from various authorities, expanding the area to its present 300 square meters.

Until late 1997, the development of Lu-Lin observatory had been supported by NSC funds. A 6-m dome was built for a homemade 76 cm reflector. The construction of the dome was finished on January 14, 1999. Discussions with the electrical power company followed. A 3.3 km underground electrical power line to the foot of Lu-Lin in the national park area was built in 2000. Another electrical power line extending from the foot to the summit of Lu-Lin was installed in early 2001 as part of the Ministry of Education's Research Excellency project. The first TAOS telescope was installed at Lu-Lin in March 2000. A 90 KW/240 VAC power line and a water pipe system was pulled up in 2001. Also a wireless network system through A-Li Shan was set up, while a faster wireless network system with 11.5 Mbit/sec bandwidth is now under development.

4.   Research Activity

The Lu-Lin observatory is being developed for both research and education. The homemade Super-Light Telescope (SLT, Figure 4, left), which includes a 76-cm diameter HEXTEK Honey Comb lightweight gas-fusion mirror, has seen the first light in the fall of 1999. The field of view (FOV) of a fast primary focus (f/1.8), Ritchey-Chretien focal ratio (f/9), and a plate scale of about 0.3"/10μm,

 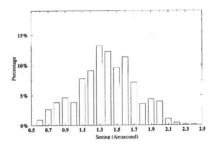

Figure 3.   **Left**: Binary star - Castor - measurements.
Right: The average seeing of Lu-Lin is about 1.39 ± 0.34 arc-second.

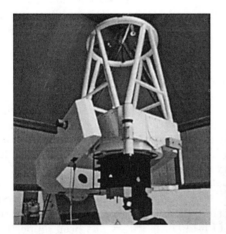 

Figure 4.   **Left**: The homemade 76 cm Super Light Telescope (SLT).
Right: The TAOS # 1 robotic telescope.

approaches half a degree square. In the first stage, only one quarter of the
FOV will be used with an Apogee AP-8 thin CCD (1K × 1K, 24μm/pixel).
A pilot program for the SLT deals with time-varying astrophysical phenomena,
exploiting the unique geophysical location (time and longitudinal coverage) of
Taiwan on the Western Pacific Rim.

The Taiwan-America Occultation Survey (TAOS, King 2001) is a collaborative project to conduct a census of comet nuclei in the outer solar system. The
TAOS experiment consists of four 50 cm, wide field (f/1.9), robotic telescopes,
each equipped with a 2K × 2K CCD camera (Figure 4, right). This experiment
provides the only means to study the cometary population in the small sized
end of the distribution. A great number of scientific products, notably variable
stars, will also derive from the huge TAOS database.

5. Future Development

Supported by the funding from the MoE Excellency project, we are going to further develop the facilities at the Lu-Lin observatory. This effort will focus on road construction, water and power supplies, communication links and other items necessary for the establishment and maintenance of the observatory as an inter-university astronomical facility for research and education. Members currently involved in this joint venture are the National Central University (which is operating the Lu-Lin observatory), Tsinghua University and Taiwan University. We expect that more universities will be included in this consortium in the near future. The proposed national infrastructure will promote Taiwan's role in many first-class astronomical projects, from solar system astronomy to cosmology; especially in view of its geographical position that beneficially complements the longitudinal coverage for joint international observing projects.

Acknowledgments. This work would not have been possible without the effort of a lot of people working on developing the site. We wish to thank the long-term support from the National Science Council, the Ministry of Education, the National Central University, the Yu-Shan National Park, and the Taiwan Forest Bureau. We also sincerely thank Mr. P.K. Chen for the first guided visit to the Lu-Lin site.

References

Ardeberg, A. 1987, in "*Identification, Optimization, and Protection of Optical Telescope Sites*", Proc. of an International Conf. Flagstaff, AZ

Fried, D.L. 1965, J. Opt. Soc. Am., 55, 1427

Fried, D.L. 1966, J. Opt. Soc. Am., 56, 1372

King, S.K. 2001, this volume

Tsay, W.S. et al. 1999, in "*The NCU Lu-Lin Observatory and its Future Direction*", Proceedings of Fourth East-Asian Meeting on Astronomy, 1999, p24

Woolf, N.J. 1982, ARA&A, 20, 367

(From left, facing camera) Nesterenko, Samus, Bochkarev (taking pictures)

Small–Telescope Astronomy on Global Scales
ASP Conference Series, Vol. 246, 2001
W.P. Chen, C. Lemme, B. Paczyński

Monitoring of AGNs at the Shanghai Astronomical Observatory

B.C. Qian & J.Tao

Shanghai Astronomical Observatory, Chinese Academy of Sciences, Shanghai, 200030

Abstract. This paper briefly introduces the program, equipment and results of monitoring of AGNs carried out at Shanghai Astronomical Observatory.

1. Introduction

A program of monitoring of AGNs has been carried out at the Shanghai Astronomical Observatory(SHAO) since 1994. We observe more than twenty AGNs with the 1.56m telescope located at She–Shan station of SHAO. The telescope is equipped with a liquid-nitrogen cooled Photometrics Series 200 CCD camera with 1024*1024 pixels. The field of view is 4'17" (1 pixel=0.25"). A focal reducer offers a field of view of about 13' (1 pixel=0.76").

The AGN monitoring program includes two parts: participating in the International AGN Watch and monitoring of blazars. The nature of AGNs is still an open question. Photometric observations of AGNs are important in order to construct their light curves and to study their variation behavior on different time scales.

By the mid-1980s, astronomers understood that their emission-line variability was closely tied to continuum variability, which strongly supported the prevailing view that the lines were driven by photoionization from the central compact continuum source, thought to be an accretion disk surrounding a supermassive black hole. It was also recognized that the emission-line variations should follow the continuum variations, but with a time delay that reflects the light-travel time across the broad-line region (BLR).

Blazars are an extreme subclass of AGNs and often show large and violent variations. We also observed the variability of many blazars on very short time scales called intraday variability or microvariability from minutes to hours which is also a common property of blazars.

2. Light Curves

Light curves of AGNs will be the direct result of monitoring. Some monitoring data have already been processed and will be published in the very near future. In late autumn 1994, the International AGN Watch started a multi-wavelength monitoring campaign on 3C390.3(Dietrich et al. 1998, O'Brien et al. 1998).

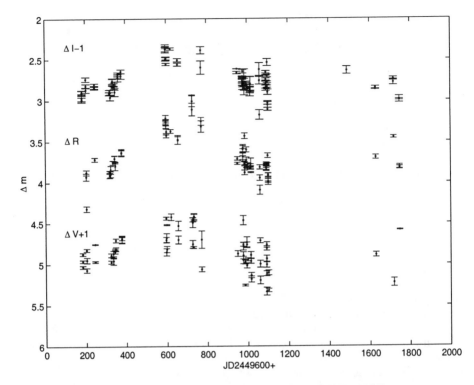

Figure 1.   Light curve of 3C390.3 for passbands V,R and I.

Figure 1 shows the light curve of 3C 390.3, obtained by the 1.56m telescope of SHAO. It includes optical observational data of 3C390.3 up to 1999.

3.   Microvariability

S5 0716+714 has exhibited significant variability. The high signal-to-noise, high time-resolution multi-frequency observations provide the most stringent constraint for physical models explaining the phenomenon of microvariability. Figure 2 shows the microvariability displayed by S5 0716+714, on March 16, 2000. The brightness varied by about 1 mag. in two hours.

4.   The Time Lag

Over the past ten years, considerable efforts have been made to get well-sampled light curves of AGNs through spectroscopic monitoring campaigns (Peterson 1993). The time lags of several AGNs have been determined (Peterson et al. 1998a; Peterson et al. 1998b). For the object BL Lac, Wagner et al (1996) found that no lag between R band and radio (5 GHz) was larger than 24 hours for S5 0716+714. Nesci et al. (1998) found for BL Lac that there is no time lag between the V and R bands within an accuracy of about 15 minutes.

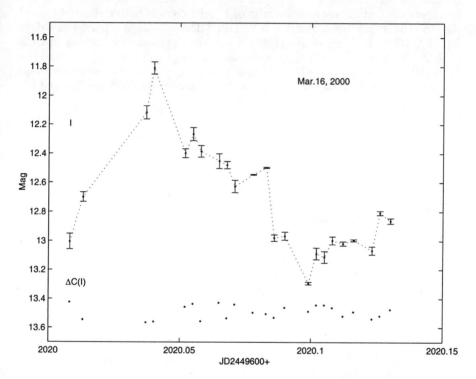

Figure 2.   Light curve of S5 0716+714 on March 16, 2000 for pass-band I. The observation was taken by the 1.56m telescope. The dots at the bottom are for two comparison stars.

In monitoring of S5 0716+714, a whole intraday variability was detected on 1995 Jan. 8. We got 80 observations in band V and I. Using cross-correlation analysis, interpolation and Discrete Correlation Function, a six minute time lag between band V and I was obtained (Qian et. al, 2000).

5. Quasiperiodic Oscillations

Quasiperiodic oscillations on short time scales have been observed only in a few BL Lac objects so far. For example, for S5 0716+714, Wagner (1991) found quasiperiodic variations on time-scales of 1 and 7 days in both radio and optical wavelengths respectively. Heidt and Wagner (1996) obtained a period of 4 days for S5 0716+714. Urry et al. (1993) found quasiperiodic oscillations in PKS 2155-304 both in UV and X-ray regions on a time-scale of one day. We used the Jurkevich method (Jurkevich 1971) to search for quasiperiodic oscillations and found quasiperiodic variation on time-scales of 10 days for S5 0716+714 and BL Lac.

References

Dietrich, M. et al. 1998, ApJS 115, 185

Heidt, J. & Wangner S.J. 1996, A&A 305, 42

Jurkevich, I. 1971, Ap&SS 13 154

Nesci, R. et al. 1998, A&A 332, L1

O'Brien, P.T. et al. 1998, ApJ 509, 163

Peterson, B.M. 1993, PASP 105, 247

Peterson, B.M. et al. 1998a, PASP 110, 660

Peterson, B.M. et al. 1998b, ApJ 501, 82

Qian, B. et al. 2000, PASJ 52, 1075

Urry, C.M. et al. 1993, ApJ 411, 614

Wagner, S.J. 1991, In: Duschl,W.J., Wagner, S.J., Camenzind, M.(eds.) Proc. Variability of Active Galaxies. Lecture Notes in Physics 377, Springer-Verlag, P. 163, 1991

Wagner, S.J., Witzel A., Heidt J., et al. 1996, AJ 111, 2187

Small–Telescope Astronomy on Global Scales
ASP Conference Series, Vol. 246, 2001
W.P. Chen, C. Lemme, B. Paczyński

Fast Drift-Scan CCD Imaging and Photometry with Small Telescopes: Lunar Occultations and Speckle Interferometry

Jorge Núñez & Octavi Fors

Departament d'Astronomia i Meteorologia, Universitat de Barcelona, Av. Diagonal 647, E-08028 Barcelona and Observatori Fabra, Barcelona, Spain. E-mail: jorge@am.ub.es

Abstract. In this paper we show how inherent features of typical CCD cameras can be used for fast (in the order of milliseconds) photometric observations (FPO), even with submeter class telescopes. This is based on a modified drift-scanning technique which we show to be advantageous for such kind of observations. In particular, we successfully carried out some lunar occultations and speckle interferometry observations using this technique. During the year 2000, we registered several occultation events. In particular, we present the occultation of SAO79031. The millisecond photometry of the event allows us subsequent data analysis for stellar diameter estimation and close binary detection. We also show an example of the application of the technique to speckle interferometry observations. Finally, we present the main characteristics of our project to build a new remotely controlled observatory near Barcelona.

1. Introduction

Since the appearance of CCDs more than two decades ago, their technical specifications have been constantly improved. Despite this rapid development, most current research grade cameras are still not able to meet the read out speed which FPO programs demand. These should be run in millisecond sampling regime with low read out noise and high digitization resolution mode. This is the main reason why CCDs have traditionally been discarded as eventual detectors in favor of photoelectric photometer systems (PEP).

With the advent of adaptive optics, a few state-of-art frame transfer CCDs (FTCCD) have been released to work as part of wavefront sensor systems (Ragazzoni et al. 1998). These devices reach 3Mpixel/sec at moderate read out noise regime, meeting most of the FPO requirements. While it is certain that they will play a key role in near future FPO programs, their cost is still far from making them accesible to most of the professional and high-end amateur community. Moreover, FTCCDs small active area makes them be dedicated detectors, which cannot be used for general imaging purposes.

In this paper we will show how standard full frame CCDs can be successfully used for FPO when operating in a modified drift-scanning mode.

2. Drift-Scanning

The conventional use of a CCD device is the operation in stare mode where the CCD chip is read out at the end of the exposure. Once the shutter is closed, the charge generated by the incident light on the surface of the CCD is converted to digital numbers, on a line per line basis, as a clocked charge moves through a serial register. This has been the usual operating mode in astronomy for years.

However, other modes can be considered since the clocking rate τ can normally be specified by the user. Typically, one has three options for the τ value: $\tau = \tau_0$, $\tau < \tau_0$ or $\tau > \tau_0$, where τ_0 is the sidereal rate at a given declination.

In the first case, acquired data appear as point-like sources provided clocking charge direction coincides with star motion over the chip and the telescope tracking system is disconnected. In the second case it is also necessary to have the camera properly aligned but, in this case, to ensure that the acquired data appear as point-like sources, it is necessary to slow the telescope tracking. These two variants in scanning mode are usually referred to as drift-scanning and time delay integration (TDI). This is the way several meridian circles (Stone et al. 1996) and Schmidt cameras (Sabbey, Coppi, & Oemler 1998) observe for fast sky coverage at moderate limiting magnitude.

Regarding the third case, τ can be chosen according to the time scale and magnitude of the event to be recorded. The detector does not need to be specifically oriented, because we keep the telescope tracking all the time. Thus, the star remains stationary over the chip while photogenerated charge is clocked through the serial register at desired rate τ. It is worth noting that a measure of the star flux is actually obtained every time a column is read out. In standard full frame CCDs this can be typically done at frequencies of 50 to 500kHz, fast enough for most FPO. We see, therefore, how a non-dedicated detector such as a full frame CCD, very common among instrumental settings in astronomical observatories, could be used for recording fast photometry events.

3. Application of Fast Drift-Scanning to Lunar Occultation Events

Fast photometry of lunar occultations (LO), with a time resolution of one millisecond, allows us to obtain an angular resolution at milliarcsecond level of object features far beyond the seeing effect, allowing, for example, stellar angular diameter measurements and close binary detection. Telescope diameter (D) and filter bandwidth ($\Delta\lambda$) are usually the two instrumental parameters that constrain the limiting angular resolution attainable by LO observations. However, this limit can be overcome with adequate data analysis if the lightcurve signal-to-noise ratio (SNR) is large enough.

Lunar occultation work has traditionally been performed using PEPs. These have been found to be extremely fast devices but show reduced quantum efficiency (QE~20%) (Kristian & Blouke 1982). On the other hand, CCDs with moderate-low read out noise would offer a better SNR due to their higher QE. Thus, as SNR plays a key role in data analysis, the selection of a detector with high QE and low read out noise turns out to be a crucial issue.

In this paper we examine the small-telescope regime for FPO. In particular, a 0.35m Schmidt Cassegrain Telescope (Celestron-14) parallelally assembled to

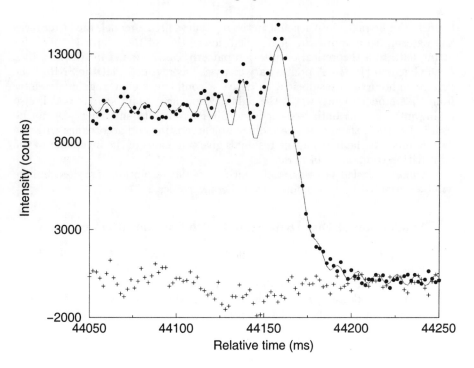

Figure 1.   SAO79031 4.0mag (4.5&4.5;0.1″) occultation at Fabra Observatory.

the Mailhat double 38cm-astrograph at Fabra Observatory was used. As detector, we employed a Texas Instruments TC-211 CCD (full frame chip, 192x164 13.75x16μ pixels, at 30KHz, QE~70% peak, 12e- rms). According to former specifications, TC-211 roughly suits FPO requirements.

Table 1.   Data for SAO79031 occultation on 19:59:29 TU 03/14/2000.

Object	Filter	$\lambda_0 \pm \Delta\lambda$(nm)	Δt(ms)	CA	PA	v($''/s$)	ϕ(mas)
SAO79031	R	641±58	2	63S	122	0.35	1.6

Table 1 shows ephemerides for the SAO79031 occultation, acquired using the scanning scheme described in Section 2.. The occultation was recorded storing a 20-pixel line every 2ms. Subsequent lightcurve shown in Figure 1 was calculated by averaging central pixels of every line and subtracting background estimated from the outer ones. A quick look reveals at least three diffraction fringes until signal vanishes into noise. This can be separated into two independent components, one due to residual periodic components in the CCD power supply and cooler and the other due to random atmospheric scintillation. While the first can be taken into account in the lightcurve fitting model, the latter lacks any deterministic approach. A detailed description of the adopted model, for fitting data to a theoretical occultation pattern, can be found in Richichi, Lisi, & Di Giacomo (1992). A preliminary analysis has revealed that the SNR of the current lightcurve is insufficient to derive a confident stellar angular diameter from that model. Thus, we fit the lightcurve using a point source model, also taking into account both telescope and filter bandwidth smearing. As can be seen in Figure 1, the model+periodic noise curve is in good accordance with the data points. At the bottom, the residuals give some idea of the behaviour of the scintillation component of noise.

A more detailed assessment of limiting angular resolution of such technique will be presented in a foregoing paper (Fors & Núñez 2001).

4. Application of Fast Drift-Scanning to Speckle Interferometry

Speckle interferometry (SI) and Speckle Imaging (Horch, 1995) provide diffraction limited resolution in small fields (isoplanatic patch) allowing, for example, to measure angular separations in close binaries far beyond the seeing. Thus, consecutive speckle campaigns allow us to trace the orbit of a close binary system. The technique consists of taking object images with a certain amount of time τ elapsing between them. This τ should be shorter than the speckle time, ranging from 20 to 40ms.

As in the case of lunar occultations, full frame CCDs are not fast enough for such τ. However, modified drift-scanning technique enables these detectors to image at this rate. In particular, drift-scanning allows us to use an observing technique similar to the charge transfer used by Horch, Ninkov, & Slawson (1997) but not limited in the strip size to the true size of the detector. This is because drift-scanning can be performed indefinitely. Thus, thousands of frames can be obtained continuously.

As a preliminary example we observed ζ Her (2.9 & 5.5mag ; 1.5″) on 07/21/2000. The CCD scale was 0.″3/pixel, just roughly above the diffraction limit of the telescope. Effective exposure time was 40ms which is right on the advisable limit for a 1.5″ seeing night with reasonable wind conditions. Figure 2 shows ten of the obtained frames. Each frame is about 4x4 arcsecs. In Figure 2 is it easy to see that we have beat the 1.5″ seeing, with both stars clearly separated. Although no detailed interferometric analysis has been carried out for angular separation, a quick visual inspection of frames shown in Figure 2 gives us $\rho=1.6″$ and $\alpha=90°$. This result seems to be at least compatible with literature values: $\rho(1980)=1.3″$, $\alpha(1980)=140°$; $\rho(1990)=1.6″$, $\alpha(1990)=83°$.

This result, although very preliminary, shows that it is possible to use the modified drift-scanning technique to perform speckle observations using common CCD cameras and small telescopes.

5. The New Observatory Project

Although this paper is primarily devoted to showing that the drift-scanning technique can be used to obtain millisecond photometric observations using small telescopes, given the characteristics of this meeting, it should be of some interest

Figure 2.   Speckle observation of ζHer (2.9&5.5;1.5″) at Fabra Observatory.

to present here some aspects of our project to build a new remotely controlled observatory near Barcelona.

The site of the new project, called *Niu de l'Aguila*, is at the top of one of the highest mountains (2531m) in the Cadí-Moixeró Regional Park, located 100 km north of Barcelona. It preserves a remarkable dark sky and presents other attractive characteristics as the surrounding area is protected by law (natural park law); the road is opened during summer and there is a ski resort during winter, it has electricity, water services, and a wireless communications antenna. We have 5 years of meteorological data for the area and there is already a usable building with live-in guard. We expect to reach a limiting magnitude to about $V = 21$ which is enough for the programs we intend to develop.

We plan to install two robotic telescopes at the new observatory. The first one will be a refurbished Baker-Nunn Camera which is presently at the Real Observatorio de la Armada (ROA) in San Fernando (Spain). This is a modified Schmidt type telescope designed for tracking Earth satellites which gives an impressive photographic field of view (FOV) of 5x30 degrees with images of less than 20 microns spot size across the entire FOV. In the first stage, a modification of the camera for CCD operation will easily give a FOV of 2x2 degs or larger with a 2kx2k CCD. To obtain a FOV of 5x5 degs, useful for a CCD, would require some additional optical modifications as described in Carter et al. (1992). We plan to fully adapt this camera for robotic and remote-controlled operation via internet using the experience of the ROA in the robotization of the Carlsberg Meridian Circle in La Palma and the San Fernando Meridian Circle located presently at El Leoncito (Argentina). Our plans for the second telescope are less definite but we expect to install a new-brand *1m-class* robotic telescope similar to the ones described in several papers of these proceedings. Both telescopes will operate complementarily, using the large FOV of the Baker-Nunn Camera mainly for discovering and the telescope with a larger collecting power mainly (but not only) for follow-up observations.

The programs we plan to carry out with both telescopes are: Near Earth Object (NEO) discovery and tracking; the same for the main belt, Trans-Neptunian asteroids and comets; finding extra-solar planets by precise photometry of transits; SN and novae discovery; location of GRBs; surveys for variable stars; normal (stare operation) and high speed (drift-scan) photometry; and any other programs suitable for this kind of small telescopes.

6.   Conclusions

We show that modified drift-scanning allows most full frame CCDs to be used for FPO as LO and SI. The instrumental settings needed to do this are very simple, with few mechanical and optical complements. This will bring the opportunity to most professional and high-end amateur observatories to do FPO. We plan to continue these kinds of observations, among others, in the new observatory we are planning to build near Barcelona.

Acknowledgments. This work was supported in part by the DGICYT Ministerio de Ciencia y Tecnología (Spain) under grant no. BP97-0903. O. Fors is supported by a fellowship from DGESIC Ministerio de Educación, Cultura i Deportes (Spain), ref. AP97 38107939.

References

Carter, B.D., Ashley, M.C.B, Sun, Y-S. & Storey, J.W.V. 1992, Proc. ASA, 10, 74.

Fors O. & Núñez J. 2001, in preparation.

Horch, E. 1995, Int. J. Im. Syst. and Tech. 6, 4, 401.

Horch, E., Ninkov, Z. & Slawson, R.W. 1997, AJ 114, 5, 2117.

Kristian J. & Blouke M. 1982, Scientific American 247, 4, 48.

Ragazzoni R., Baruffolo A., Farinato J. 1998, Ghedina A., Mallucci S., Marchetti E. & Niero T., 1998, SPIE 3353, 132.

Richichi A., Lisi F. & Di Giacomo A. 1992, A&A 254, 149.

Sabbey C.N., Coppi P. & Oemler A. 1998, PASP 110, 1067.

Stone R.C., Monet D.G., Monet A.K.B., Walker, R.L. & Ables, H.D. 1996, AJ 111, 4, 1721.

Small–Telescope Astronomy on Global Scales
ASP Conference Series, Vol. 246, 2001
W.P. Chen, C. Lemme, B. Paczyński

The Automated Telescope of Novosibirsk State University

A. Nesterenko, M. Nikulin, D. Vyprentsev, A. Zaytsev

Novosibirsk State University, Pirogova 2, 630090, Novosibirsk, Russia

I. Nesterenko, V. Prosvetov, V. Tsukanov, A. Valishev

Budker Institite of Nuclear Physics, Lavrenteyva av.11, 630090, Novosibirsk, Russia

Abstract. The current status of the 30 cm automated telescope project at Novosibirsk State University (NSU) observatory with Internet access is presented.

1. Introduction

The automated telescope facility was designed basically for photometric observations. The 30 cm reflector is intended to be used as a photometric monitor of variable stars and other objects at visible bands. A piggy-back 10 cm refractor will serve as a monitor of the solar surface. It is also planned to use the facility in the University education process. The NSU observatory location is 54°50′40″ north latitude and 83°05′42″ east longitude.

2. The Optical Systems of the Telescopes and the *UBVR* Photometric System

The facility will consist of two telescopes, a 30 cm Newtonian system (f/4.9) with a meniscus coma corrector and a 10 cm achromatic refractor (f/5.0). Calculations demonstrate, that the image is close to diffraction quality at the size of the CCD area sensors (ICX084AL and ICX075AL). For observations of the Sun it is necessary to use additional colour filters with transmission band at about 100 nm. In this case the lens chromatism does not restrict the field of view in high quality images. For the sunlight reduction we use a full aperture filter with a transmission of about 1/10000.

At the present time the *UBVR* system is used for photometric measurements. Measured spectral response curves of the CCD camera with our filters have acceptable deviation from curves of the *UBVR* system.

3. Hardware and Network Scheme

The telescope control devices are presented in Fig. 1. The main 30 cm Newtonian and the piggy-back 10 cm achromatic refractor telescopes are placed inside a 4

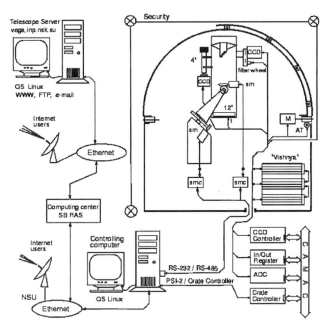

Figure 1. The hardware and network scheme

m diameter dome. We use an equatorial fork mount equipped with two stepper motors to guide the telescope to the selected area in the sky. The motor control system consists of two independent stepper motor controllers SMC-485, which are connected to the standard computer RS-232 port via the RS-485 interface bus and the RS-232/RS-485 converter. The RS-485 interface allows us to connect up to 32 devices by a monochannel line with up to 1200 m length. The dome guiding system is based on two collector motors and potentiometric azimuth transmitter, which provides a rotation synchronism accuracy with the telescope of about 1°.

The control over all the automated telescope facility and the security system is performed by a computer powered by the GNU/Linux OS via the CAMAC and "Vishnya" (cherry in English) modules. Internet users can gain access to the telescopes after filling a registration form. This form and observation requests can be submitted by e-mail or filled interactively with a web browser.

4. Conclusion

The basic elements of the future automated telescope have already been built. The infrastructure to access this telescope through the Internet is basically formed. In the next year it is planned to complete the installation of the equipment and to initiate its commissioning.

The website of the NSU observatory is *http://vega.inp.nsk.su/*

Small–Telescope Astronomy on Global Scales
ASP Conference Series, Vol. 246, 2001
W.P. Chen, C. Lemme, B. Paczyński

Some Aspects of Astronomy at Maidanak Observatory

Alisher S. Hojaev

Astronomical Institute, Astronomicheskaya 33, Tashkent, Uzbekistan

Abstract. A brief overview for some aspects of astronomical investigations at Maidanak is given. An excursus is made into its history and past research. Future plans and their requirements are described.

1. Introduction

By now it has been generally recognized that Central Asia, especially its south-western part, may be one of the most promising regions for ground-based astronomy. The necessity to choose a place for the 6.0 m telescope (BTA) have stimulated astroclimatic (astronomical site testing) campaigns in the region of the Tian-Shan, Pamir and Alay mountains in the early 1960s. It was found that the relatively dry air and the availability of suitable sites at altitudes of about 3 km above sea level provide high transparency in the optical, as well as in the far-IR and submillimeter range, with a maximum number of clear nights (at least in the area of the FSU), maximum sky stability, and a low amount of precipitation.

2. The Observatory

An analysis of the geographic, climatic, and atmospheric conditions for about a few dozen different sites in this region revealed Mt. Maidanak (E 66°56'; N 38°41') to be the best for an astronomical observatory. This isolated summit is located on one of the parallel ridges belonging to the Pamir and Alay mountain region in the south-eastern part of Uzbekistan with an altitude of about 2600 m. Fig. 1 presents the location of Maidanak on a world map. The first astroclimatic measurements started in August 1969 and became regular in April 1970 (Hetselius 1972). The long-term explorations have clearly shown that Maidanak has a seeing in the sub-arcsecond range (Shevchenko 1973, Shcheglov & Gur'yanov 1991, Gur'yanov et al. 1992), a large number of clear nights (Novikova 1970, Gladyshev & Shirokova 1987), a low night sky background (Kardopolov & Filip'ev 1979), and a high optical transparency (Zheleznyakova 1984), thus making it one of the most preferable sites worldwide for ground-based astronomy. Final tests by the ESO Differential Image Motion Monitor (Sarazin & Roddier 1990), which was used for site testing in Chile, and a Generalized Seeing Monitor (Martin et al. 1994) confirmed these estimations (e.g. Sarazin 1999, Egamberdiev et al. 2000).

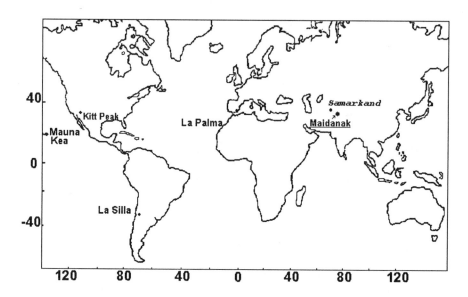

Figure 1.   Location of the Maidanak observatory

In fact there are two summits – Eastern and Western Maidanak, about 5km apart with the eastern peak being a bit higher. After ensuring the high quality of atmospheric conditions, a satellite laser ranging station was established on the Eastern summit (Scott 1995). Researchers from the Astronomical Institutes of Uzbekistan, Ukraine, and Lithuania carried out observations with a few small telescopes (two 60 cm, one 48 cm, two 40 cm and some smaller ones) installed on the same summit. The Western summit hosted the observational facilities of the Sternberg Astronomical Institute of the Moscow State University, Astronomical observatories of Lithuania, Ukraine, St.Petersburg University and others. All these astronomical facilities including their infrastructure are now under the jurisdiction of the Astronomical Institute of Uzbekistan and are operated according to agreements between the astronomical institutions of Uzbekistan, Russia, Ukraine and Lithuania. Steps are being made to arrange an international observatory at Maidanak. Presently the observatory facilities consist of a 1.5 m and a 1.0 m telescope as well as 9 other small telescopes with apertures of 60 cm or less. A general view of the observatory on the Western summit is given in Fig. 2.

3.   Research : A Brief Overview

The first astronomical investigations at Eastern Maidanak started in the early 1970s. In 1971 a small Schmidt telescope was installed on the summit and since 1972 this telescope has been used for observations of T-associations (Isakov 1975). In the same year, a 16" Cassegrain reflector for photometry received its first light there (Shevchenko 1974). Later on, two 60 cm Zeiss600 telescopes, a

Figure 2.   A bird's-eye view of Western Maidanak

48 cm LOMO AZT-14, a 40 cm and different other smaller telescopes started their operation.

Some of these telescopes, equipped with photomultiplier tube photon counting photometers, which have been adapted to the international UBVRI system, are used for monitoring of variable stars, such as T Tauri stars, Herbig Ae-Be stars and other PMS-stars, Cepheids, eclipsing binaries, patrolling of stellar flares and surveying of other unique objects. A large photometric data bank has been collected since the mid 1970s. Ukrainian astronomers also carried out photometric and polarimetric observations of solar system planets (especially Jupiter and Saturn), their satellites and Saturn's rings using one of the Zeiss600 telescopes. Baltic colleagues in cooperation with visiting scientists carried out stellar photometry of star-forming regions as well as other regions in the Vilnius and VilGen systems by the 40 cm telescope. In the late 1970s and in the early 1980s, researchers from the Astronomical Institutes of Uzbekistan and Byurakan Astrophysical observatory of Armenia have used the local network of observatory telescopes (usually 3) for simultaneous multiband patrol of flare stars (UV Cet, EV Lac, Pleiades campaigns). Such local telescope networks were also used to observe X-ray sources (Cyg X-1, Cyg X-2, Her X-1 etc).

During that period, successful observations in the infrared range were also made: near-IR spectroscopy of PMS-stars, mainly HAeBe stars and T Tauri stars, using an image intensifier with an oxygen-silver-cesium photocatode (e.g. Kotyshev 1988) and JHK photometric measurements of IR-sources, based on lead-sulfide cells, both cooled by dry ice. Moreover, spectroscopy of HAeBe stars in both optical and UV wavelengths was attempted using the image intensifier with a multi-alkaline photocatode.

Figure 3.   A speckle frame synthesized image (with R = 0.11″) of the asteroid 4 Vesta; for details, see Tsvetkova et al. (1991).

The installation of the 1.0 m Zeiss1000 telescope on the Western summit has opened new possibilities, particularly in high-resolution imaging of celestial bodies. As an example an image of the asteroid 4 Vesta (Tsvetkova et al. 1991) is presented in Fig. 3, which reveals the excellent seeing conditions of the site. This telescope had also an equipment for Vilnius system photometry, used by the Baltic astronomers for surveying the Galaxy stellar content.

Meanwhile a 1.5 m AZT-22 telescope, constructed by LOMO and installed at Western Maidanak, obtained better images due to the high optical quality of its main mirror (near diffraction limit). In July 1994 most of our telescopes were used to observe the crash of comet Shoemaker-Levy 9/P on Jupiter. Polarimetric measurements carried out by the 1.5 m and 1.0 m telescopes in addition to multicolor imaging, revealed significant differences in the polarization of the impact sites (Dudinov et al. 1995, Fig. 4). Based on these observations a consistent physical model of the phenomena was developed (Shevchenko et al. 1996). In addition, using the 1.0 m Zeiss telescope with the special image intensifying camera we observed young stellar objects, such as V1331 Cyg (Hojaev & Zheleznyak 1998). The star and its circumstellar environment are shown in Fig. 5. An outflow phenomenon in this relatively massive star detected in the optical range (Hojaev 1999) was corroborated by other authors. An absolute elements and physical models were defined for close binaries, e.g. IT Cas (Lacy et al. 1997), V364 Lac (Torres et al 1999), WZ Cep (Djuracevic et al. 1997), see also Zakirov et al. (1996).

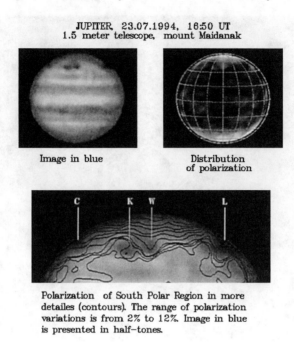

JUPITER, 23.07.1994, 16:50 UT
1.5 meter telescope, mount Maidanak

Image in blue   Distribution
of polarization

Polarization of South Polar Region in more
details (contours). The range of polarization
variations is from 2% to 12%. Image in blue
is presented in half–tones.

Figure 4.   Jupiter during the CSL9/P impact

4.   Future Exploration

Its specific geographical position covering the gap between observatories in Chile, Hawaii and the Canary islands (see Fig. 1), makes Maidanak very favorable for projects requiring uninterrupted observations of celestial bodies. The excellent seeing conditions render images with high angular resolution.

New developments demand that we update our telescopes by new back-illuminated chips, digital photometers, effective high-resolution multidimensional spectrographs and other new techniques in focal plane instrumentation. It is essential to establish a broad-band internet connection using a satellite link at the observatory to obtain necessary information in time, to receive and send data on-line, to join international networks of telescopes, to arrange for appropriate ground based support for space born observations etc.

Maidanak is worthy of installing a new generation of telescopes with larger apertures as well as automatic and robotic telescopes, demanding high capability internet links for remote observations. Existing telescopes can be transfered to automatic or at least semiautomatic ones. The site also provides the conditions for optical interferometry, and steps are being made to gradually improve the instrumentation and infrastructure. Lately a few CCD cameras including LN-cooled ones have started their operations mainly for monitoring GLS, AGNs and QSOs within the framework of international cooperation. Our aim is to make such collaboration more intensive and extensive as well.

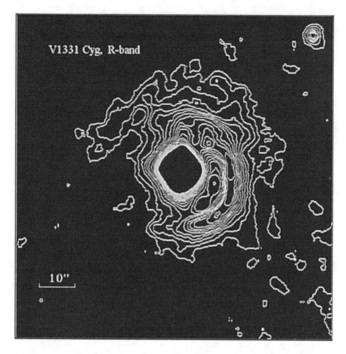

Figure 5.   The PMS-star V1331 Cyg and its environment

5. Conclusion

We have described the outstanding position of Maidanak for ground-based astronomy and have presented some previous and ongoing research as well as our future plans for improving the research capabilities at this site. It is not possible here to give a complete and detailed description, but we hope that we provided some idea about this observatory which deserves a place among the best ones. A close collaboration with astronomers and astronomical institutions world-wide is desired and their possible support by, for instance, the Maidanak Foundation, will be mutually beneficial.

Acknowledgments. The author is thankful to Prof. W.P. Chen for his generous support, discussions and interest in the paper. We are also grateful to Mr. Denis Ustimenko for his kind software assistance in preparing the manuscript.

References

Djurasevic, G. et al. 1998, A&AS, v.131, 17

Dudinov, V.N. et al. 1995, in ESO Conf. & Workshop Proceed., No52, 323

Egamberdiev, S.A. et al. 2000, ESO Preprint No1381

Gladyshev, S.A.& Shirokova, M.G. 1987 in Methods for Increasing the Efficiency
    of Optical Telescopes (Moscow: MSU), 45

Gur'yanov, A.E. et al. 1992, A&A, 262, 373

Hetselius, V.G. 1972, in Young Stellar Complexes and Astroclimate (Tashkent: Fan), 137

Hojaev, A.S. 1999, New Astronomy Reviews, 43, 431

Hojaev, A.S. & Zheleznyak, A.P. 1997 in Low Mass Star Formation from Infall to Outflow (Grenoble: IAU), 218

Isakov, I.S. 1975, in Study of Extremely Young Stellar Complexes (Tashkent: Fan), 143

Kardopolov, V.I. & Filip'ev, G.K. 1979, Pisma v AZh, 5, 106

Kotyshev, V.V. 1988, Tsirk. AI, N127(474), 18

Lacy, C.H.S. et al. 1997, AJ, 114, 1206

Martin, F. et al. 1994, A&AS, 108, 173

Novikova, G.V. 1970, in Atmospheric Optics (Moscow: Nauka), 10

Sarazin, M. & Roddier, F. 1990, A&A, 227, 294

Sarazin, M. 1999, http://www.eso.org/gen-fac/pubs/astclim/espas/maidanak/

Scott, W.B. 1995, Av.Week Space Tech., 142, N20 (15 May), 68

Shcheglov, P.V. & Gur'yanov, A.E. 1991, Soviet Ast., 68, 632

Shevchenko, V.S. 1973, AZh, 50, N3, 632

Shevchenko, V.S. 1974, in 100 years for Astronomical institute of Uzbek Academy of Sciences (Tashkent: Fan), 113

Shevchenko, V.V. et al. 1996, Astron.Vestnik (Solar System Research), 30, No 2, 101

Torres, G. et al. 1999, AJ, 118, 1831

Tsvetkova, V.S. et al. 1991, Icarus, 92, 342

Zakirov, M.M. et al. 1996, Journal Korean Astron. Soc., 29, S245

Zheleznyakova, A.I. 1984, in Astroclimate and Efficiency of Telescopes (Moscow: Nauka), 55

Small–Telescope Astronomy on Global Scales
ASP Conference Series, Vol. 246, 2001
W.P. Chen, C. Lemme, B. Paczyński

Observational Results with the 1 Meter Telescope at Yunnan Observatory During 1990-2000

P.S. Chen & W.Y. Zhang

Yunnan Observatory & United Laboratory of Optical Astronomy, Chinese Academy of Sciences, Kunming 650011, China

Abstract. The specification and instruments for the 1-m telescope at Yunnan Observatory of China are briefly described in this paper. Scientific results observed during 1990-2000 are also summarized.

1. Telescope Specification and Instrument Equipment

1.1. Telescope Specification

The telescope is a Cassegrain–Ritchey-Chretien–Coude system with a 1016 mm primary mirror and F/13. It is fully computer controlled for pointing, tracking and dome servo. The HST Guide Star Catalog is used as the reference to guide the program stars.

1.2. Instrument Equipped

(a) CCD Camera Format: 1024x1024 imaging pixels with a size of 24x24μm. Field of view in the Cassegrain focus: 6.5' x 6.5' (0."38/pixel).

(b) Cassegrain Spectrometer (with the CCD receiver above). Focal length: 150 mm. Spectrum range: 330-900 nm. Dispersion: 34-195 Å/mm.

(c) Coude spectrograph (with the CCD receiver above). Focal length: 1900 mm. Spectrum range: 330-900 nm. Dispersion:2.8-8.4 Å/mm.

2. Main Scientific Results

2.1. AGNs and Blazars

a) Optical monitoring in BVRI for over 70 BL Lac objects and Blazars. Short-term variations (hours) have been found for 28 BL Lac objects and 18 Blazars (Jia et al. 1998; Zhang et al. 1998; Xie et al. 1999).

b) Determination of optical positions for 30 extragalactic radio sources with a mean standard error of better than 0."2 (Tang et al. 2000).

c) Long-term optical monitoring for some BL Lac objects that have shown no short-term, only long-term periodic variations. Periods have been determined for ON 231, Mkr 421 and OJ 287 (Zhang et al. 1998).

d) Optical monitoring of OJ 287 from 1993-1998. An outburst was observed in 1994-1995 (Bai et al. 1999).

e) Photometry in BVRI and JHK for the infrared quasar IRAS 00275-2859 to determine its spectral energy distribution and absolute magnitude (Bao & Chen 1992).

2.2.   Binary Systems and Variable Stars

a) Light curves in B and V for many binary systems: AH Tau, FG Hya, SS Ari, AU Ser, XY Leo, BV Eri and BL Eri, YY Eri etc. (e.g. Yang et al. 1999; Liu, Q. et al. 1993,1996; Gu 1999).

b) Photometric monitoring of AU Ser has shown variation not only in the light curve but also in the color curve and found possible short period oscillations (Li et al. 1998).

c) Observations have shown the period change for some binary stars, like XY Leo and BL Eri (Liu, Q. et al. 1996; Pan et al. 1998).

d) Activity on the classical T Tau star BP Tau and its short-term variability in BVRI (Gullbring & Chen et al. 1996).

2.3.   Hα Observations for Chromo-spherically Active Binaries

a) High-resolution Hα profiles, equivalent widths and radial velocities were determined for 20 chromo-spherically active binaries (Liu, X. et al. 1993, 1994).

b) High-resolution Hα observations were performed for Aur in 1989-1992. Hα profile significantly varied outside the eclipse (Cha et al. 1994).

c) Spectroscopic observations for Eri to study the magnetic field (Liu, X. et al. 1994).

2.4.   Star Rotations

Rotational parameters for 50 bright stars were determined (Pan, K. et al. 1992).

2.5.   Objects in the Solar System

a) Positions of 14 fragments and the brightness of 8 fragments of comet Shoemaker-Levy 9 were measured three months before its collision with Jupiter (Wu et al. 1996).

b) Determination of positions for comets Hale-Bopp and Hyakutake were performed in the spring of 1996. 56 Positions derived on Feb. 28 - March 18, 1996 for Hale-Bopp were adopted by the "Minor Planets and Comets" (Wu & Chen 1996).

c) Determination of positions of the Uranian satellites: Ariel, Titania, Oberon and Umbriel with a single measurement error of better than 0.06 (Peng et al. 1998).

References

Bai, J.M. et al. 1999, A&AS, 136, 455

Bao, M.X., & Chen, P.S. 1992, Acta Astronomica Sinica, 33, 337

Cha, G.W. et al. 1994, A&A, 284, 874

Gu, S.H. 1999, A&A, 346, 437

Gullbring, E., Chen, P.S. et al. 1996, A&A, 307, 791

Jia, G.B. et al. 1998, A&AS, 128, 315

Li, Z.Y. et al. 1998, A&AS, 131, 115

Liu, Q.Y. et al. 1993, A&AS, 101, 253

Liu, Q.Y. et al. 1996, A&AS, 118, 453

Liu, X.F. et al. 1994, Acta Astrophisica Sinica, 14, 53

Liu, X.F. et al. 1994, Acta Astronomica Sinica, 37, 225

Pan, K.K. et al. 1992, Acta Astronomica Sinica, 33, 32

Peng, Q.Y. et al. 1998, Chinese Journal of Space Science, 18, 91

Tang, Z.H. et al. 2000, MNRAS, 319, 717

Wu, G.J., & Chen,P.S. et al. 1997, Acta Astronomica Sinica, 38, 183

Xie, G.Z. et al. 1999, ApJ, 522, 846

Yang, Y.L., & Liu, Q.Y. 1999, A&AS, 136, 139

Zhang, X. et al. 1998, Pub. Yunnan Obs., 3, 5

Small–Telescope Astronomy on Global Scales
ASP Conference Series, Vol. 246, 2001
W.P. Chen, C. Lemme, B. Paczyński

Systematic Spectroscopic Observations on Small Telescopes: Past and Future Research of Stellar Kinematics

M.E. Sachkov

Institute of Astronomy, 48 Pyatnitskaya Str., Moscow 109017, Russia

E.V.Glushkova, & A.S.Rastorguev

Sternberg Astronomical Institute, 13 Universitetsky Ave., Moscow 119899, Russia

Abstract. We have used a CORAVEL-type spectrometer on small telescopes (70-cm, Moscow observatory, Russia; 60- and 100-cm, Simeiz observatory, Crimea, Ukraine and some others) for systematic observations of northern sky stars, since 1987 (Rastorguev & Glushkova, 1997; Gorynya et al., 1998). The typical accuracy of a single measurement is about 0.5 km/s. The limiting magnitude is 14 in V-band. The young Galaxy disk population, metal-poor stars and Kaptein selected areas are the important parts of our program. In this paper we present new results on galactic kinematics and our future program.

The young disk population, which includes neutral hydrogen, HII-regions, OB-associations and supergiants, classical Cepheids and young open clusters, is characterized by small velocity dispersions (6-15 km/s) and, consequently, by small rotational lag relative to the LSR. This explains why such objects are used to study the rotational law of the galactic disk. Most studies of Population-I kinematics are based on radial-velocity data for objects with well-established and homogeneous distance scales.

We have summarized the information about our own observations in Table 1. Our measurements of radial velocities, taken together with precise absolute proper motion measurements (HIPPARCOS, 1997), have enabled the detailed investigation of the space velocity field of young objects. The most reliable results have been derived for open clusters and Cepheids:

1. The Galactic rotation curve and the kinematical parameters:

$(\Omega_0, \Omega'_0, \Omega''_0) = (28.0 \pm 1.3, -4.67 \pm 0.24\mathrm{kpc}^{-1}, 1.16 \pm 0.27\mathrm{kpc}^{-2})\mathrm{km/s/kpc}$ - the disk angular speed and its derivatives;

$V_0 = (6.9 \pm 3.2, 11.1 \pm 2.2, 6.9 \pm 1.3)\mathrm{km/s}$ - the solar velocity with respect to the local centroid, including the noncircular motion of the centroid;

$(\sigma_U, \sigma_V, \sigma_W) = (12.1 \pm 1.4, 8.0 \pm 1.0, 7.2 \pm 1.3)\mathrm{km/s}$ - the components of the velocity dispersion tensor;

$(f_R, f_\theta) = (-6.8 \pm 2.7, 1.4 \pm 1.9)\mathrm{km/s}$ - the amplitudes of systematic deviations;

$(i, \chi_0) = (5.8 \pm 0.6, 86 \pm 22)^0$ - pitch angle and phase angle of the Sun respectively.

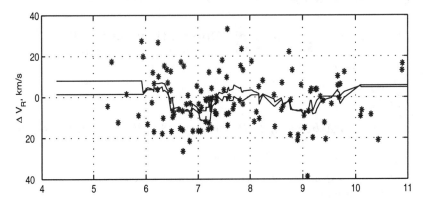

Figure 1.   Radial residual velocity, ΔV_R, from purely circular motions (in km/s) of young open clusters and Cepheids as a function of galactocentric distance R (kpc) (positive from the galactic center).

2. The radial systematic deviation with an amplitude of the order 7 km/s can be considered as an argument for the wave nature of the galactic spiral arms (see Fig.1).

3. The statistical-parallax method applied to classical Cepheids with periods $> 10^d$ leads to an LMC distance modulus of less than 18.40, which agrees, within the errors, with the short distance scale.

Table 1.   Radial velocity measurements

Objects	Number of measurements	Number of objects	Aim of research
Open clusters	1400	642 for 43 clusters	Kinematics
Cepheids	7211	144	Kinematics
Late type supergiants	1080	300	Kinematics
Stars in			
Kaptein selected areas	1300	1150	$[Fe/H]$
Metal-poor stars	120	100	$[Fe/H]$

Acknowledgments.   We are grateful to M.V.Zabolotskikh for useful discussions and comments, to Drs. N.N.Samus and N.A.Gorynya for taking part in our observations and to Mrs. Sachkova for the assistance. This study was supported, in part, by the Russian Foundation for Basic Research.

References

ESA, 1997, The Hipparcos and Tycho Catalogues, SP-1200.

Rastorguev, A.S., Glushkova, E.V. 1997, Astronomy Letters, V. 23, P.1.

Gorynya, N.A., Samus, N.N., Sachkov, M.E., Rastorguev, A.S., Glushkova, E.V., and Antipin, S.V. 1998, Astronomy Letters, V. 24, P. 815.

Small–Telescope Astronomy on Global Scales
ASP Conference Series, Vol. 246, 2001
W.P. Chen, C. Lemme, B. Paczyński

Advantages and Drawbacks of the ISM Method in Globular Cluster Photometry

József M. Benkő

*Konkoly Observatory of the Hungarian Academy of Sciences,
P.O. Box 67. H-1525 Budapest, Hungary*

Abstract. In the framework of our survey of variable stars in galactic globular clusters, we tested the promising new photometric technique, the Image Subtraction Method (Alard & Lupton 1998, Alard 2000). We found that the standard deviations of light curves are much better compared to those of the more common PSF methods (e.g. DAOPHOT), especially in the most crowded regions. Moreover, many additional variable stars have become measurable with this method. Unlike previously published data we have transformed our results into conventional magnitudes. The accuracies of the derived magnitudes are discussed.

All previously published data where ISM photometry was applied to globular clusters (Olech et al. 1999, Kopacki 2000), were in relative units (e.g. e^-/ADU) and this is not an accident! The big strength of ISM – that it is a relative method – is at the same time its most serious drawback. The method serves to determine differential fluxes for the variables relative to their fluxes in the reference frame, that is: $\Delta I_i^{(j)} = I_0^{(j)} - I_i^{(j)}$, where $I_0^{(j)}$ and $I_i^{(j)}$ are the fluxes of the jth variable in the reference image and in the ith frame, respectively. There are two possible ways to convert these into the usual instrumental magnitudes.

[1] If both magnitudes $(m_0^{(j)})$ and fluxes $(I_0^{(j)})$ for each variable can be *measured* in the reference frame, we simply apply the formula (see also Woźniak 2000)

$$m_i^{(j)} = 2.5 \log_{10} \frac{I_0^{(j)}}{I_0^{(j)} - \Delta I_i^{(j)}} + m_0^{(j)}.$$

[2] If we know the magnitude values of two arbitrary points (e.g., $m_1^{(j)} = m_{\max}^{(j)}$, $m_2^{(j)} = m_{\min}^{(j)}$) of the light curve, and we have $\Delta I_{1,2}$ values corresponding to the same phase, we can *calculate* the values I_0 and m_0. For each star

$$I_0 = \frac{\Delta I_1 10^{(m_1 - m_2)/2.5} - \Delta I_2}{10^{(m_1 - m_2)/2.5} - 1},$$

$$\text{and} \qquad m_0 = m_k - 2.5 \log_{10} \frac{I_0}{I_0 - \Delta I_k}, \text{ where k=1 or 2.}$$

These transformations provide excellent results if the star is located in a relatively sparse region, where the traditional PSF methods provide accurate enough reference values $(m_0^{(j)}, I_0^{(j)})$. In other cases the transformed light curves will be strongly distorted. To obtain reliable results the reference magnitudes have to be known within some hundredths of a magnitude.

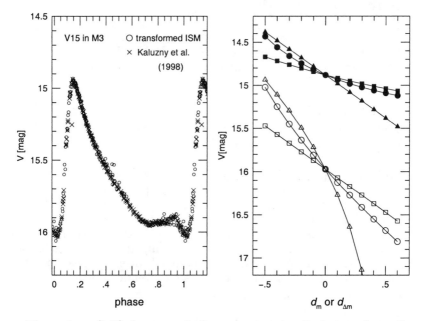

Figure 1.   (left) An example for accurate magnitude transformation
compared to previous DoPhot photometric results. (right) The effects
of errors in magnitude zero points and amplitudes on the shape of the
transformed light curve of an RR Lyrae star with $< m_V >= 15.5$ and
1^m total amplitude. d_m and $d_{\Delta m}$ represent the errors of the reference
magnitudes and amplitude, respectively. Empty and filled symbols
respectively signal the transformed magnitudes of minima and maxima.
Triangles and squares show the effect of d_m errors in $m_0 = m_{max}$ and
$m_0 = m_{min}$ respectively. If we know the total amplitude ($\Delta m = m_{min} - m_{max}$) with an error of $d_{\Delta m}$, the transformed light curve will be located
between the lines denoted by circles.

Acknowledgments.   I would like to thank J. Jurcsik for her continuous
help. This work is partially supported by OTKA, Grants No. T-30954 and
T-24022.

References

Alard C. 2000 A&AS, 144, 363

Alard C. and Lupton R.H. 1998 ApJ, 503, 325

Kaluzny J., Hilditch R.W., Clement C., Rucinski S.M. 1998 MNRAS, 296, 347

Kopacki G. 2000 A&A, 358, 547

Olech A., Woźniak P.R., Alard C., Kaluzny J., Thompson I.B. 1999 MNRAS,
        310, 759

Woźniak P.R., 2000 Acta Astron., 50, 421

Small–Telescope Astronomy on Global Scales
ASP Conference Series, Vol. 246, 2001
W.P. Chen, C. Lemme, B. Paczyński

Spatial Structure of Star Clusters by the 2MASS Database

J. W. Chen and W. P. Chen

Institute of Astronomy, National Central University, Chung-Li 32054, Taiwan, Email: m899008@astro.ncu.edu.tw

Abstract. We present some results of a pilot program to study star clusters with the Two-Micron All-Sky Survey (2MASS) observations. While 2MASS cannot resolve the cores or detect much of the main sequence of globular clusters, the homogeneity and large angular coverages make the database suitable to study young star clusters. We show that member stars are centrally concentrated in open clusters, with a density distribution markedly shallower than that for globular clusters. In NGC2506 (age 3 Gyr) giant stars appear to occupy a smaller region than main sequence stars—a natural consequence of mass segregation.

1. Spatial Distribution in a Star Cluster

The stellar distribution in a young star cluster is dictated by the conditions in the molecular cloud from which the cluster was formed and, as the cluster evolves, the dynamical interaction among member stars, and the disruption by Galactic tidal force or differential rotation. An investigation of spatial structure in young clusters may shed light on the role each effect plays during the cluster evolution. The youngest clusters may bear more imprints of star formation, or even the molecular cloud structure, than the subsequent dynamical history.

Stars in a globular cluster are known to concentrate progressively toward the center. The King model—successfully applied to globular clusters, some open clusters and dwarf elliptical galaxies (King 1962)— is understood as a combination of an isothermal sphere (i.e., dynamically relaxed system) in the inner part of a cluster, and tidal truncation by the Milky Way in the outer part.

2. Star Clusters in the 2MASS Database

Fig. 1 shows an example of the (projected) number density distribution — estimated by counting numbers of stars detected by 2MASS per square arcminute in concentric annuli—of the globular cluster M55. The radius of a cluster, R_s is defined to be where the number density drops to 3-σ of the background. The background is found to be rather uniform out to large angular extents, demonstrating the advantages of the wide sky coverage with the 2MASS data. A large background coverage is essential as distinction between cluster members and background stars is addressed in a statistical sense. While the central part of M55 is too crowded to resolve, the outer part is well fitted by the King model (Fig. 1). With its sensitivity, the 2MASS in general would not detect much of the main sequence in globular clusters.

In each of the two open clusters presented here, NGC2506 (1.9 Gyr, Twarog et al 1999) and IC348 (∼6 Myr, Lada & Lada 1995), the density profile is shallower than that for globular clusters. Fig 2 compares the cumulative density distributions of these two open clusters and the globular clusters, M55 and M13, within their respective R_s. In NGC2506, the bright giant stars (presumably more massive and evolved) seem to concentrate more toward the center, lending evidence of mass segregation in this open cluster.

The 2MASS database is proven a powerful tool for our application, and we are in the process of extending our pilot study to the analysis of a sample of young star clusters to determine their spatial structure. We plan to address the mass segregation history, and for the youngest clusters, the comparison with the density profiles or mass distribution in molecular cloud fragments.

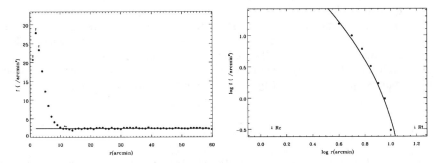

Figure 1.   Linear (left) and logarithmic (right) star density profile.

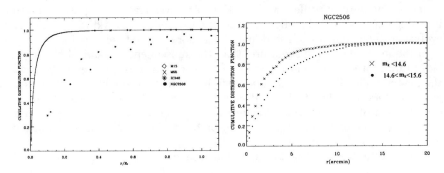

Figure 2.   Cumulative density profiles (left) of globular and open clusters, and (right) of NGC2506.

References

King, I. 1962, AJ, 67, 471

Twarog, B. A., Anthony-Twarog, B. J. & Bricker, A. R. 1999, AJ, 117, 1816

Lada, E. A. & Lada, C. J. 1995, AJ, 109, 1682

Small–Telescope Astronomy on Global Scales
ASP Conference Series, Vol. 246, 2001
W.P. Chen, C. Lemme, B. Paczyński

The Kinematics of Globular Cluster NGC 288

Chan-Kao Chang

Institute of Astronomy, National Central University, 32054 Chung-li, Taiwan; rex@phy.ncu.edu.tw

Alfred B. Chen, Wean-Shun Tsay, Wen-Ping Chen

Institute of Astronomy, National Central University, 32054 Chung-li, Taiwan

Phillip K. Lu

Department of Astronomy, Yale University, P.O. Box 208101, New Haven, CT 06520, USA

Abstract. The mean radial velocity of NGC 288 (accuracy 5.5 km/s) is determined to be -56.3 ± 20.1 km/s which, when combined with the mean proper motion (Guo, 1995), yields a peculiar velocity with respect to the LSR of $(u, v, w) = (29.7 \pm 18.1, -258.6 \pm 18.3, 62.3 \pm 20.3)$ km/s. This implies that NGC 288 moves in a retrograde sense with the Galactic rotation. We also derived the effective temperatures for stars in our sample and, as a corroborative effort, compared with those estimated previously from the BATC data (Tsai 1998) by spectral energy distribution fitting. We demonstrate that the BATC/SED fitting is an appropriate and efficient way to estimate the effective temperature of a star.

1. Introduction

The globular cluster NGC 288 has an obvious blue horizontal branch with intermediate metallicity (the second parameter problem). Its location near the South Galactic Pole makes it a useful probe to study the gravitational force and the local mass density of the Milky Way. In the past two decades, however, investigation on the metal abundances and the age of NGC 288 has been mainly by photometry. Kinematics information has been scarce. To remedy this, spectra of NGC 288 were obtained, and radial velocities of known member stars were measured. In addition, stellar effective temperature was derived for each star.

The WIYN/Hydra Multi-Fiber positioner was used in 1988 to obtain the spectra for radial velocity determination and spectral classification. Typical integration times were between 1000 and 4000 seconds, with a spectral resolution of 2.2Å. The internal accuracy of our data, as determined by cross analysis of about 100 solar spectra in a single sky flat exposure, is about 5.5 km/s.

2. Results

The mean radial velocity of NGC 288, as determined from 23 member stars, is -56.3 ± 20.1 km/s which, when combined with the proper motion previously determined by Guo (1995), yields a space velocity with respect to the solar system $(U, V, W) = (38.7 \pm 18.1, -270.6 \pm 18.3, 55.3 \pm 20.3)$ km/s. Adopting a peculiar velocity for the sun $(u, v, w) = (-9, 12, 7)$ km/s (Mihalas, 1981), we obtained the peculiar velocity for NGC 288 to be $(u, v, w) = (29.7 \pm 18.1, -258.6 \pm 18.3, 62.3 \pm 20.3)$ km/s. The total velocity of NGC 288 with respect to the LSR is thus 267.7 ± 32.8 km/s toward $l = 261.3^\circ$, $b = 13.5^\circ$. If the rotational velocity of the LSR with respect to the Galactic center is -220 km/s (Mihalas, 1981), NGC 288 would be rotating about the Galactic center with -38 ± 18.3 km/s, that is, in a retrograde sense.

We also derived the effective temperatures, $T_{\rm eff}$, of each star from its spectral type and, as a corroborative effort, compared with that derived from the BATC data by the SED fitting technique. The agreement is in general reasonable (Fig.1), suggesting that the intermediate-band multi-color fitting technique is appropriate in temperature determination, with higher efficiency than spectral observations.

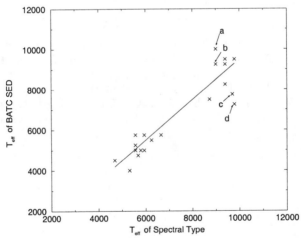

Figure 1.   Comparison between the effective temperature determined from spectral classification (x-axis) and from BATC/SED fitting (y-axis). The agreement is reasonable, except at high temperatures for which the BATC data seem less sensitive to.

References

Chen, A.B.-C., Tsay, W.-S., Tsai, W.-S., Lu, P. K. 2000, AJ, 120, 2569

Guo, X. 1995, Ph.D dissertation, Yale University

Mihalas, D., Binney, J. 1981, Galactic Astronomy: Structrue and Kinematics, 2nd ed., (Freeman)

Small–Telescope Astronomy on Global Scales
ASP Conference Series, Vol. 246, 2001
W.P. Chen, C. Lemme, B. Paczyński

Bright Young Star Candidates in the Rosette Nebula

P. S. Chiang, W. P. Chen, J. Z. Li

Graduate Institute of Astronomy, National Central University, Chung-Li 32054, Taiwan

Y.-H. Chu

Department of Astronomy, University of Illinois, Urbana-Champaign, IL, 61801, USA

Abstract. We identified possible optical young stars in the Rosette Nebula by using the IRAS and the ROSAT databases. The previously reported "diffuse" X-ray emission has been resolved into point sources. Forty-seven point sources are detected above 3-sigma of the X-ray background, and twenty-seven of these have stellar counterparts. Some of the young star candidates appear to be associated Herbig-Haro nebular features.

1. The Rosette Nebula

The Rosette Nebula (NGC 2237) is a prominent H II region currently expanding due to the radiation pressure and stellar winds from the OB association in NGC 2244 located near the center. The Nebula has an angular extent of ~1.5 degrees, which at a distance of 1.5 kpc (Pérez 1991), corresponds to a physical size of more than about 40 pc.

2. The Young Star Candidates

The Rosette Nebular has been observed with the ROSAT PSPC and HRI instruments. IRAS sources in the region have been reported by Cox et al (1990). To identify possible pre-main sequence stars, we compiled a list of sources detected by both ROSAT and IRAS in the vicinity of the Rosette Nebula. The combination of X-ray and infrared observations to search for young stars has been successful in nearby molecular clouds, such as in Taurus, Ophiuchus, Chamaeleon, and Lupus (see the review by Feigelson & Montmerle 1999). Because Rosette Nebula is some 10 times further away than these molecular clouds, only the brightest IRAS and ROSAT objects in Rosette Nebula could have been detected. Our analysis has resolved previous reported "diffuse" X-ray emission into point sources. A total of 47 point sources are detected at 3-sigma above the background. Here we report only sources detected in *both* catalogs. Optical counterparts have been sought within 1' of the nominal IRAS and ROSAT positions in the H-alpha image taken by T. Rector et al. (private comm.) and the Digitized Sky Survey images. Infrared and X-ray sources without optical counterparts are possibly embedded sources (Feigelson & Montmerle 1999).

Table 1 lists 5 IRAS/ROSAT sources thus identified. These should be studied to confirm their pre-main sequence nature. Fig. 1 shows an example of such an IRAS/ROSAT source, for which the H_α image is compared with the

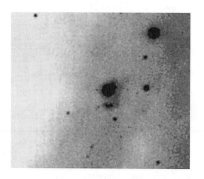 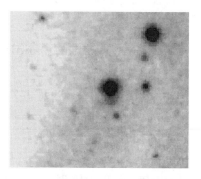

Figure 1.   The H-alpha (left) and Digitized Sky Survey (right) images of Object 1 in Table 1. Each image is about 90" on a side, with north to the top and east to the left.

Digitized Sky Survey image. An emission nebulosity, possibly of Herbig-Haro object nature, is readily seen to the south of, but may or may not be associated with, the bright optical star (No. 1 in Table 1). A full papaer of our results will be published soon.

Table 1: Bright Young Star Candidates in the Rosette Nebula

No.	IRAS	ROSAT/RA DEC (2000)	USNO/RA DEC (2000)	m_B m_R
1	06288+0456	06 31 29.7 +04 54 50	06 31 29.8 +04 54 49	12.4 16.4
			06 31 28.4 +04 54 50	19.2 19
2	06288+0452	06 31 33.2 +04 50 52	06 31 31.5 +04 51 00	10.5 16.2
		06 31 33.3 +04 50 47	06 31 34.9 +04 51 00	20 19.7
		06 31 33.7 +04 50 35	06 31 36.1 +04 51 00	16.3 18.3
			06 31 30.0 +04 50 59	18.4 18.9
			06 31 33.4 +04 50 38	9.4 12.3
3	06289+0504	06 31 38.2 +05 01 31	06 31 38.4 +05 01 37	8.4 11.1
		06 31 38.4 +05 01 42	06 31 40.5 +05 01 36	20.4 19.9
			06 31 40.7 +05 01 36	20.2 20.1
			06 31 41.7 +05 01 36	20.5 19.2
4	06290+0508	06 31 39.7 +05 05 58	06 31 41.8 +05 06 06	19.5 19.4
		06 31 39.9 +05 05 48	06 31 43.5 +05 05 49	12.8 15.4
		06 31 40.3 +05 05 46	06 31 38.9 +05 05 40	20.9 19.7
			06 31 40.0 +05 05 57	12.5 15.1
			06 31 41.8 +05 05 50	18.9 19.1
			06 31 42.7 +05 05 45	20.1 19.6
			06 31 43.5 +05 05 49	12.8 15.4
5	06298+0444	06 32 31.0 +04 42 37	06 32 31.4 +04 42 34	16.4 17.5

References

Feigelson, E. D. & Montmerle, T. 1999, ARA&A, 37, 363

Cox, P., Deharveng, L., & Leene, A. 1990, A&A, 230,181

Pérez, M. R., 1991, RMxAA, 22, 99

Small–Telescope Astronomy on Global Scales
ASP Conference Series, Vol. 246, 2001
W.P. Chen, C. Lemme, B. Paczyński

Deprojection of Planetary Nebulae

Z.W. Zhang and W.H. Ip

Institute of Astronomy, National Central University, Chung-Li 32054, Taiwan, Email: m889002@astro.ncu.edu.tw

Abstract. Useful information can be obtained by deprojection of the two-dimensional images of extended objects such as planetary nebulae or supernova remnants. Three-dimensional distributions (assumed to be axially symmetric) of emissivity, electron number density and temperature might be derived from the deprojected structures. As a test, we follow the iterative algorithm of Leahy & Volk (1994) to project the images of several PNe taken at different wavelengths. This method can also be applied to the images of supernova remnants and galaxy clusters.

1. Introduction

Planetary Nebulae have many different shapes. Their intriguing configurations might be explained in terms of the projection of axially symmetric ionization structures viewed at different look angles. Simple theoretical models have been found to be successful in fitting the observed morphologies (Aaquist and Kwok, 1996). The real structures of PNe as revealed by the deprojection method turn out to be far more complex (See Volk and Leahy, 1993). We have therefore built a program to investigate the deprojected three-dimensional structures of a large sample of PNe and eventually FLIERS, cometary knots and cometary globules in HII regions surrounding OB cluster.

The basic idea of deprojection is to derive the three-dimensional configuration of an object from the two-dimensional image (Fig.1). This is somewhat similar to tomography if it could be assumed that the structure of the object itself is axially symmetric. Mathematical treatments such as the Fourier Slice theorem have been used to deal with the deprojection of astrophysical objects. This is a very powerful technique with wide applications.

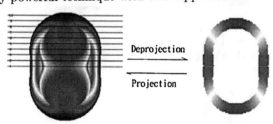

Figure 1. Projection and deprojection can derive the density or emissivity distribution.

2. Matrix Method for Deprojection

The algorithm of the Leahy method is based on the iterative deconvolution technique given by Lucy (1974). The basic idea of matrix computation is derived from the integration of emissivity along the line of sight which can be written as (Leahy & Volk 1994)

$$I_j^k = 3D \sum_{i=3D1}^{N} K_{ji}\varepsilon_i^k \tag{1}$$

Here, I_j is the intensity distribution on the sky plane from projected radius r_{j-1} to r_j. ε_i is a spherically symmetric emissivity distribution $\varepsilon(r)$. K_{ji} is the kernel matrix to transfer the coordinates from the spherical shell i to the sky plane j. The emissivity is then derived by using the inverse kernel Q_{ij} and iteration method

$$Q_{ij}^k = 3D K_{ji}\varepsilon_i^k / I_j^k \tag{2}$$

$$\varepsilon_i^{k+1} = 3D \frac{\sum_{j=3D1}^{N} I_j^{obs} Q_{ij}^k}{\sum_{j=3D1}^{N} K_{ij}} \tag{3}$$

As an example, the result for the line of IC 418 from WFPC2 (left of Fig.2) are shown in the right-bottom of Figure 2.

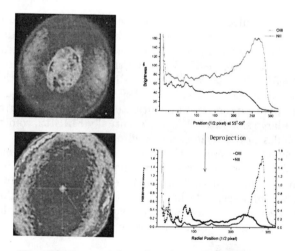

Figure 2.   Numerical examples of emissivity distributions of [N II] and [O III] from the Leahy deprojection of the WFPC2 images of IC 418(left-top is [N II] and left-bottom is [O III]).

References

Aaquist, O. B.,& Kwok, S. 1996, APJ, 426, 813

Leahy, L. A., & Volk, K. 1994, A&A, 282, 561

Volk, K., & Leahy, L. A. 1993, AJ, 106, 1954

Small–Telescope Astronomy on Global Scales
ASP Conference Series, Vol. 246, 2001
W.P. Chen, C. Lemme, B. Paczyński

Chemical Abundances of the Planetary Nebulae NGC 2392 and NGC 3242

C.H. Wu[1], J.Z. Li[1,2], Z.W. Chang[1], C.Y. Lin[1], J.Y. Hu[2] and W.H. Ip[1]

1. *Institute of Astronomy, National Central University, Chung-Li 32054,Taiwan(Email: m879008@astro.ncu.edu.tw)*
2. *Beijing Astronomical Observatory, National Astronomical Observatory, Chinese Academy of Sciences, Beijing 100012, China*

Abstract. Planetary nebulae represent the end product of the evolution of low mass stars with $M < 8M_\odot$. The central stars have masses between 0.55 and $0.85M_\odot$. This means a large amount of material must have been distributed in the interiors of the PNe. The observed variations of the chemical compositions and mass distributions therefore carry important information about the nature of the associated AGB superwinds and Red Giant outflows.(Kwok et al,1978) A program for comprehensive multiwavelength study of PNe has been initiated at NCU. One recent study has to do with the spectrographic observations of the planetary nebulae NGC 2392 and NGC 3242 using the 2.16 m telescope of the Beijing Astronomical Observatory. Some preliminary results are presented here.

1. NGC 2392 (The Eskimo Nebula) and NGC 3242

The Eskimo Nebula (NGC 2392) is unique in the sense that its elliptical structure is infiltrated by filamentary structures reminiscent of the cometary knots seen also in the Helix Nebula. There is also the interesting question on the ionization structures and dynamical histories of its different layers. It is therefore important to obtain spectra with high spatial resolution covering key features. The bipolar nebula NGC 3242 has prominent brightness structures at its two ansas (Balick et al. 1993). The origin of these so-called Fast Low Ionization Emission Regions (FLIERs) is still unclear.

2. Preliminary Results

The observations were performed with the 2.16 m telescope of the Beijing Astronomical Observatory on Dec 24 2000. The slit is about 4' long and 2.5" wide. The spectral resolution is 200Å/6mm, see Figure 1. We are particularly interested in NGC 2392 because in a previous study by Wu (2000) the filamentary structures surrounding the central region were found to display strong symmetry. This effect could not be explained by the occurrence of Rayleigh-Taylor instability as proposed previously by Capriotti (1973). It is more likely that pairs of polar jets were responsible for such structures. Phillips and Cuesta

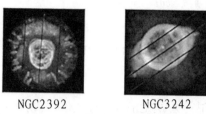

NGC2392   NGC3242

Figure 1.   Slit positions and orientations for the BAO observation of
NGC2392 and NGC3242.

(1999) made similar observations in their analysis of their narrowband images
of NGC 2392 in [O III] and [SII]. The whirling features in the central nebula are
also puzzling. It might be that the wind-brown bubble mode for NGC 6543 (see
Balick and Preston, 1987) could be applied to the Eskimo Nebula. We hope to
use the present spectrographic data to shed further light on this issue.

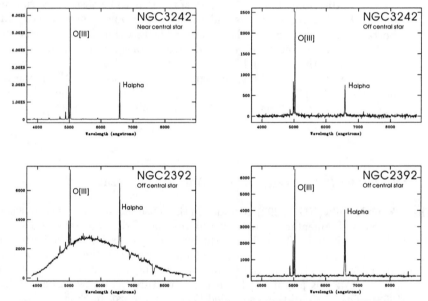

Figure 2.   Preliminary spectrographic of NGC3242 and NGC2392.
Left: spectrum taken near the central star. Right: spectrum taken off
the central star.

References

Balick, B. and Preston, H.L. 1987, AJ, 94, 958

Balick B., M. Rugers, Terzian Y., Chengalur J.N. 1993, ApJ 411,778

Capriotti, E.R. 1973, ApJ 179, 495

Kwok S., Purton C.R., Fitzgerald P.M. 1978, ApJ 219, L125

Phillips, J.P. and Cuesta, L. 1999, AJ 118, 2929

Small–Telescope Astronomy on Global Scales
ASP Conference Series, Vol. 246, 2001
W.P. Chen, C. Lemme, B. Paczyński

Former Soviet Union / West Europe Consortium for AGN Monitoring

N.G. Bochkarev

Sternberg Astronomical Institute, University of Moscow, Universitetskij Prospect 13, Moscow 119899, Russia

A.I. Shapovalova and A.N. Burenkov

Special Astrophysical Observatory of the Russian AS, Nizhnij Arkhyz, Karachaevo-Cherkesia, 369167, Russia

In the same year as Blandford and McKee (1982), Prof. N. Bochkarev and Dr. I. Antokhin (Bochkarev and Antokhin, 1982) published an article on the reverberation method of AGN BLR structure analysis. In 1983, testing observations were carried out with the 6 m telescope.

From 1986 up to now spectral monitoring of several AGNs has been carried out at SAO RAS (North Caucasus). Since 1988 we have been participating in AGN Watch Program. Since 1997 we have been carrying on an international program that is realised at several observatories of CIS (former Soviet Union) in collaboration with several West-European scientific groups (Bochkarev and Shapovalova, 1999). Some spectra of 3C390.3, one of the dimmest target objects of our program, $m = 15.6$ mag, taken with different telescopes are presented in Fig. 1(left).

It should be noted (see Fig. 1) that the 1 m telescope is too small for obtaining spectra of good quality (S/N$\approx$30), unless long exposure times are used. Therefore the most usable spectra should be taken with telescopes $\geq$2 m.

In Fig. 1 (right) the light curves for the emission Hβ line and the continuum of 3C 390.3 obtained in 1995-2000 are shown. The BVRI broad-band monitoring was carried out at 4 CIS observatories with small telescopes (60 cm and 70 cm). The errors of the photometric data are about 0.01-0.03 mag. From Fig. 1 (right) we see that the monitoring using photometric data (for continuum) and spectral data (for continuum and Hβ emission line) agrees each other well.

Presently our consortium on AGN monitoring includes 12 teams from 9 countries (about 50 participants, 19 of them younger than 35). We have observational bases located at different hour zones, and teams with vast experience in observations of AGN: optical broad-band photometry - at 5 CIS observatories (three other observatories, including one situated in Siberia, one in the Central Asia in a place with very good astroclimate, and one of the Trans-Caucasus observatories, are ready to join the program); CCD-spectroscopy - at 4 observatories (one of them situated in Mexico); near-IR-photometry - at one observatory; 4 groups involved in X-ray observations (mainly in Western Europe); 5 groups of interpretators, including 4 using photoionization codes CLOUDY and others, as well as groups of theoretical researches.

Besides, our consortium possesses unique archival data: 10-30 year long dense time series of observations of many objects: photoelectric UBV-photometry

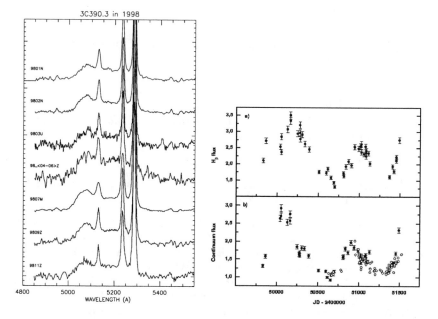

Figure 1.   Left: The spectra of 3C390.3 in Hβ region observed in 1998, each labeled with year (98), month (01=Jan, 02=Feb etc.), and telescope (N= 6 m telescope Nasmyth focus; U = Prime focus; Z= 1 m SAO telescope; M=2.1-m GHO telescope (México)). From top (9801N) to bottom (9811Z) exposure times in minutes are 30, 10, 15, 60, 120, 540, 270 and S/N are 43, 32, 13, 11, 56, 34, 34.

Right: Light curves for the emission Hβ line and the optical continuum of 3C 390.3 from 1995-2000. Filled circles show the spectral data. Those calculated out of V-band photometry are represented by open circles. Units are in 10^{-13} ergs cm^{-2}s^{-1} for Hβ fluxes, and 10^{-15} ergs cm^{-2}s^{-1}A^{-1} for the continuum.

with an accuracy of 1% and low S/N spectra of many AGN since the beginning of 1970 up to now obtained at the Crimean Observatory, Kazakhstan and SAO (Caucasus). Most of the archival spectra remain unprocessed.

This paper has had financial support from INTAS (grant N96-0328), RFBR (grants: N94-02-4885; N97-02-17625; N00-02-16272a), scientific-technical programme "Astronomy" (Russia), RFBR+CHINA (grant 99-02-39120) and CONACYT grants GXXXX-E and 32106-E (México).

References

Blandford, R.D., McKee, C.F. 1982, ApJ, 255, 419
Bochkarev, N. G., Antokhin, I. 1982, Astron. Tsirk., 1238
Bochkarev N.G., Shapovalova A.I. 1999, 75 (ASPConf.Ser.175)

Small-Telescope Astronomy on Global Scales
ASP Conference Series, Vol. 246, 2001
W.P. Chen, C. Lemme, B. Paczyński

Russian/Former Soviet Union Experience in Small Telescope Usage for Investigation of Interstellar Matter (ISM) and Nebulae

N.G. Bochkarev

Euro-Asian Astronomical Society, Sternberg Astronomical Institute
(SAI), Universitetskij Prospect 13, Moscow 119899, Russia
e-mail: boch@sai.msu.ru

The deepest tradition in ISM study in the optical range was built in Russia/FSU by V.Fessenkov, the founder of Fessenkov Astrophysical (Aph) Institute (AFIF, Kazakhstan) and G.Shain (Crimean Aph.Obs. - CrAO, Ukraine). The tradition was handed over to SAI (Moscow) by I.Shklovski and S.Pikelner, to Abastumani Aph. Obs. (AAO, Georgia), where a catalogue of dark nebulae (Khavtassi, 1960) was produced, and to Byurakan Aph. Obs. (BAO, Armenia).

For a long time 0.3-0.7 m telescopes were used for determination of interstellar extinction in the Galaxy by the standard technique (SAI; Engelhart Astron. Obs. of Kazan Univ., Russia; AAO; BAO and others. The most sophisticated investigations were carried out in Lithuania (e.g. Straizys, 1977; Sudzius, 1974).

The Catalogue of Reflection Nebulae was completed using the data obtained with the AFIF 0.5 m meniscus telescope (Rozkovsky & Kurchakov, 1968). Some nebulae were studied more deeply (e.g. Kurchakov, 1968; Sabitov, 1968; Rozkovsky, 1981). Research of polarization in dim reflection nebulae and imbedded stars was also carried out at AFIF (e.g. Sabitov, 1981; Pavlova & Rspaev, 1985). D.Rozkovsky measured polarization in many dim dust nebulae, including those outside the galactic plane, thus getting information on the nebulae dust properties, and the dust outside the galactic disk. Diffuse galactic radiation was measured by Rozkovsky et al. (1968). The observations were carried out with a Schmidt camera (focal ratio 1:1, focal distance 17 cm).

L.Kondratyeva performed, with a 0.7 m meniscus telescope of AFIF, a large number of spectroscopic and photometric observations of a large sample of planetary nebulae (PN), which allowed to determine their physical characteristics (e.g. Kondratyeva, 1975, 1978, 1985, 1992, 1998). The formation of the nebulae Th4-4 was traced: before 1970, it was a Be star with $T_{eff} = 22000$ K. Presently, it is a compact PN with a 54000 K hot nucleus (Kondratyeva, 1989, 1993).

Characteristics of PN (nuclei variability, binary nature etc.) were studied in detail based on a 25 year long data set obtained at SAI Crimean Lab. (0.6 m and 1.25 m telescopes) by Kostyakova (1999), Arkhipova et al., (2001).

Physical characteristics of many diffuse HII regions and their inner dispersion have been determined; nebulae structure was investigated (e.g. Glushkov et al., 1975). One of the most important results was the discovery of dust rims (bright and dark) around young hot stars (e.g. Glushkov, 1968, 1995). The inner stucture of diffuse nebulae has lately been investigated at BAO with a 1 m Schmidt camera (e.g. Parsamian & Petrosian, 1984).

SAI has a many years long tradition and experience of Fabry-Perot interferometers (FPI) usage for nebulae investigation. T.Lozinskaya and her group traced, with the 0.48 m, 0.6 m and 1.25 m telescopes of SAI Cr. Lab., the radial velocities distribution pattern in expanding nebulae (all the bright SNR of the Northern hemisphere (e.g. Sitnik et al., 1982) with radial velocities dispersion $\geq$ 15-20 km/s). Lozinskaya (1986) studied nebulae blown by WR-stellar wind and giant bubbles around stellar associations (Lozinskaya et al., 1994), discovered and studied wind-blown bubbles around Of stars.

P.Shcheglov (SAI) and his pupils were the first to use, in the 1960s-1970s, FPI for observations of very faint Hα, Hβ BG emission of the Galaxy. The observations were performed with FPIs 5-15 cm in diameter using no telescope but a tube of a 1-2 degrees field of view. Zhitkov (1970, 1971) discovered and Kutyrev (1985), Shestakova et al. (1988), and, since the 1970s, R.Reinolds (Kutyrev & Reinolds, 1989) all studied very faint Galaxy Hα, Hβ emission (EM = 0.1 - 10 $cm^{-6}pc$); i.e, low density HII regions were discovered in LISM and other regions.

References

Arkhipova, V. P. et al. 2001, Pisma AZh, 27, 122 (As. Lett., 27, 99)

Glushkov, Yu. I. 1968, Trudy AFIF, 11, 84 & 126

Glushkov, Yu. I. 1995, AApTr, 8, 105

Glushkov, Yu. I., Denisyuk, E. A. & Karyagina, Z. V. 1975, A & A, 39, 481

Khavtassi, J.Sh. 1960, The Atlas of Dark Galactic Nebulae. (Tbilisi: Acad.Sci.)

Kondratyeva, L. N. 1975, Pisma AZh., 1, N.10, 14; 1978, AZh, 55, 334; 1985, Astrofizika, 22, 153; 1992, AZh., 69, 1219; 1998, Izvest. AFIF, 4, 93

Kondratyeva, L. N. 1989, As. Lett., 15, 29; 1993, Pisma AZh., 19, 811

Kostyakova, E. B. 1999, Pisma AZh., 25, 457

Kurchakov, A. V. 1968, Trudy AFIF, 11, 56

Kutyrev, A. S., & Reinolds, R. 1989, ApJ, 344, L9

Kutyrev, A. S. 1985, Astr. Tsirk., 1369, 1; 1396, 3

Lozinskaya, T. A. 1986, Supernovae and stellar wind: interaction with interstellar matter (Moscow: Nauka)

Lozinskaya, T. A., et al. 1994, Pisma AZh, 71, 515

Parsamian, E. S. & Petrosian, V. M. 1985, Astrofizika, 20, 495

Pavlova, L. A. & Rspaev, F. K. 1985, Astrofizika, 22, 145

Rozkovsky, D. A. 1981, Astr.Tsirk., 1149, 4

Rozkovsky, A. D. et al. 1968, Trudy AFIF, 11, 66

Rozkovsky, D. A. & Kurchakov, A. V. 1968, Trudy AFIF, 11, 3

Sabitov, Sh. N. 1968, Trudy AFIF, 11, 79

Sabitov, Sh. N. 1981, Astr.Tsirk., 1154, 6

Shestakova, L. I. et al. 1988, Pisma AZh, 14, 60

Sitnik, T. G. et al. 1982, Pisma AZh, 8, 286

Straizys, V. 1977, Multicolor Stellar Photometry (Vilnius: Mokslas Publ.)

Sudzius, J. 1974, Bull. Vilnius Obs., 39, 3 & 18

Zhitkov, V. F. 1970, Astr. Tsirk., 578, 5; 1971, 655, 4

Small–Telescope Astronomy on Global Scales
ASP Conference Series, Vol. 246, 2001
W.P. Chen, C. Lemme, B. Paczyński

Russian/Former Soviet Union Experience in Professional Small Telescope Usage

N.G. Bochkarev

Euro-Asian Astr.Soc., Sternberg Astr.Inst.(SAI), Universitetskij Prospect 13, Moscow 119899, Russia; boch@sai.msu.ru

FSU astronomers traditionally use small telescopes ($\oslash$ $\leq$1.5 m, hereafter STs) for both science and education. Russian/FSU experience here is among the largest world-wide. There are only 2 large and moderate-sized facilities in whole Russia: the 6 m telescope of SAO RAS and Russian-Ukrainian 2 m one on the 3100 m high peak Terskol in Central Caucasus.

Equipped with good light receivers and handled by skilled observers, STs can produce first class scientific data. Important results are typically yielded by long-time sequences of observations and/or new observational "know how": good instrument/receiver design, appropriate selection of objects and moments, etc. Examples of what has been done with STs in FSU, within my memory, (in the last $\simeq$1/3 century) are listed below, without a list of references, because of lack of space. The author plans to publish a larger article on this subject in Astr.&Aph.Trans.

1. Solar System Objects

Meteors. S.Mukhamednazarov obtained a unique sequence of >200 spectra of 2-6^m meteors with a small ($\approx$5 cm) aperture wide-field camera (20x20^o) supplied with a slitless spectrograph and a TV light receiver. The observations, carried on at Turkmenistan since 1972, trebled the number of spectra obtained during the previous 75 yrs, moved the observational limit to 3-4^m dimmer events, and contributed considerably to meteor chemical composition research.

Comets. Monitoring of comet tails (their size, shape, brightness, structure: non-stationary processes, e.g. tail-breaks) with STs to search for interplanetary magnetic field and solar activity influence is yet an important task for amateurs and professionals. High-quality images come from places with good seeing: Mt. Maidanak Obs. (Uzbekistan), Mt. Sanglok Obs. (Tajikistan), Mt.Dushak-Eregdag Obs. (Turkmenistan) and others.

Spectroscopy is easily done with STs for bright comets. K.Churyumov (AO of Kiev State Univ.(SU), Ukraine) & colleagues have obtained good spectra (resolution of $\approx$50000) with 1 m ST + CCD-camera. This is a valuable input into the gas chemical and isotopic composition and outflow parameter study.

Positional astronomy. Missions to planet satellites and asteroids required a considerable improvement of ephemerides accuracy. Examples of ST usage for this purpose are:

1. S.Novikov & A.Shokin (SAI) performed, in 1987, >1000 positional measurements of Phobos and Daimos from Mt. Maidanak (a place with very good

seeing - see A.Hojaev in this volume), with a 1 m ST equipped with a screen for Mars light eclipsing, and a 0.23 m wide-field astrograph designed in 1847(!) and moved to Mt. Maidanak in the 1970s. Thanks to this research the resulting pointing deviation for the ship "Phobos" was only 2 km.

2. Ephemerides for Jupiter and Saturn satellites have been improved through photometric observation campaigns of their rare mutual occultations and eclipses. The 1990s campaign supervised by N. Emel'yanov (SAI) involved about 10 teams of FSU observers that used 0.3-0.8 m STs. Even visual observations by an SAI student, A.Berezhnoj, with a monocular ($\oslash$=6 cm; f=30 cm) were a great input similar to those by a ST+CCD-camera.

3. Stellar occultations by asteroids.

Asteroids. In 1965-1995 N.&L.Chernykh, at Crimean Aph.Obs. (CrAO, Ukraine), using a wide-field 40 cm astrograph made the world-wide largest input in asteroids discovery: $\approx 50\%$ of the asteroid discoveries in the world and a great impact on the asteroid hazard problem. They named about 800 asteroids.

V.Prokof'eva's group (CrAO) using a TV 2D-receiver (I-Isocon) with a 50 cm ST and 5 min. exposure measures objects of 19-20^m with a 5% accuracy. Such observations yield important data: besides rotation and shape determination, binary asteroids were discovered, and precession of one of them detected.

Asteroid stellar occultations are another important field of ST application: knowledge of asteroid orbits, shapes, and the accuracy of some physical parameters may be increased.

B.Artamonov (SAI) with V.Dudinov (AO of Kharkov SU, AO KhSU, Ukraine) et al., obtained at Mt. Maidanak, using speckle-interferometry and a 1 m ST, a direct image of Vesta (angular size $\approx$0.5") with a resolution of 0.13".

The mineral composition of some asteroids was estimated by S.Omarov et al. from spectra obtained with a 0.7 m ST of Shemakha Aph. Obs. (Azerbaijan) in 1980s.

Planets. The great and rare opportunity to observe the Shoemaker-Levy comet impact on Jupiter was successfully used by many observers with STs. On June 20, 1994, K.Churyumov with the 0.5 m ST of Lesniki Obs. near Kiev and V.Prokof'eva (CrAO) with 50 and 20 cm ST measured an $\approx$1 s long flash ($\approx$0.1^m) produced by the fragment Q2 crashing into an invisible part of the planet's atmosphere and reflected by Io. V.Prokof'eva obtained, on Jul. 16 - Aug. 30, 1994, >500 spectral and UBV measurements of the fireballs produced by the falling comet fragments. NaI doublet emission allowed investigations of the interaction of comet matter with the Jovian magnetosphere plasma. B.Artamonov et al., at Mt. Maidanak, obtained high angular resolution images of the fireballs.

V.Prokof'eva performed, with a 0.5 m ST and a TV-receiver, spectroscopic observations of NaI in the Jovian atmosphere. She and V.Lyuty & V.Tarashchuk, with a 60 cm ST, studied in detail Pluto's UBV variation with its rotational period.

Moon. SAI, Abastumani Aph.Obs. (AAO, Georgia) & AO KhSU (Ukraine) have great experience in the Moon surface structure and mineralogy analysis by polarimetric mapping with STs.

2. "Classic" Fields of Stellar Astronomy

Optical sky patrol has been carried on for decades at many observatories with 5-50 cm STs. The largest photoplate archives belong to SAI, AO of Odessa Univ. (OAO, Ukraine), and the Inst. for Astr. of Tajik Acad.Sci (IA TAS). The plates are intensely used for research in the history of brightness of variable objects: CV-stars, X-ray binaries, AGN, QSO, gamma-ray bursts (GRB), etc. (see below), the largest experience here being that of SAI & OAO. Presently, plate collecting and wide-field sky patrols are canceled, but D.Tsvetkov (SAI) continues SN patrol using wide-field cameras and plates or CCD-cameras.

Proper motion of stars was measured with photoplates exposed at distant epochs at SAI, Pulkovo, OAO, Main AO of NAS Ukraine (MAO, Kiev) etc.

Photometry of variable stars is a rich field for both professionals and amateur work with STs. With the amateurs, AAVSO experience is the largest.

Many scientists—from SAI MSU (pupils of P.P.Parenago, B.V.Kukarkin, D.Ya.Martynov: A.Cherepashchuk, V.Goranskii, R.i Khaliullin, V. Lyuty, N. Samus, S.i Shugarov et al.), OAO (pupils of V.P. Tsessevich: I. Andronov, D. Mkrtichian, others), and Kazan SU (pupils of A.Dubyago: M. Lavrov, E. Zhukov, others) got numerous brilliant results partially listed below.

Cepheids. 55000 photometric observations of 369 Cepheids (>50% of the world data archive) have been taken by L.Berdnikov (Saratov SU/SAI) at Mt. Maidanak, Las Campanas, Cerro Tololo & South Africa, with the 0.48-1 m ST. Prior to this work, about 150 Cepheides had never been observed photoelectrically; data on 100 Cepheids were poor. The results have had an important impact on the determination of the Galaxy distance scale and structure, and in the investigation of Cepheid period variations, i.e. stellar evolution.

To specify globular clusters distance scale, V.Goransky et al. observed the clusters RR Lyr type stars (B = $19-20^m$) light curves with a 1 m ST at Tian-Shan Mt. Obs. The >30000 observations of RR Lyr type stars have been collected in OAO.

Orbital period variations. Long-time sequences of data on variable stars by M.A.Svechnikov and his pupils (Ural SU) have resulted in the production of a catalogue submitted to Strasbourg Data Center. They found impressive examples of fast changes of orbital periods, P_o, in binaries which indicate episodes of intensive mass eruptions.

Relativistic effects in classical binaries (periastron shifting) were detected and investigated by late D.Martynov & R.Khaliullin et al. with ST at SAI Cr.Lab.

Novae and nova-like stars. The brightest Nova, N Cygni 1975 (V1500 Cyg) discovered visually by S.Shugarov (SAI) was studied at CrAO by E.Pavlenko et al. with a 50 cm ST. Its post-nova stable state (V=17.2^m) light curve was analyzed in detail; after outburst the orbital period increased on a time-scale of 230 yrs.

Using the 60-cm SAI Cr.Lab. ST, I.Voloshina found the low state photometric periods for the dwarf novae V795 Her, AH Her, and AB Dra. Observations of the dwarf nova, SS Cyg, including outbursts fast photometry have been carried on for 10 years. Quasi-periodic oscillations have been registered and independently studied by Yu.Gnedin et al. (Pulkovo AO, St.Peterburg) using AAO ST observations.

Symbiotic stars are studied in CrAO starting with the cycle of papers by A.Boyarchuk & T.Belyakina in the 1960s which showed their physical nature. We should also mention the detailed search of CH Cyg by L.Luud (IAPA, Estonia), IR photometric observations of a sample of the objects and monitoring of V1016 Cyg symbiotic nova and other peculiar objects carried on by O.Taranova's group (SAI) with InSb receiver at 1.25 m ST in SAI Cr.Lab. as well as spectral and photometric observations by V.Arkhipova, V.Esipov et al. (SAI) using the same ST.

Young stars. More than 70000 UBVR observations (≈230 stars) of T Tau type and Herbig Ae/Be young stars and FUORs were taken at Mt. Maidanak with the 48 cm and 60 cm STs over 20 yrs (long-term program) by the group of late V.Shevchenko (K.Grankin, O.Ezhkova, S.Mel'nikov). Star spots were studied, new subclasses of proto-Algols and quasi-Algols discovered, information on proto-planetary matter around stars were gathered.

Fine results for UV Ceti stars were obtained at BAO (Armenia), CrAO (R.Gershberg et al.), and Mt. Maidanak (V.Shevchenko et al., Astr.Inst., Uzbekistan).

Lunar occultations fast photometry is a rich field for STs even for ⊘ =0.3-0.4 m. The observations can yield star diameters (or their upper limits), and help to discover/study close binaries (distance between components ≈0.005 -1"). In Russia the best results have been obtained by E.Trunkovsky (SAI).

Double stars. A.Tokovinin has determined, through observations with STs of 0.6-1 m and an original hand-made photoelectric phase-grating interferometer positional angles and separations of several hundreds of double stars. The rms measurement error is ≈0.007" (at 1 m ST). The results form a considerable input into statistics of individual double system orbits, masses and mass ratios, i.e. for the determination of fundamental properties and leading to an understanding of the formation process of stars.

Spectrophotometric catalogues and robotic telescopes. Bright star absolute spectrophotometry for catalogue creation has carefully been carried out at SAI (I.Glushneva, V.Kornilov et al.), AFIF (A.Kharitonov et al.), and OAO (N.Komarov et al.) etc.

Twin 1 m robotic telescopes were installed in the 1980s by SAI at Tian-Shan Mt. Obs. (2700 m) for such catalogue production and for studying optical counterparts of radio/X-ray sources. These STs do not presently function.

Radial velocities. Exciting results were achieved with 0.5 - 1 m STs in stellar radial velocity study with COROVEL-spectrometer designed by A.Tokovinin even with as bad seeing conditions as in Moscow. See the paper by N.Samus in this volume.

High precision star photometry, even with a relatively bad seeing/weather has been successfully realized at the 70 cm ST of Ural SU Kourovskaya AO with a guided four-beam CCD photometer. The device yields a precision of 0.005^m or better. An accuracy of 0.0015^m was achieved in observations of the *planetary* transit across the disc of the star HD209458.

The device permits to separate a flux of close double star components, e.g. allowed to get, for the 1st time, parameters of the triple variable AM Leo.

One example of the study of very low amplitude variability ($\approx 0.008^m$) is V.Goransky et al.'s (1999) study of V1674 Cyg used as a standard for observations of Cyg X-1. High accuracy observations were achieved thanks to the highly stable atmosphere transparency at Tian-Shan and Maidanak Mt. Obs.

Stellar polarimetry. Famous results were achieved with STs in FSU in this area. Many researchers are/were involved: N. Shakhovskoy (who discovered intrinsic variable stellar polarization in 1962) & Yu.Efimov in CrAO; Yu. Gnedin et al. (Pulkovo AO); A. Kurchakov et al. (AFIF, Kazakhstan); V. Dombrovskij's pupils in SPbSU (Russia); M. Vashakidze's pupils in AAO (Georgia), V. Oskanian's pupils in BAO (Armenia), IA TAS (Tajikistan) etc.

3. Modern Stellar Astronomy

Optical monitoring of X-ray binaries. Since X-ray binaries were discovered by UHURU at the beginning of the 1970s, their optical identification on archive photoplates and photometric history study were started at SAI. Since then V.Lyuty has performed optical monitoring of many such objects with a 60 cm ST. Even the preliminary results allowed to determine the nature of optical variability, which differs for different objects. The lower mass limit of the invisible X-ray component of the first black hole candidate, Cyg X-1, proved that the object could not be a neutron star. Further data gathering and analysis revealed many fine details in the structure and light curves and lead to an understanding of physical processes within the objects.

A six month long set of observations of a number of southern X-ray binaries by A.Cherepashchuk in Australia with a 60 cm ST proved that long-period variability, the so called precession effects in accretion disks ($P_{pr} = 35^d$ for Her X-1 and 164^d for SS 433) is quite typical for X-ray binaries.

Detailed study of the light curves allowed to determine the fundamental parameters of the system: the componentent masses, the orbit size and inclination. A large contribution to the understanding of the nature of SS 433 was made by A.Cherepashchuk et al. First of all, the light curve was studied in detail using data from the Mt. Maidanak 0.6 and 1 m STs. Dense sequences of multi-color observations over several 164^d cycles allowed to trace the orbital light curve variations with the cycle phase and set limits on the geometry of the system.

Deviations from average light curves (first of all, flashes) allow to study non-stationary accretion processes and gas behavior at the accretion disks outer rim. Presently the best studied case is Cyg X-1. Intensive high-accuracy UBV monitoring with the 0.6-1 m ST at Maidanak and Tian-Shan Mt. has permitted E.Karitskaya (INASAN) et al. to find and analyze optical flashes ($\leq 0.04^m$) and, for the first time, compare them with X-ray data from RXTE/ASM and find time lag of X-ray events relative to the optical ones ($7-12^d$) which is the time interval of the matter flow from the outer to the inner rim of the disk. At SAI Cr.Lab. (0.6 m ST) the largest flash in the last 30 yrs was registered ($\approx 0.12^m$) pointing to explosion-type gas outflow from the disk's outer rim (N.Bochkarev & V.Lyuty, SAI, 1998).

Cooperative programs of Cyg X-1 observations in different spectral ranges set in action by V.Lyuty (UBVR data from 0.6 m ST) detected P_{pr} in radio-, optic and X-ray ranges, which do not contradict to earlier optic data of J.Kemp

(USA) with E.Karitskaya, V.Lyuty et al. and others, and permit to include Cyg X-1 in the list of so-called mini-quasars.

In the 1980s detailed precision (0.002^m) photometry of Her X-1 carried out by N.Kilyachkov & V.Shevchenko (AI, Uzbekistan) with the Mt. Maidanak 60 cm ST revealed, during the X-ray source occultation by the optical component, a series of its light curve peculiarities in the minimal brightness stage. N.Shakhovskoy & Yu.Efimov with CrAO 1.25 m ST detected, during that same period, flashes of polarization. Peculiarities of the Her X-1 light curve near the minima are being studied with STs of SAI Cr.Lab. The data is still being discussed by theorists.

Stimulating UBV observations of X-ray binaries were carried out with the 1 m ST at Sanglok Mt. Obs. IA TAS by R.Rakhimov in the 1980s.

Polarimetry. N.Shakhovskoy & Yu.Efimov performed 5-color polarimetry of a number of X-ray binaries with CrAO 1.25 m ST: SS 433, Cyg X-1, polars, etc. Magnetic fields of magnetized white dwarfs (WD) were measured and accretion on WD investigated.

GRB. Late E.Moskalenko (SAI) together with OAO, before the BeppoSAX launch in 1997, searched for optical counterparts of GRBs using the archive photoplates, mainly, of the OAO 7-cameras ST and, a larger number of artifacts were excluded, revealing non-trivial events that could be short ($\leq$40 min.) optical flashes.

Gravitational microlensing is now being widely discussed. Unfortunately, in Russia, the work on the topic is mainly theoretical.

Asteroseismology is a very prospective trend of ST usage. As high precision 24^h-round monitoring is required, investigations need to be conducted with wide international cooperation. An example of this type of network initiated and put into action, in particular, by FSU astronomers is the Central Asian Network (CAN) organized by D.Mkrtichian (OAO) with co-investigators from Russia, Georgia, Kazakhstan, and Turkmenistan; another branch includes West-European and US colleagues. In 1996, 29 Cyg observation global campaign 10 oscillation modes were detected.

For interstellar matter/nebulae see Bochkarev's paper in this volume.

4. Extragalactic Objects

Photometry. Since the late 1960s V.Lyuty (SAI) using a 60 cm ST, has been carrying on a long term program of UBV-photometry of many AGN. He investigated many outbursts (characteristic time 10-30^d) re-occurring about every 100^d and discovered quasi-periodic brightness variations (characteristic time 1-3 yrs). By comparison with archive data from SAI plate collection, AGN variation cycles were traced over a time interval of up to 80 years. He studied minor brightness variations over one night. This had an important impact on research of the nature of activity, first of all, stability and variability of accretion disc parameters.

Polarimetry in UBVRI bands of BL Lac type objects was performed by N.Shakhovskoy (CrAO) & Yu.Efimov with a 1.25 m ST. They showed that the linear polarization of some objects may reach 40% in some cases and is highly variable.

Spectra. A unique low S/N ratio spectra archive was collected by E.Denissiuk (AFIF, Kazakhstan) with a 70 cm ST and an image tube. In 1968-2000, thousands of spectra of 1515 galaxies were taken, emission lines in 48 galaxies and 42 new SyG were discovered. Vast amounts of material on spectral variability of 50 galaxies (>3000 spectra) has been collected over 30 years. This archive is unique in the world.

Moderate to high S/N (20-100) AGN spectral monitoring is being carried out by N.Bochkarev (SAI), A.Shapovalova (SAO) & others with a 1 m telescope of SAO RAS and a CCD-spectrograph (see N.Bochkarev & A.Shapovalova, this volume). For the first time, an attempt to determine the lag of broad-line variations with respect to the continuum was made by A.Cherepashchuk and V.Lyuty in 1973 with the 0.6 m ST of SAI Cr.Lab.

High resolution images of cosmological gravitational lensing were obtained by B.P.Artamonov (SAI), V.Dudinov (AO KhSU) et al. at Mt. Maidanak Obs. The images with deconvolution allow to obtain Einstein Cross images with a resolution higher than 0.5", which is the world-wide best ground-based result, matching those by the HST. B.Artamonov & V.Oknyansky (SAI) studied variations of its components brightness over ≈5 yrs with the aim of finding the time lag of one component of the image with respect to the others.

(Front from left) Fu, Warner, A.-L. Chen, J. H. Wang

Small-Telescope Astronomy on Global Scales
ASP Conference Series, Vol. 246, 2001
W.P. Chen, C. Lemme, B. Paczyński

Concluding Remarks

Brian Warner

Department of Astronomy, University of Cape Town, Rondebosch 7700, South Africa

This is the first IAU conference devoted to research with small telescopes since the Symposium on Instrumentation & Research Programs for Small Telescopes, held in Christchurch, New Zealand, almost exactly fifteen years ago. Both the climate for, and the definition of, small telescopes has changed since then. Now, with the construction and operation of 8m class telescopes, anything less than about 3m is considered small. But whereas fifteen years ago there were concerns whether instrumentation could be improved in order to make small telescopes competitive, and whether there was any ultimate future for these outside of education, it is now clear that an extensive niche has been found, with remodelling, refurbishment and construction of new small telescopes well under way.

Many of the research programs proposed in 1985 have been carried through (e.g. using the multiplex advantage in radial velocity meters) and have in some ways been completed successfully (e.g. lunar occultations for positional measurements of the Moon and for stellar angular diameters). Some of the instrumentation has passed into oblivion, but CCD detectors have proved to be all that been hoped for. (A state-of-the-art CCD can turn a 20-inch telescope into the equivalent of the 200-inch telescope of 50 years ago - but there are many more 20-inch telescopes than there ever were 200-inch telescopes.) The development of the Internet and its advantages to collaborations were also not fully foreseen.

In this Colloquium we have seen that large parts of variable star research are conducted entirely on small telescopes; wide fields of view are helpful for surveys; searches for moving objects, almost undreamt of in 1985, are producing numbers of NEOs and EKBOs; gravitational microlensing has been specifically a small telescope industry; there has been an increased interest on the part of telescope manufacturers - as witnessed by the presence of some of them at the Colloquium; and the best indicator of all of the good health of the use of small telescopes is that there is a looming data handling and archiving crisis. It is perhaps ironic that whereas most theoreticians and modellers wait impatiently for computers that are orders of magnitude faster than those presently available, excellent science is being done with telescopes that are an order of magnitude smaller than the largest currently coming into operation.

Some of the obvious strengths of small telescopes are their wide field of view - useful for all-sky searches and searches within clusters; the fact that there are so many already in existence and often underused; the flexibility of scheduling and the concomitant rapid response to alerts; the absence of TACs to get in the way of the enthusiastic researcher; the expertise in technology that is rapidly growing among the manufacturers; the affordability for smaller and poorer countries and institutions - which as a result can participate in unique and often cutting-edge science; the ease of networking and organising, made possible by the Internet; the possibility of designing telescopes to be dedicated, specialised and hence efficient; the relative ease of using robotic and/or autonomous systems.

A few definitions are required here: there used to be no doubt about what a financially poor institution is, but with small telescopes within financial reach of most institutions, there has been a reversal of meaning - now a financially disadvantaged observatory is one with a Director and 8m class telescope to support. A robotic telescope is one which does what it is told to do (parallels were drawn with the perfect student); an autonomous telescope is able to make decisions and adjust the observing program according to prevailing conditions and history (here parallels were drawn with faculty members).

Currently, small telescopes, in addition to carrying out programs not possible on large telescopes (see below), act as a Discovery Service for larger telescopes - providing the All-Sky searches to find objects to be studied in more detail, providing triggers for immediate study by larger telescopes (e.g. supernovae) and providing general follow-up opportunities for long-term behaviour. One obvious area in which small is bigger is the possibility of coordinated round-the-world observations (e.g. WET, CBA) which generate the multi-day, almost continuous, coverage that is essential for asteroseismology and the study of fast non-periodic variable stars. Such distributed observing is rarely possible even with telescopes of intermediate size. The practical demonstrations possible with small telescope do more to promote interest in astronomy in particular and science in general than does the often mere existence and bulk of large telescopes.

In the Discovery area, small telescopes are generating catalogues and light curves of tens of thousands of new variables stars - as by-products of the MACHO and other gravitational lensing experiments. But, as Bohdan Paczynski pointed out, our knowledge even of variable stars brighter than 12th magnitude is still very incomplete. The All-Sky surveys (rather than selected areas, as in the lensing projects) needed to complete knowledge of variable stars down even to 20th are within reach of present technology. These variables include myriads of eclipsing and contact systems, as well as intrinsic variables. In particular, various arguments show that there should be a population of short period detached white dwarfs and M dwarfs, of which only a few have so far been found.

Almost all the Colloquium was devoted to variability - of brightness, position or radial velocity (an exception was John Gaustad's beautiful H alpha survey of the southern sky, made with one of the smallest of lenses). This is where the strengths of small telescopes lie - in repeated observations on a variety of time scales, which reveal changes. Although large telescopes may technically be able to do everything that small telescopes can, it is not practicably or politically feasible for them to do so. This leaves a large amount of science undone, unless small telescopes come to the rescue. In Figure 1 I show an Apparent Brightness/Time scale diagram which is partitioned according to what is feasible for very large telescopes (VLTs) to explore (both intentionally and serendipitously) and what is left for very little telescopes (vlts) to concentrate on. There is a fuzzy region of overlap, but no one will doubt that VLTs are not going to be employed to find or study large numbers of Mira variables, or follow dwarf novae through entire outburst cycles, or be used on bright stars, or for any All-Sky surveys to find NEOs. Also, it does not make sense for smaller telescopes to try

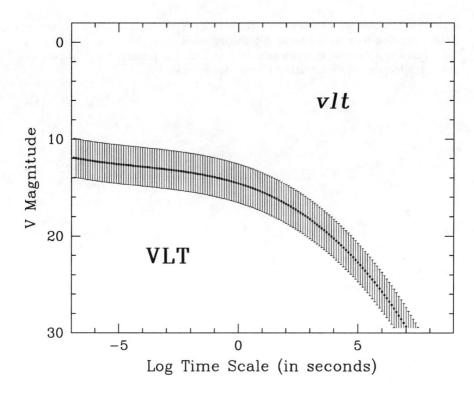

Figure 1.   The Apparent Brightness – Variability Time Scale diagram.

to compete where large numbers of photons are indispensable (e.g., very high resolution light curves of optical pulsars, or in the study of excessively faint objects). Figure 1 becomes even more interesting if one uses a linear rather than logarithmic time scale axis - there is plenty of phase space for vlts to occupy.

The evident successes of vlts are shown in the rapidly growing data archives. But it should not be overlooked that the small telescope users can participate in High Energy Astrophysics through the discovery of optical Gamma Ray bursts and high-z supernovae; this is Big Science and New Science, and it does not require large accelerators nor VLTs to join in.

There are also some evident needs of the small telescopes community. Good Public Relations, showing the successes and uniqueness of the research output, are needed, to maintain the flow of funds for continuation of the present programs, and to generate funds for the next generations of specialised small telescopes and their software. Coordinated, standardised archiving is already an urgent need - this should be addressed through international cooperation and not left to a form of survival of the fittest (in fact, it could well result in survival of the largest, which is not necessarily the best - viz. the UBV system of photometry). The overwhelming number of light curves of variable stars that are being generated surpasses what can be digested by human inspection - there is a need here for development of software, perhaps using Artificial Intelligence techniques, where self-learning software can sieve through the light curves and draw attention only to the really interesting ones.

There is also another evident need - and that is for another conference on small telescopes before a further fifteen years passes by.

Small–Telescope Astronomy on Global Scales
ASP Conference Series, Vol. 246, 2001
W.P. Chen, C. Lemme, B. Paczyński

APPENDIX: a List of Robotic Telescopes

Frederic V. Hessman

Universitäts-Sternwarte, Geismarlandstr. 11, 37083 Göttingen, Germany

Abstract. In order to help the reader keep somewhat abreast of the rapid development in the field of robotic telescopes, the current list on the web available at the Göttingen MONET website[1] is included here as an appendix.

The selection of sites in the following tables is probably not exhaustive – I apologize to any colleagues who feel left out. I have not included systems which are nearly robotic (i.e. which need a local operator to get the system started each night) but have included proposed (i.e. unfunded) projects (e.g. ROBONET and NOT) in the hopes that they will be funded and eventually of more immediate interest. The following abbreviations are used:

APT : (usually) Automatic Photoelectric Telescope

B&C : Boller & Chivens

C14 : Celestron 14-inch Schmidt-Cassegrain

GRB : Gamma-Ray Bursts

NEO : Near-Earth Objects

TNO : Trans-Neptunian Objects = Kuiper-Belt Objects

---

[1] http://www.astro.physik.uni-goettingen.de/~hessman/MONET

Table 1: List of Robotic Telescopes

	Name & Institution	Telescope(s) & Site	Purpose(s)	Status & WWW
1	All-Sky Astronomical Survey (ASAS) [Warsaw]	2cm, Las Campanas/Chile	All-sky survey of stars < 12 mag	In operation from 1997-2000 (www.astrouw.edu.pl/~gp/asas/asas.html)
2	All-Sky Astronomical Survey III (ASAS-3) [Warsaw, Princeton]	2x7cm, 1x25cm, Las Campanas/Chile	All-sky survey of stars < 16 mag	Being commissioned (www.astrouw.edu.pl/~gp/asas/asas.asas3.html)
3	Antipodal Transient Observatory (ATO) [Washington Univ., IIA Bangalore]	2x Torus 50cm, Sonoita/Arizona, Mt. Saraswati/India	GRB	Being commissioned (jelley.wustl.edu/pages/?def+ato)
4	Apogee Robotic Observatory (ARO) [Apogee Inc.]	C14, Tucson/Arizona	Commercial test-bed	Under construction (www.apogee-ccd.com/aro.html)
5	Automated Patrol Telescope [Univ. New South Wales]	Modified 50cm Baker-Nunn, Siding Spring/Australia	GRB, service obs.	In operation (newt.phys.unsw.edu.au/~mcba/apt.html)
6	Automatic Imaging Telescope [Univ. Perugia]	40cm Newtonian, Perugia/Italy	Service obs.	In operation since 1994 (wwwospg.pg.infn.it/osserv.htm)
7	Automatic Radio-Linked Telescope [Western New Mexico Univ.]	44cm, Silver City/New Mexico	Nova, supernova searches	In operation (www.nfo.edu/nfo.html)
8	Automatic Telescope [Novosibirsk State Univ.]	Novosibirsk/Russia		(vega.inp.nsk.su/index.php3)
9	Bisei Spaceguard Center [Japan Space Forum]	Torus 50cm, 100cm, Bisei/Japan	NEOs	Being commissioned (www.torusoptics.com/japan.htm)
10	Bradford Robotic Telescopes [Bradford Univ., IAC, Manchester]	46cm, Tenerife	Education, GRB searches	Replacement of original telescope underway (www.telescope.org/rti)
11	Burst Observer & Optical Transient Exploring System (BOOTES) [Spain, Czech Republic]	2x5cm, 2x30cm, Mazagon/Spain	GRB followups	Being commissioned (www.laeff.esa.es/~ajct/BOOTES)
12	Carlberg Meridian Telescope [Copenhagen Obs.]	18cm, La Palma	Astrometry	In operation since 1984 (www.ast.cam.ac.uk/~dwe/SRF/camc.html)
13	Carl Sagan Observatory [Univ. Sonora]	Coronado Inst. Group 55cm, Cerro Azul/Mexico	SN, Education	Under construction (cosmos.cifus.uson.mx/Infraestructura/ocs/ocs.htm)
14	Catania Automatic Photoelectric Telescope [Catania Obs.]	80cm APT, Mt. Aetna/Sicily	Photometry	In operation since 1992 (w3.ct.astro.it/aptweb/aptweb.html)
15	Centro de Astrobiologica [CAB]	Torus 40cm, 2x50cm, Spain	Exoplanets	Under construction (www.cab.inta.es)
16	El Enano [Swarthmore, Las Cumbres Obs., Illinois]	50mm, Cerro Tololo/Chile	Hα survey	In operation since 1998 (www.swarthmore.edu/Home/News/Astronomy)
17	Explosive Transient Camera (ETC) [MIT]	16x 25mm, Kitt Peak/Arizona	GRB	In operation since 1991 (space.mit.edu/ETC)
18	Fairborn Observatory [Fairborn Obs.]	Home of many, Patagonia/Arizona	Mostly photoelec. service obs.	In operation (24.1.225.36)
19	Four-College Automatic Photoelectric Telescope [Villinova,Cidadel,Charleston College, University of Nevada/Las Vegas]	80cm APT, Patagonia/Arizona, Patagonia/Arizona	Photometry	In operation (24.1.225.36/t5.html) (24.1.225.36/t5.html)
20	Gamma Ray Binocular (GRAB) [NCU Taipei]	C14, Meade 40cm, Kenting/Taiwan	GRB	Being commissioned

Table 1: List of Robotic Telescopes (continued)

	Name & Institution	Telescope(s) & Site	Purpose(s)	Status & WWW
21	Global Network of Automatic Telescopes (GNAT) [GNAT]	SciTech 50cm Tucson/Arizona	Service Obs.	Being commissioned (www.gnat.org/~ida/gnat)
22	Hungarian Automatted Telescope (HAT) [Konkoly Obs.]	6cm Hungary	All-sky monitoring	In operation (www.konkoly.hu/staff/bakos/HAT)
23	Iowa Robotic Telescope Facility (IRTF) [Univ. Iowa]	18cm, 50cm Iowa City/Iowa, Sonoita/Arizona	Education	In operation (denali.physics.uiowa.edu)
24	Italian Robotic Antarctic Infrared Telescope (IRAIT) [Perugia Univ.]	80cm Montone/Italy	IR service obs.	Being commissioned (wwwospg.pg.infn.it/irait.htm)
25	IUCAA 2-m [IUCAA]	TTL 2.0m India	Potential robotic use	Under construction (www.iucaa.ernet.in/~hkdas/telescope.html)
26	Katzman Automated Imaging Telescope (KAIT) [UC Berkeley]	76cm Lick Obs./California	Supernova searches	In operation since 1989 (astro.berkeley.edu/~bait/kait.html)
27	KAO Robotic Telescope [Korean Astronomical Observatory]	ACE 1.0m Arizona	Service obs.	Being commissioned
28	Livermore Optical Transit Imaging System (LOTIS) [LLNL]	4x11cm Livermore/California	GRB	In operation since 1996 (hubcap.clemson.edu/~ggwilli/LOTIS)
29	Liverpool Telescope [John Moores Univ.]	TTL 2.0m La Palma	Service obs.	Under construction (telescope.livjm.ac.uk)
30	Multicolor Active Galactic NUclei Monitoring (MAGNUM) [Univ. of Tokyo, NAOJ, ANU]	EOST 2.0m Haleakala/Hawaii	AGN	Being commissioned (merope.mtk.nao.ac.jp/~yuki/mage.html)
31	MOnitoring NEtwork of Telescopes (MONET) [Univ. Göttingen, McDonald, SAAO]	80-120cm Mt. Locke/Texas, Sutherland/South Africa	Service obs., Education	Funded (www.uni-sw.gwdg.de/~hessman/MONET)
32	Nassau Station Robotic Telescope [Case Western Reserve Univ.]	90cm Chardon/Ohio	Education, Service obs.	In operation since 1999 (astrwww.cwru.edu/nassau/nassau.html)
33	National School Observatories (NSO) [John Moores Univ., Liverpool]	2x TTL 2.0m Haleakala/Hawaii, Siding Spring/Australia	Education only	Under construction (www.schoolsobservatory.org.uk)
34	Nepean Astronomy Center [Univ. West. Sydney]	60cm Werrington/Australia	Education	Currently being roboticized (www.uws.edu.au/observatory/)
35	Network of Oriental Telescopes (NOT)	1.5-2.0m North Africa, Near East	Education, service obs.	Proposed (www.saao.ac.za/~wgssa/as2/nort.html)
36	Ondrejov Robotic Telescope [Astron. Inst., Ondrejov]	Meade 30cm Ondrejov/Czech Republic	GRB	In operation since 1997 (altamira.asu.cas.cz/ort.html)
37	Phoenix 10" [Franklin äMarshal College]	25cm Patagonia/Arizona	Photometry	In operation since 1983 (www.fandm.edu/Departments/Astronomy)
38	Riken-Bisei Optical Transient Survey (RIBOTS) [Riken, Bisei Obs.]	Meade 30cm Bisei/Japan	GRB	In operation (www.astro.indiana.edu/RoboScope.html)
39	ROBONET [St. Andrews Univ.]	6x 2.0m global network	Service obs.	Proposed (polaris.st-and.ac.uk/~kdh1/jitpage.html)
40	RoboScope [Univ. Indiana]	B&C 41cm Morgan-Monroe/Indiana	Monitoring	In operation since 1990 (www.astro.indiana.edu/RoboScope.html)

Table 1: List of Robotic Telescopes (continued)

	Name & Institution	Telescope(s) & Site	Purpose(s)	Status & WWW
41	Robotic Optical Transient Search Experiment I (ROTSE I) [Los Alamos, LLNL, Univ. Michigan]	4x11cm, Los Alamos/New Mexico	GRB, all-sky survey	In operation since 1998 (www.umich.edu/~rotse)
42	Robotic Optical Transient Search Experiment III (ROTSE III) [Univ. Michigan, Univ. New South Wales, MPI Physik]	4x50cm, Los Alamos/New Mexico, Siding Spring/Australia, Namibia	GRB, all-sky survey	Under construction (www.umich.edu/~rotse/rotse-iii/rotseiii.htm)
43	Robotically Controlled Telescope (RCT) [Western Kentucky Univ., S. Carolina State Univ., Planetary Sci. Inst., Berkeley]	1.3m, Kitt Peak/Arizona	Service Obs.	Being refurbished (www.psi.edu/rct)
44	Robotisches Observatorium der Theoretische Astrophysik Tübingen (ROTAT) [Univ. Tübingen]	Meade 30cm, refurbished 60cm Tübingen/Germany, Haute-Provence/France	Service obs.	Being commissioned (www.tat.physik.uni-tuebingen.de/~blum/ROTAT)
45	SAAO APT [SAAO]	75cm, Sutherland/South Africa	Photometry	In operation since 2000
46	SpectraBot [Univ. Indiana]	1.25m, Morgan-Monroe/Indiana	Spectroscopy	In operation since 1999 (www.astro.indiana.edu/spectrabot.html)
47	STELLA [Hamburg, AIP, IAC]	Halfmann 1.2m, Tenerife	Spectroscopy	Under construction (www.aip.de/groups/turbulence/stella.html)
48	Strömgren Automatic Telescope [Copenhagen Obs.]	50cm, La SIlla/Chile	Photometry	In operation (www.astro.ku.dk/~jjaf/0p5m.html)
49	StromloTNO [MSSSO]	1.3m, Mt. Stromlo/Australia	TNOs	Being fully roboticized (wwwmacho.anu.edu.au)
50	Super-LOTIS [LLNL, Clemson]	Upgraded B&C 60cm, Kitt Peak/Arizona	GRB	Being commissioned (hubcap.clemson.edu/~ggwilli/LOTIS/super.html)
51	Taiwan-America-Occultation-Survey (TAOS) [Academia Sinica, NCU Taiwan, LLNL]	3x Torus 50cm, Yu Shan/Taiwan	NEOs	Under construction (taos.asiaa.sinica.edu.tw)
52	Telescope a Action Rapide pour les Objets Transitoires (TAROT) [Toulouse]	Valmeca 25cm, Plateau du Calern/France	GRB	In operation since 1998 (www.cesr.fr/~boer/tarot)
53	TENAGRA Observatory [private]	C14, OCS 50cm, SciTech 78cm Oregan, Patagonia/Arizona	SN searches	In operation, 78cm being commissioned (www.tenagraobservatories.com)
54	Tenessee Automatic Photoelectric Telescopes [Tennessee State Univ.]	30,60,76cm APT, Patagonia/Arizona	Photometry	In operation (schwab.tsuniv.edu)
55	Tenessee Spectroscopic Survey Telescope [Tennessee State Univ.]	2.1m, Patagonia/Arizona	Spectroscopy	Being commissioned (schwab.tsuniv.edu/t13.html)
56	Torus Observatory [private]	Torus 60cm, Nevada	Service obs.	Replacement under construction (www.torusoptics.com/torus.htm)
57	Wolfgang-Amadeus [Univ. Vienna]	2x76cm APT, Patagonia/Arizona	Photometry	In operation since 1996 (www.astro.univie.ac.at/~kgs/APT)
58	Yonsei Survey Telescopes for Astron. Research (YSTAR) [Yonsei Univ./Korea]	2+x Torus 0.5m, Korea	All-sky survey	Commissioning of first telescopes (csaweb.yonsei.ac.kr/~byun/Ystar)

ASTRONOMICAL SOCIETY OF THE PACIFIC CONFERENCE SERIES

and

INTERNATIONAL ASTRONOMICAL UNION VOLUMES

Published
by

The Astronomical Society of the Pacific
(ASP)

ASP CONFERENCE SERIES VOLUMES

Published by the Astronomical Society of the Pacific

PUBLISHED: 1988 (* asterisk means OUT OF STOCK)

Vol. CS -1   PROGRESS AND OPPORTUNITIES IN SOUTHERN HEMISPHERE
OPTICAL ASTRONOMY: CTIO 25TH Anniversary Symposium
eds. V. M. Blanco and M. M. Phillips
ISBN 0-937707-18-X

Vol. CS-2   PROCEEDINGS OF A WORKSHOP ON OPTICAL SURVEYS FOR QUASARS
eds. Patrick S. Osmer, Alain C. Porter, Richard F. Green, and Craig B. Foltz
ISBN 0-937707-19-8

Vol. CS-3   FIBER OPTICS IN ASTRONOMY
ed. Samuel C. Barden
ISBN 0-937707-20-1

Vol. CS-4   THE EXTRAGALACTIC DISTANCE SCALE:
Proceedings of the ASP 100th Anniversary Symposium
eds. Sidney van den Bergh and Christopher J. Pritchet
ISBN 0-937707-21-X

Vol. CS-5   THE MINNESOTA LECTURES ON CLUSTERS OF GALAXIES
AND LARGE-SCALE STRUCTURE
ed. John M. Dickey
ISBN 0-937707-22-8

PUBLISHED: 1989

Vol. CS-6   SYNTHESIS IMAGING IN RADIO ASTRONOMY: A Collection of Lectures
from the Third NRAO Synthesis Imaging Summer School
eds. Richard A. Perley, Frederic R. Schwab, and Alan H. Bridle
ISBN 0-937707-23-6

PUBLISHED: 1990

Vol. CS-7   PROPERTIES OF HOT LUMINOUS STARS: Boulder-Munich Workshop
ed. Catharine D. Garmany
ISBN 0-937707-24-4

Vol. CS-8*   CCDs IN ASTRONOMY
ed. George H. Jacoby
ISBN 0-937707-25-2

Vol. CS-9   COOL STARS, STELLAR SYSTEMS, AND THE SUN: Sixth Cambridge Workshop
ed. George Wallerstein
ISBN 0-937707-27-9

Vol. CS-10*   EVOLUTION OF THE UNIVERSE OF GALAXIES:
Edwin Hubble Centennial Symposium
ed. Richard G. Kron
ISBN 0-937707-28-7

Vol. CS-11   CONFRONTATION BETWEEN STELLAR PULSATION AND EVOLUTION
eds. Carla Cacciari and Gisella Clementini
ISBN 0-937707-30-9

Vol. CS-12   THE EVOLUTION OF THE INTERSTELLAR MEDIUM
ed. Leo Blitz
ISBN 0-937707-31-7

PUBLISHED: 1991

Vol. CS-13   THE FORMATION AND EVOLUTION OF STAR CLUSTERS
ed. Kenneth Janes
ISBN 0-937707-32-5

ASP CONFERENCE SERIES VOLUMES
Published by the Astronomical Society of the Pacific

PUBLISHED: 1991 (* asterisk means OUT OF STOCK)

Vol. CS-14   ASTROPHYSICS WITH INFRARED ARRAYS
ed. Richard Elston
ISBN 0-937707-33-3

Vol. CS-15   LARGE-SCALE STRUCTURES AND PECULIAR MOTIONS IN THE UNIVERSE
eds. David W. Latham and L. A. Nicolaci da Costa
ISBN 0-937707-34-1

Vol. CS-16   Proceedings of the 3rd Haystack Observatory Conference on ATOMS, IONS,
AND MOLECULES: NEW RESULTS IN SPECTRAL LINE ASTROPHYSICS
eds. Aubrey D. Haschick and Paul T. P. Ho
ISBN 0-937707-35-X

Vol. CS-17   LIGHT POLLUTION, RADIO INTERFERENCE, AND SPACE DEBRIS
ed. David L. Crawford
ISBN 0-937707-36-8

Vol. CS-18   THE INTERPRETATION OF MODERN SYNTHESIS OBSERVATIONS
OF SPIRAL GALAXIES
eds. Nebojsa Duric and Patrick C. Crane
ISBN 0-937707-37-6

Vol. CS-19   RADIO INTERFEROMETRY: THEORY, TECHNIQUES, AND APPLICATIONS,
IAU Colloquium 131
eds. T. J. Cornwell and R. A. Perley
ISBN 0-937707-38-4

Vol. CS-20   FRONTIERS OF STELLAR EVOLUTION:
50th Anniversary McDonald Observatory (1939-1989)
ed. David L. Lambert
ISBN 0-937707-39-2

Vol. CS-21   THE SPACE DISTRIBUTION OF QUASARS
ed . David Crampton
ISBN 0-937707-40-6

PUBLISHED: 1992

Vol. CS-22   NONISOTROPIC AND VARIABLE OUTFLOWS FROM STARS
eds. Laurent Drissen, Claus Leitherer, and Antonella Nota
ISBN 0-937707-41-4

Vol CS-23   ASTRONOMICAL CCD OBSERVING AND REDUCTION TECHNIQUES
ed. Steve B. Howell
ISBN 0-937707-42-4

Vol. CS-24   COSMOLOGY AND LARGE-SCALE STRUCTURE IN THE UNIVERSE
ed. Reinaldo R. de Carvalho
ISBN 0-937707-43-0

Vol. CS-25   ASTRONOMICAL DATA ANALYSIS, SOFTWARE AND SYSTEMS I - (ADASS I)
eds. Diana M. Worrall, Chris Biemesderfer, and Jeannette Barnes
ISBN 0-937707-44-9

Vol. CS-26   COOL STARS, STELLAR SYSTEMS, AND THE SUN:
Seventh Cambridge Workshop
eds. Mark S. Giampapa and Jay A. Bookbinder
ISBN 0-937707-45-7

Vol. CS-27   THE SOLAR CYCLE: Proceedings of the
National Solar Observatory/Sacramento Peak 12th Summer Workshop
ed. Karen L. Harvey
ISBN 0-937707-46-5

ASP CONFERENCE SERIES VOLUMES
Published by the Astronomical Society of the Pacific

PUBLISHED: 1992 (asterisk means OUT OF STOCK)

Vol. CS-28   AUTOMATED TELESCOPES FOR PHOTOMETRY AND IMAGING
eds. Saul J. Adelman, Robert J. Dukes, Jr., and Carol J. Adelman
ISBN 0-937707-47-3

Vol. CS-29   Viña del Mar Workshop on CATACLYSMIC VARIABLE STARS
ed. Nikolaus Vogt
ISBN 0-937707-48-1

Vol. CS-30   VARIABLE STARS AND GALAXIES
ed. Brian Warner
ISBN 0-937707-49-X

Vol. CS-31   RELATIONSHIPS BETWEEN ACTIVE GALACTIC NUCLEI
AND STARBURST GALAXIES
ed. Alexei V. Filippenko
ISBN 0-937707-50-3

Vol. CS-32   COMPLEMENTARY APPROACHES TO DOUBLE
AND MULTIPLE STAR RESEARCH, IAU Colloquium 135
eds. Harold A. McAlister and William I. Hartkopf
ISBN 0-937707-51-1

Vol. CS-33   RESEARCH AMATEUR ASTRONOMY
ed. Stephen J. Edberg
ISBN 0-937707-52-X

Vol. CS-34   ROBOTIC TELESCOPES IN THE 1990's
ed. Alexei V. Filippenko
ISBN 0-937707-53-8

PUBLISHED: 1993

Vol. CS-35*   MASSIVE STARS: THEIR LIVES IN THE INTERSTELLAR MEDIUM
eds. Joseph P. Cassinelli and Edward B. Churchwell
ISBN 0-937707-54-6

Vol. CS-36   PLANETS AROUND PULSARS
ed. J. A. Phillips, S. E. Thorsett, and S. R. Kulkarni
ISBN 0-937707-55-4

Vol. CS-37   FIBER OPTICS IN ASTRONOMY II
ed. Peter M. Gray
ISBN 0-937707-56-2

Vol. CS-38   NEW FRONTIERS IN BINARY STAR RESEARCH: Pacific Rim Colloquium
eds. K. C. Leung and I.-S. Nha
ISBN 0-937707-57-0

Vol. CS-39   THE MINNESOTA LECTURES ON THE STRUCTURE
AND DYNAMICS OF THE MILKY WAY
ed. Roberta M. Humphreys
ISBN 0-937707-58-9

Vol. CS-40   INSIDE THE STARS, IAU Colloquium 137
eds. Werner W. Weiss and Annie Baglin
ISBN 0-937707-59-7

Vol. CS-41   ASTRONOMICAL INFRARED SPECTROSCOPY:
FUTURE OBSERVATIONAL DIRECTIONS
ed. Sun Kwok
ISBN 0-937707-60-0

ASP CONFERENCE SERIES VOLUMES
Published by the Astronomical Society of the Pacific

PUBLISHED: 1993 (* asterisk means OUT OF STOCK)

Vol. CS-42 GONG 1992: SEISMIC INVESTIGATION OF THE SUN AND STARS
ed. Timothy M. Brown
ISBN 0-937707-61-9

Vol. CS-43 SKY SURVEYS: PROTOSTARS TO PROTOGALAXIES
ed. B. T. Soifer
ISBN 0-937707-62-7

Vol. CS-44 PECULIAR VERSUS NORMAL PHENOMENA IN A-TYPE AND RELATED STARS,
IAU Colloquium 138
eds. M. M. Dworetsky, F. Castelli, and R. Faraggiana
ISBN 0-937707-63-5

Vol. CS-45 LUMINOUS HIGH-LATITUDE STARS
ed. Dimitar D. Sasselov
ISBN 0-937707-64-3

Vol. CS-46 THE MAGNETIC AND VELOCITY FIELDS OF SOLAR ACTIVE REGIONS,
IAU Colloquium 141
eds. Harold Zirin, Guoxiang Ai, and Haimin Wang
ISBN 0-937707-65-1

Vol. CS-47 THIRD DECENNIAL US-USSR CONFERENCE ON SETI --
Santa Cruz, California, USA
ed. G. Seth Shostak
ISBN 0-937707-66-X

Vol. CS-48 THE GLOBULAR CLUSTER-GALAXY CONNECTION
eds. Graeme H. Smith and Jean P. Brodie
ISBN 0-937707-67-8

Vol. CS-49 GALAXY EVOLUTION: THE MILKY WAY PERSPECTIVE
ed. Steven R. Majewski
ISBN 0-937707-68-6

Vol. CS-50 STRUCTURE AND DYNAMICS OF GLOBULAR CLUSTERS
eds. S. G. Djorgovski and G. Meylan
ISBN 0-937707-69-4

Vol. CS-51 OBSERVATIONAL COSMOLOGY
eds. Guido Chincarini, Angela Iovino, Tommaso Maccacaro, and Dario Maccagni
ISBN 0-937707-70-8

Vol. CS-52 ASTRONOMICAL DATA ANALYSIS SOFTWARE AND SYSTEMS II - (ADASS II)
eds. R. J. Hanisch, R. J. V. Brissenden, and Jeannette Barnes
ISBN 0-937707-71-6

Vol. CS-53 BLUE STRAGGLERS
ed. Rex A. Saffer
ISBN 0-937707-72-4

PUBLISHED: 1994

Vol. CS-54* THE FIRST STROMLO SYMPOSIUM: THE PHYSICS OF ACTIVE GALAXIES
eds. Geoffrey V. Bicknell, Michael A. Dopita, and Peter J. Quinn
ISBN 0-937707-73-2

Vol. CS-55 OPTICAL ASTRONOMY FROM THE EARTH AND MOON
eds. Diane M. Pyper and Ronald J. Angione
ISBN 0-937707-74-0

Vol. CS-56 INTERACTING BINARY STARS
ed. Allen W. Shafter
ISBN 0-937707-75-9

ASP CONFERENCE SERIES VOLUMES
Published by the Astronomical Society of the Pacific

PUBLISHED: 1994 (* asterisk means OUT OF STOCK)

Vol. CS-57   STELLAR AND CIRCUMSTELLAR ASTROPHYSICS
eds. George Wallerstein and Alberto Noriega-Crespo
ISBN 0-937707-76-7

Vol. CS-58*   THE FIRST SYMPOSIUM ON THE INFRARED CIRRUS
AND DIFFUSE INTERSTELLAR CLOUDS
eds. Roc M. Cutri and William B. Latter
ISBN 0-937707-77-5

Vol. CS-59   ASTRONOMY WITH MILLIMETER AND SUBMILLIMETER WAVE
INTERFEROMETRY,
IAU Colloquium 140
eds. M. Ishiguro and Wm. J. Welch
ISBN 0-937707-78-3

Vol. CS-60   THE MK PROCESS AT 50 YEARS: A POWERFUL TOOL FOR ASTROPHYSICAL
INSIGHT, A Workshop of the Vatican Observatory --Tucson, Arizona, USA
eds. C. J. Corbally, R. O. Gray, and R. F. Garrison
ISBN 0-937707-79-1

Vol. CS-61   ASTRONOMICAL DATA ANALYSIS SOFTWARE AND SYSTEMS III - (ADASS III)
eds. Dennis R. Crabtree, R. J. Hanisch, and Jeannette Barnes
ISBN 0-937707-80-5

Vol. CS-62   THE NATURE AND EVOLUTIONARY STATUS OF HERBIG Ae/Be STARS
eds. Pik Sin Thé, Mario R. Pérez, and Ed P. J. van den Heuvel
ISBN 0-9837707-81-3

Vol. CS-63   SEVENTY-FIVE YEARS OF HIRAYAMA ASTEROID FAMILIES:
THE ROLE OF COLLISIONS IN THE SOLAR SYSTEM HISTORY
eds. Yoshihide Kozai, Richard P. Binzel, and Tomohiro Hirayama
ISBN 0-937707-82-1

Vol. CS-64*   COOL STARS, STELLAR SYSTEMS, AND THE SUN:
Eighth Cambridge Workshop
ed. Jean-Pierre Caillault
ISBN 0-937707-83-X

Vol. CS-65*   CLOUDS, CORES, AND LOW MASS STARS:
The Fourth Haystack Observatory Conference
eds. Dan P. Clemens and Richard Barvainis
ISBN 0-937707-84-8

Vol. CS-66*   PHYSICS OF THE GASEOUS AND STELLAR DISKS OF THE GALAXY
ed. Ivan R. King
ISBN 0-937707-85-6

Vol. CS-67   UNVEILING LARGE-SCALE STRUCTURES BEHIND THE MILKY WAY
eds. C. Balkowski and R. C. Kraan-Korteweg
ISBN 0-937707-86-4

Vol. CS-68*   SOLAR ACTIVE REGION EVOLUTION:
COMPARING MODELS WITH OBSERVATIONS
eds. K. S. Balasubramaniam and George W. Simon
ISBN 0-937707-87-2

Vol. CS-69   REVERBERATION MAPPING OF THE BROAD-LINE REGION
IN ACTIVE GALACTIC NUCLEI
eds. P. M. Gondhalekar, K. Horne, and B. M. Peterson
ISBN 0-937707-88-0

Vol. CS-70*   GROUPS OF GALAXIES
eds. Otto-G. Richter and Kirk Borne
ISBN 0-937707-89-9

ASP CONFERENCE SERIES VOLUMES
Published by the Astronomical Society of the Pacific

PUBLISHED: 1995 (* asterisk means OUT OF STOCK)

Vol. CS-71   TRIDIMENSIONAL OPTICAL SPECTROSCOPIC METHODS IN ASTROPHYSICS, IAU Colloquium 149
eds. Georges Comte and Michel Marcelin
ISBN 0-937707-90-2

Vol. CS-72   MILLISECOND PULSARS: A DECADE OF SURPRISE
eds. A. S Fruchter, M. Tavani, and D. C. Backer
ISBN 0-937707-91-0

Vol. CS-73   AIRBORNE ASTRONOMY SYMPOSIUM ON THE GALACTIC ECOSYSTEM: FROM GAS TO STARS TO DUST
eds. Michael R. Haas, Jacqueline A. Davidson, and Edwin F. Erickson
ISBN 0-937707-92-9

Vol. CS-74   PROGRESS IN THE SEARCH FOR EXTRATERRESTRIAL LIFE:
1993 Bioastronomy Symposium
ed. G. Seth Shostak
ISBN 0-937707-93-7

Vol. CS-75   MULTI-FEED SYSTEMS FOR RADIO TELESCOPES
eds. Darrel T. Emerson and John M. Payne
ISBN 0-937707-94-5

Vol. CS-76   GONG '94: HELIO- AND ASTERO-SEISMOLOGY FROM THE EARTH AND SPACE
eds. Roger K. Ulrich, Edward J. Rhodes, Jr., and Werner Däppen
ISBN 0-937707-95-3

Vol. CS-77   ASTRONOMICAL DATA ANALYSIS SOFTWARE AND SYSTEMS IV - (ADASS IV)
eds. R. A. Shaw, H. E. Payne, and J. J. E. Hayes
ISBN 0-937707-96-1

Vol. CS-78   ASTROPHYSICAL APPLICATIONS OF POWERFUL NEW DATABASES:
Joint Discussion No. 16 of the 22nd General Assembly of the IAU
eds. S. J. Adelman and W. L. Wiese
ISBN 0-937707-97-X

Vol. CS-79*   ROBOTIC TELESCOPES: CURRENT CAPABILITIES, PRESENT DEVELOPMENTS, AND FUTURE PROSPECTS FOR AUTOMATED ASTRONOMY
eds. Gregory W. Henry and Joel A. Eaton
ISBN 0-937707-98-8

Vol. CS-80*   THE PHYSICS OF THE INTERSTELLAR MEDIUM AND INTERGALACTIC MEDIUM
eds. A. Ferrara, C. F. McKee, C. Heiles, and P. R. Shapiro
ISBN 0-937707-99-6

Vol. CS-81   LABORATORY AND ASTRONOMICAL HIGH RESOLUTION SPECTRA
eds. A. J. Sauval, R. Blomme, and N. Grevesse
ISBN 1-886733-01-5

Vol. CS-82*   VERY LONG BASELINE INTERFEROMETRY AND THE VLBA
eds. J. A. Zensus, P. J. Diamond, and P. J. Napier
ISBN 1-886733-02-3

Vol. CS-83*   ASTROPHYSICAL APPLICATIONS OF STELLAR PULSATION, IAU Colloquium 155
eds. R. S. Stobie and P. A. Whitelock
ISBN 1-886733-03-1

ATLAS   INFRARED ATLAS OF THE ARCTURUS SPECTRUM, 0.9 - 5.3 μm
eds. Kenneth Hinkle, Lloyd Wallace, and William Livingston
ISBN: 1-886733-04-X

ASP CONFERENCE SERIES VOLUMES
Published by the Astronomical Society of the Pacific

PUBLISHED: 1995 (* asterisk means OUT OF STOCK)

Vol. CS-84   THE FUTURE UTILIZATION OF SCHMIDT TELESCOPES, IAU Colloquium 148
              eds. Jessica Chapman, Russell Cannon, Sandra Harrison, and Bambang Hidayat
              ISBN 1-886733-05-8

Vol. CS-85*   CAPE WORKSHOP ON MAGNETIC CATACLYSMIC VARIABLES
              eds. D. A. H. Buckley and B. Warner
              ISBN 1-886733-06-6

Vol. CS-86   FRESH VIEWS OF ELLIPTICAL GALAXIES
              eds. Alberto Buzzoni, Alvio Renzini, and Alfonso Serrano
              ISBN 1-886733-07-4

PUBLISHED: 1996

Vol. CS-87   NEW OBSERVING MODES FOR THE NEXT CENTURY
              eds. Todd Boroson, John Davies, and Ian Robson
              ISBN 1-886733-08-2

Vol. CS-88*   CLUSTERS, LENSING, AND THE FUTURE OF THE UNIVERSE
              eds. Virginia Trimble and Andreas Reisenegger
              ISBN 1-886733-09-0

Vol. CS-89   ASTRONOMY EDUCATION: CURRENT DEVELOPMENTS,
              FUTURE COORDINATION
              ed. John R. Percy
              ISBN 1-886733-10-4

Vol. CS-90   THE ORIGINS, EVOLUTION, AND DESTINIES OF BINARY STARS
              IN CLUSTERS
              eds. E. F. Milone and J. -C. Mermilliod
              ISBN 1-886733-11-2

Vol. CS-91   BARRED GALAXIES, IAU Colloquium 157
              eds. R. Buta, D. A. Crocker, and B. G. Elmegreen
              ISBN 1-886733-12-0

Vol. CS-92*   FORMATION OF THE GALACTIC HALO .... INSIDE AND OUT
              eds. Heather L. Morrison and Ata Sarajedini
              ISBN 1-886733-13-9

Vol. CS-93   RADIO EMISSION FROM THE STARS AND THE SUN
              eds. A. R. Taylor and J. M. Paredes
              ISBN 1-886733-14-7

Vol. CS-94   MAPPING, MEASURING, AND MODELING THE UNIVERSE
              eds. Peter Coles, Vicent J. Martinez, and Maria-Jesus Pons-Borderia
              ISBN 1-886733-15-5

Vol. CS-95   SOLAR DRIVERS OF INTERPLANETARY AND TERRESTRIAL DISTURBANCES:
              Proceedings of 16th International Workshop National Solar
              Observatory/Sacramento Peak
              eds. K. S. Balasubramaniam, Stephen L. Keil, and Raymond N. Smartt
              ISBN 1-886733-16-3

Vol. CS-96   HYDROGEN-DEFICIENT STARS
              eds. C. S. Jeffery and U. Heber
              ISBN 1-886733-17-1

Vol. CS-97   POLARIMETRY OF THE INTERSTELLAR MEDIUM
              eds. W. G. Roberge and D. C. B. Whittet
              ISBN 1-886733-18-X

ASP CONFERENCE SERIES VOLUMES
Published by the Astronomical Society of the Pacific

PUBLISHED: 1996 (* asterisk means OUT OF STOCK)

Vol. CS-98
FROM STARS TO GALAXIES: THE IMPACT OF STELLAR PHYSICS
ON GALAXY EVOLUTION
eds. Claus Leitherer, Uta Fritze-von Alvensleben, and John Huchra
ISBN 1-886733-19-8

Vol. CS-99
COSMIC ABUNDANCES:
Proceedings of the 6th Annual October Astrophysics Conference
eds. Stephen S. Holt and George Sonneborn
ISBN 1-886733-20-1

Vol. CS-100
ENERGY TRANSPORT IN RADIO GALAXIES AND QUASARS
eds. P. E. Hardee, A. H. Bridle, and J. A. Zensus
ISBN 1-886733-21-X

Vol. CS-101
ASTRONOMICAL DATA ANALYSIS SOFTWARE AND SYSTEMS V – (ADASS V)
eds. George H. Jacoby and Jeannette Barnes
ISBN 1080-7926

Vol. CS-102
THE GALACTIC CENTER, 4th ESO/CTIO Workshop
ed. Roland Gredel
ISBN 1-886733-22-8

Vol. CS-103
THE PHYSICS OF LINERS IN VIEW OF RECENT OBSERVATIONS
eds. M. Eracleous, A. Koratkar, C. Leitherer, and L. Ho
ISBN 1-886733-23-6

Vol. CS-104
PHYSICS, CHEMISTRY, AND DYNAMICS OF INTERPLANETARY DUST,
IAU Colloquium 150
eds. Bo Å. S. Gustafson and Martha S. Hanner
ISBN 1-886733-24-4

Vol. CS-105
PULSARS: PROBLEMS AND PROGRESS, IAU Colloquium 160
ed. S. Johnston, M. A. Walker, and M. Bailes
ISBN 1-886733-25-2

Vol. CS-106
THE MINNESOTA LECTURES ON EXTRAGALACTIC NEUTRAL HYDROGEN
ed. Evan D. Skillman
ISBN 1-886733-26-0

Vol. CS-107
COMPLETING THE INVENTORY OF THE SOLAR SYSTEM:
A Symposium held in conjunction with the 106th Annual Meeting of the ASP
eds. Terrence W. Rettig and Joseph M. Hahn
ISBN 1-886733-27-9

Vol. CS-108
M.A.S.S. -- MODEL ATMOSPHERES AND SPECTRUM SYNTHESIS:
5th Vienna - Workshop
eds. Saul J. Adelman, Friedrich Kupka, and Werner W. Weiss
ISBN 1-886733-28-7

Vol. CS-109
COOL STARS, STELLAR SYSTEMS, AND THE SUN: Ninth Cambridge Workshop
eds. Roberto Pallavicini and Andrea K. Dupree
ISBN 1-886733-29-5

Vol. CS-110
BLAZAR CONTINUUM VARIABILITY
eds. H. R. Miller, J. R. Webb, and J. C. Noble
ISBN 1-886733-30-9

Vol. CS-111
MAGNETIC RECONNECTION IN THE SOLAR ATMOSPHERE:
Proceedings of a Yohkoh Conference
eds. R. D. Bentley and J. T. Mariska
ISBN 1-886733-31-7

ASP CONFERENCE SERIES VOLUMES
Published by the Astronomical Society of the Pacific

PUBLISHED: 1996 (* asterisk means OUT OF STOCK)

Vol. CS-112   THE HISTORY OF THE MILKY WAY AND ITS SATELLITE SYSTEM
eds. Andreas Burkert, Dieter H. Hartmann, and Steven R. Majewski
ISBN 1-886733-32-5

PUBLISHED: 1997

Vol. CS-113   EMISSION LINES IN ACTIVE GALAXIES: NEW METHODS AND TECHNIQUES,
IAU Colloquium 159
eds. B. M. Peterson, F.-Z. Cheng, and A. S. Wilson
ISBN 1-886733-33-3

Vol. CS-114   YOUNG GALAXIES AND QSO ABSORPTION-LINE SYSTEMS
eds. Sueli M. Viegas, Ruth Gruenwald, and Reinaldo R. de Carvalho
ISBN 1-886733-34-1

Vol. CS-115   GALACTIC CLUSTER COOLING FLOWS
ed. Noam Soker
ISBN 1-886733-35-X

Vol. CS-116   THE SECOND STROMLO SYMPOSIUM:
THE NATURE OF ELLIPTICAL GALAXIES
eds. M. Arnaboldi, G. S. Da Costa, and P. Saha
ISBN 1-886733-36-8

Vol. CS-117   DARK AND VISIBLE MATTER IN GALAXIES
eds. Massimo Persic and Paolo Salucci
ISBN-1-886733-37-6

Vol. CS-118   FIRST ADVANCES IN SOLAR PHYSICS EUROCONFERENCE:
ADVANCES IN THE PHYSICS OF SUNSPOTS
eds. B. Schmieder. J. C. del Toro Iniesta, and M. Vázquez
ISBN 1-886733-38-4

Vol. CS-119   PLANETS BEYOND THE SOLAR SYSTEM
AND THE NEXT GENERATION OF SPACE MISSIONS
ed. David R. Soderblom
ISBN 1-886733-39-2

Vol. CS-120   LUMINOUS BLUE VARIABLES: MASSIVE STARS IN TRANSITION
eds. Antonella Nota and Henny J. G. L. M. Lamers
ISBN 1-886733-40-6

Vol. CS-121   ACCRETION PHENOMENA AND RELATED OUTFLOWS, IAU Colloquium 163
eds. D. T. Wickramasinghe, G. V. Bicknell, and L. Ferrario
ISBN 1-886733-41-4

Vol. CS-122   FROM STARDUST TO PLANETESIMALS:
Symposium held as part of the 108th Annual Meeting of the ASP
eds. Yvonne J. Pendleton and A. G. G. M. Tielens
ISBN 1-886733-42-2

Vol. CS-123   THE 12th 'KINGSTON MEETING': COMPUTATIONAL ASTROPHYSICS
eds. David A. Clarke and Michael J. West
ISBN 1-886733-43-0

Vol. CS-124   DIFFUSE INFRARED RADIATION AND THE IRTS
eds. Haruyuki Okuda, Toshio Matsumoto, and Thomas Roellig
ISBN 1-886733-44-9

Vol. CS-125   ASTRONOMICAL DATA ANALYSIS SOFTWARE AND SYSTEMS VI
eds. Gareth Hunt and H. E. Payne
ISBN 1-886733-45-7

ASP CONFERENCE SERIES VOLUMES
Published by the Astronomical Society of the Pacific

PUBLISHED: 1997 (* asterisk means OUT OF STOCK)

Vol. CS-126   FROM QUANTUM FLUCTUATIONS TO COSMOLOGICAL STRUCTURES
eds. David Valls-Gabaud, Martin A. Hendry, Paolo Molaro, and Khalil Chamcham
ISBN 1-886733-46-5

Vol. CS-127   PROPER MOTIONS AND GALACTIC ASTRONOMY
ed. Roberta M. Humphreys
ISBN 1-886733-47-3

Vol. CS-128   MASS EJECTION FROM AGN (Active Galactic Nuclei)
eds. N. Arav, I. Shlosman, and R. J. Weymann
ISBN 1-886733-48-1

Vol. CS-129   THE GEORGE GAMOW SYMPOSIUM
eds. E. Harper, W. C. Parke, and G. D. Anderson
ISBN 1-886733-49-X

Vol. CS-130   THE THIRD PACIFIC RIM CONFERENCE ON
RECENT DEVELOPMENT ON BINARY STAR RESEARCH
eds. Kam-Ching Leung
ISBN 1-886733-50-3

PUBLISHED: 1998

Vol. CS-131   BOULDER-MUNICH II: PROPERTIES OF HOT, LUMINOUS STARS
ed. Ian D. Howarth
ISBN 1-886733-51-1

Vol. CS-132   STAR FORMATION WITH THE INFRARED SPACE OBSERVATORY (ISO)
eds. João L. Yun and René Liseau
ISBN 1-886733-52-X

Vol. CS-133   SCIENCE WITH THE NGST (Next Generation Space Telescope)
eds. Eric P. Smith and Anuradha Koratkar
ISBN 1-886733-53-8

Vol. CS-134   BROWN DWARFS AND EXTRASOLAR PLANETS
eds. Rafael Rebolo, Eduardo L. Martin, and Maria Rosa Zapatero Osorio
ISBN 1-886733-54-6

Vol. CS-135   A HALF CENTURY OF STELLAR PULSATION INTERPRETATIONS:
A TRIBUTE TO ARTHUR N. COX
eds. P. A. Bradley and J. A. Guzik
ISBN 1-886733-55-4

Vol. CS-136   GALACTIC HALOS: A UC SANTA CRUZ WORKSHOP
ed. Dennis Zaritsky
ISBN 1-886733-56-2

Vol. CS-137   WILD STARS IN THE OLD WEST: PROCEEDINGS OF THE 13th NORTH
AMERICAN WORKSHOP ON CATACLYSMIC VARIABLES
AND RELATED OBJECTS
eds. S. Howell, E. Kuulkers, and C. Woodward
ISBN 1-886733-57-0

Vol. CS-138   1997 PACIFIC RIM CONFERENCE ON STELLAR ASTROPHYSICS
eds. Kwing Lam Chan, K. S. Cheng, and H. P. Singh
ISBN 1-886733-58-9

Vol. CS-139   PRESERVING THE ASTRONOMICAL WINDOWS:
Proceedings of Joint Discussion No. 5 of the 23rd General Assembly of the IAU
eds. Syuzo Isobe and Tomohiro Hirayama
ISBN 1-886733-59-7

ASP CONFERENCE SERIES VOLUMES
Published by the Astronomical Society of the Pacific

PUBLISHED: 1998 (* asterisk means OUT OF STOCK)

Vol. CS-140   SYNOPTIC SOLAR PHYSICS --18th NSO/Sacramento Peak Summer Workshop
eds. K. S. Balasubramaniam, J. W. Harvey, and D. M. Rabin
ISBN 1-886733-60-0

Vol. CS-141   ASTROPHYSICS FROM ANTARCTICA:
A Symposium held as a part of the 109th Annual Meeting of the ASP
eds. Giles Novak and Randall H. Landsberg
ISBN 1-886733-61-9

Vol. CS-142   THE STELLAR INITIAL MASS FUNCTION: 38th Herstmonceux Conference
eds. Gerry Gilmore and Debbie Howell
ISBN 1-886733-62-7

Vol. CS-143*   THE SCIENTIFIC IMPACT OF THE GODDARD HIGH RESOLUTION
SPECTROGRAPH (GHRS)
eds. John C. Brandt, Thomas B. Ake III, and Carolyn Collins Petersen
ISBN 1-886733-63-5

Vol. CS-144   RADIO EMISSION FROM GALACTIC AND EXTRAGALACTIC COMPACT
SOURCES, IAU Colloquium 164
eds. J. Anton Zensus, G. B. Taylor, and J. M. Wrobel
ISBN 1-886733-64-3

Vol. CS-145   ASTRONOMICAL DATA ANALYSIS SOFTWARE AND SYSTEMS VII – (ADASS VII)
eds. Rudolf Albrecht, Richard N. Hook, and Howard A. Bushouse
ISBN 1-886733-65-1

Vol. CS-146   THE YOUNG UNIVERSE GALAXY FORMATION
AND EVOLUTION AT INTERMEDIATE AND HIGH REDSHIFT
eds. S. D'Odorico, A. Fontana, and E. Giallongo
ISBN 1-886733-66-X

Vol. CS-147   ABUNDANCE PROFILES: DIAGNOSTIC TOOLS FOR GALAXY HISTORY
eds. Daniel Friedli, Mike Edmunds, Carmelle Robert, and Laurent Drissen
ISBN 1-886733-67-8

Vol. CS-148   ORIGINS
eds. Charles E. Woodward, J. Michael Shull, and Harley A. Thronson, Jr.
ISBN 1-886733-68-6

Vol. CS-149   SOLAR SYSTEM FORMATION AND EVOLUTION
eds. D. Lazzaro, R. Vieira Martins, S. Ferraz-Mello, J. Fernández, and C. Beaugé
ISBN 1-886733-69-4

Vol. CS-150   NEW PERSPECTIVES ON SOLAR PROMINENCES, IAU Colloquium 167
eds. David Webb, David Rust, and Brigitte Schmieder
ISBN 1-886733-70-8

Vol. CS-151   COSMIC MICROWAVE BACKGROUND
AND LARGE SCALE STRUCTURES OF THE UNIVERSE
eds. Yong-Ik Byun and Kin-Wang Ng
ISBN 1-886733-71-6

Vol. CS-152   FIBER OPTICS IN ASTRONOMY III
eds. S. Arribas, E. Mediavilla, and F. Watson
ISBN 1-886733-72-4

Vol. CS-153   LIBRARY AND INFORMATION SERVICES IN ASTRONOMY III -- (LISA III)
eds. Uta Grothkopf, Heinz Andernach, Sarah Stevens-Rayburn,
and Monique Gomez
ISBN 1-886733-73-2

ASP CONFERENCE SERIES VOLUMES
Published by the Astronomical Society of the Pacific

PUBLISHED: 1998 (* asterisk means OUT OF STOCK)

Vol. CS-154   COOL STARS, STELLAR SYSTEMS AND THE SUN: Tenth Cambridge Workshop
eds. Robert A. Donahue and Jay A. Bookbinder
ISBN 1-886733-74-0

Vol. CS-155   SECOND ADVANCES IN SOLAR PHYSICS EUROCONFERENCE:
THREE-DIMENSIONAL STRUCTURE OF SOLAR ACTIVE REGIONS
eds. Costas E. Alissandrakis and Brigitte Schmieder
ISBN 1-886733-75-9

PUBLISHED: 1999

Vol. CS-156   HIGHLY REDSHIFTED RADIO LINES
eds. C. L. Carilli, S. J. E. Radford, K. M. Menten, and G. I. Langston
ISBN 1-886733-76-7

Vol. CS-157   ANNAPOLIS WORKSHOP ON MAGNETIC CATACLYSMIC VARIABLES
eds. Coel Hellier and Koji Mukai
ISBN 1-886733-77-5

Vol. CS-158   SOLAR AND STELLAR ACTIVITY: SIMILARITIES AND DIFFERENCES
eds. C. J. Butler and J. G. Doyle
ISBN 1-886733-78-3

Vol. CS-159   BL LAC PHENOMENON
eds. Leo O. Takalo and Aimo Sillanpää
ISBN 1-886733-79-1

Vol. CS-160   ASTROPHYSICAL DISCS: An EC Summer School
eds. J. A. Sellwood and Jeremy Goodman
ISBN 1-886733-80-5

Vol. CS-161   HIGH ENERGY PROCESSES IN ACCRETING BLACK HOLES
eds. Juri Poutanen and Roland Svensson
ISBN 1-886733-81-3

Vol. CS-162   QUASARS AND COSMOLOGY
eds. Gary Ferland and Jack Baldwin
ISBN 1-886733-83-X

Vol. CS-163   STAR FORMATION IN EARLY-TYPE GALAXIES
eds. Jordi Cepa and Patricia Carral
ISBN 1-886733-84-8

Vol. CS-164   ULTRAVIOLET–OPTICAL SPACE ASTRONOMY BEYOND HST
eds. Jon A. Morse, J. Michael Shull, and Anne L. Kinney
ISBN 1-886733-85-6

Vol. CS-165   THE THIRD STROMLO SYMPOSIUM: THE GALACTIC HALO
eds. Brad K. Gibson, Tim S. Axelrod, and Mary E. Putman
ISBN 1-886733-86-4

Vol. CS-166   STROMLO WORKSHOP ON HIGH-VELOCITY CLOUDS
eds. Brad K. Gibson and Mary E. Putman
ISBN 1-886733-87-2

Vol. CS-167   HARMONIZING COSMIC DISTANCE SCALES IN A POST-HIPPARCOS ERA
eds. Daniel Egret and André Heck
ISBN 1-886733-88-0

Vol. CS-168   NEW PERSPECTIVES ON THE INTERSTELLAR MEDIUM
eds. A. R. Taylor, T. L. Landecker, and G. Joncas
ISBN 1-886733-89-9

ASP CONFERENCE SERIES VOLUMES
Published by the Astronomical Society of the Pacific

PUBLISHED: 1999 (* asterisk means OUT OF STOCK)

Vol. CS-169   11th EUROPEAN WORKSHOP ON WHITE DWARFS
eds. J.-E. Solheim and E. G. Meištas
ISBN 1-886733-91-0

Vol. CS-170   THE LOW SURFACE BRIGHTNESS UNIVERSE, IAU Colloquium 171
eds. J. I. Davies, C. Impey, and S. Phillipps
ISBN 1-886733-92-9

Vol. CS-171   LiBeB, COSMIC RAYS, AND RELATED X- AND GAMMA-RAYS
eds. Reuven Ramaty, Elisabeth Vangioni-Flam, Michel Cassé, and Keith Olive
ISBN 1-886733-93-7

Vol. CS-172   ASTRONOMICAL DATA ANALYSIS SOFTWARE AND SYSTEMS VIII
eds. David M. Mehringer, Raymond L. Plante, and Douglas A. Roberts
ISBN 1-886733-94-5

Vol. CS-173   THEORY AND TESTS OF CONVECTION IN STELLAR STRUCTURE:
First Granada Workshop
ed. Álvaro Giménez, Edward F. Guinan, and Benjamín Montesinos
ISBN 1-886733-95-3

Vol. CS-174   CATCHING THE PERFECT WAVE: ADAPTIVE OPTICS AND
INTERFEROMETRY IN THE 21st CENTURY,
A Symposium held as a part of the 110th Annual Meeting of the ASP
eds. Sergio R. Restaino, William Junor, and Nebojsa Duric
ISBN 1-886733-96-1

Vol. CS-175   STRUCTURE AND KINEMATICS OF QUASAR BROAD LINE REGIONS
eds. C. M. Gaskell, W. N. Brandt, M. Dietrich, D. Dultzin-Hacyan,
and M. Eracleous
ISBN 1-886733-97-X

Vol. CS-176   OBSERVATIONAL COSMOLOGY: THE DEVELOPMENT OF GALAXY SYSTEMS
eds. Giuliano Giuricin, Marino Mezzetti, and Paolo Salucci
ISBN 1-58381-000-5

Vol. CS-177   ASTROPHYSICS WITH INFRARED SURVEYS: A Prelude to SIRTF
eds. Michael D. Bicay, Chas A. Beichman, Roc M. Cutri, and Barry F. Madore
ISBN 1-58381-001-3

Vol. CS-178   STELLAR DYNAMOS: NONLINEARITY AND CHAOTIC FLOWS
eds. Manuel Núñez and Antonio Ferriz-Mas
ISBN 1-58381-002-1

Vol. CS-179   ETA CARINAE AT THE MILLENNIUM
eds. Jon A. Morse, Roberta M. Humphreys, and Augusto Damineli
ISBN 1-58381-003-X

Vol. CS-180   SYNTHESIS IMAGING IN RADIO ASTRONOMY II
eds. G. B. Taylor, C. L. Carilli, and R. A. Perley
ISBN 1-58381-005-6

Vol. CS-181   MICROWAVE FOREGROUNDS
eds. Angelica de Oliveira-Costa and Max Tegmark
ISBN 1-58381-006-4

Vol. CS-182   GALAXY DYNAMICS: A Rutgers Symposium
eds. David Merritt, J. A. Sellwood, and Monica Valluri
ISBN 1-58381-007-2

Vol. CS-183   HIGH RESOLUTION SOLAR PHYSICS: THEORY, OBSERVATIONS,
AND TECHNIQUES
eds. T. R. Rimmele, K. S. Balasubramaniam, and R. R. Radick
ISBN 1-58381-009-9

ASP CONFERENCE SERIES VOLUMES

Published by the Astronomical Society of the Pacific

PUBLISHED: 1999 (* asterisk means OUT OF STOCK)

Vol. CS-184   THIRD ADVANCES IN SOLAR PHYSICS EUROCONFERENCE:
MAGNETIC FIELDS AND OSCILLATIONS
eds. B. Schmieder, A. Hofmann, and J. Staude
ISBN 1-58381-010-2

Vol. CS-185   PRECISE STELLAR RADIAL VELOCITIES, IAU Colloquium 170
eds. J. B. Hearnshaw and C. D. Scarfe
ISBN 1-58381-011-0

Vol. CS-186   THE CENTRAL PARSECS OF THE GALAXY
eds. Heino Falcke, Angela Cotera, Wolfgang J. Duschl, Fulvio Melia,
and Marcia J. Rieke
ISBN 1-58381-012-9

Vol. CS-187   THE EVOLUTION OF GALAXIES ON COSMOLOGICAL TIMESCALES
eds. J. E. Beckman and T. J. Mahoney
ISBN 1-58381-013-7

Vol. CS-188   OPTICAL AND INFRARED SPECTROSCOPY OF CIRCUMSTELLAR MATTER
eds. Eike W. Guenther, Bringfried Stecklum, and Sylvio Klose
ISBN 1-58381-014-5

Vol. CS-189   CCD PRECISION PHOTOMETRY WORKSHOP
eds. Eric R. Craine, Roy A. Tucker, and Jeannette Barnes
ISBN 1-58381-015-3

Vol. CS-190   GAMMA-RAY BURSTS: THE FIRST THREE MINUTES
eds. Juri Poutanen and Roland Svensson
ISBN 1-58381-016-1

Vol. CS-191   PHOTOMETRIC REDSHIFTS AND HIGH REDSHIFT GALAXIES
eds. Ray J. Weymann, Lisa J. Storrie-Lombardi, Marcin Sawicki,
and Robert J. Brunner
ISBN 1-58381-017-X

Vol. CS-192   SPECTROPHOTOMETRIC DATING OF STARS AND GALAXIES
ed. I. Hubeny, S. R. Heap, and R. H. Cornett
ISBN 1-58381-018-8

Vol. CS-193   THE HY-REDSHIFT UNIVERSE:
GALAXY FORMATION AND EVOLUTION AT HIGH REDSHIFT
eds. Andrew J. Bunker and Wil J. M. van Breugel
ISBN 1-58381-019-6

Vol. CS-194   WORKING ON THE FRINGE:
OPTICAL AND IR INTERFEROMETRY FROM GROUND AND SPACE
eds. Stephen Unwin and Robert Stachnik
ISBN 1-58381-020-X

PUBLISHED: 2000

Vol. CS-195   IMAGING THE UNIVERSE IN THREE DIMENSIONS:
Astrophysics with Advanced Multi-Wavelength Imaging Devices
eds. W. van Breugel and J. Bland-Hawthorn
ISBN 1-58381-022-6

Vol. CS-196   THERMAL EMISSION SPECTROSCOPY AND ANALYSIS OF DUST,
DISKS, AND REGOLITHS
eds. Michael L. Sitko, Ann L. Sprague, and David K. Lynch
ISBN: 1-58381-023-4

Vol. CS-197   XVth IAP MEETING DYNAMICS OF GALAXIES:
FROM THE EARLY UNIVERSE TO THE PRESENT
eds. F. Combes, G. A. Mamon, and V. Charmandaris
ISBN: 1-58381-24-2

ASP CONFERENCE SERIES VOLUMES
Published by the Astronomical Society of the Pacific

PUBLISHED: 2000 (* asterisk means OUT OF STOCK)

Vol. CS-198   EUROCONFERENCE ON "STELLAR CLUSTERS AND ASSOCIATIONS: CONVECTION, ROTATION, AND DYNAMOS"
eds. R. Pallavicini, G. Micela, and S. Sciortino
ISBN: 1-58381-25-0

Vol. CS-199   ASYMMETRICAL PLANETARY NEBULAE II: FROM ORIGINS TO MICROSTRUCTURES
eds. J. H. Kastner, N. Soker, and S. Rappaport
ISBN: 1-58381-026-9

Vol. CS-200   CLUSTERING AT HIGH REDSHIFT
eds. A. Mazure, O. Le Fèvre, and V. Le Brun
ISBN: 1-58381-027-7

Vol. CS-201   COSMIC FLOWS 1999: TOWARDS AN UNDERSTANDING OF LARGE-SCALE STRUCTURES
eds. Stéphane Courteau, Michael A. Strauss, and Jeffrey A. Willick
ISBN: 1-58381-028-5

Vol. CS-202   PULSAR ASTRONOMY – 2000 AND BEYOND, IAU Colloquium 177
eds. M. Kramer, N. Wex, and R. Wielebinski
ISBN: 1-58381-029-3

Vol. CS-203   THE IMPACT OF LARGE-SCALE SURVEYS ON PULSATING STAR RESEARCH, IAU Colloquium 176
eds. L. Szabados and D. W. Kurtz
ISBN: 1-58381-030-7

Vol. CS-204   THERMAL AND IONIZATION ASPECTS OF FLOWS FROM HOT STARS: OBSERVATIONS AND THEORY
eds. Henny J. G. L. M. Lamers and Arved Sapar
ISBN: 1-58381-031-5

Vol. CS-205   THE LAST TOTAL SOLAR ECLIPSE OF THE MILLENNIUM IN TURKEY
eds. W. C. Livingston and A. Özgüç
ISBN: 1-58381-032-3

Vol. CS-206   HIGH ENERGY SOLAR PHYSICS – *ANTICIPATING HESSI*
eds. Reuven Ramaty and Natalie Mandzhavidze
ISBN: 1-58381-033-1

Vol. CS-207   NGST SCIENCE AND TECHNOLOGY EXPOSITION
eds. Eric P. Smith and Knox S. Long
ISBN: 1-58381-036-6

ATLAS   VISIBLE AND NEAR INFRARED ATLAS OF THE ARCTURUS SPECTRUM 3727-9300 Å
eds. Kenneth Hinkle, Lloyd Wallace, Jeff Valenti, and Dianne Harmer
ISBN: 1-58381-037-4

Vol. CS-208   POLAR MOTION: HISTORICAL AND SCIENTIFIC PROBLEMS, IAU Colloquium 178
eds. Steven Dick, Dennis McCarthy, and Brian Luzum
ISBN: 1-58381-039-0

Vol. CS-209   SMALL GALAXY GROUPS, IAU Colloquium 174
eds. Mauri J. Valtonen and Chris Flynn
ISBN: 1-58381-040-4

Vol. CS-210   DELTA SCUTI AND RELATED STARS: Reference Handbook and Proceedings of the 6th Vienna Workshop in Astrophysics
eds. Michel Breger and Michael Houston Montgomery
ISBN: 1-58381-043-9

PUBLISHED: 2000 (* asterisk means OUT OF STOCK)

Vol. CS-211   MASSIVE STELLAR CLUSTERS
eds. Ariane Lançon and Christian M. Boily
ISBN: 1-58381-042-0

Vol. CS-212   FROM GIANT PLANETS TO COOL STARS
eds. Caitlin A. Griffith and Mark S. Marley
ISBN: 1-58381-041-2

Vol. CS-213   BIOASTRONOMY `99: A NEW ERA IN BIOASTRONOMY
eds. Guillermo A. Lemarchand and Karen J. Meech
ISBN: 1-58381-044-7

Vol. CS-214   THE Be PHENOMENON IN EARLY-TYPE STARS, IAU Colloquium 175
eds. Myron A. Smith, Huib F. Henrichs and Juan Fabregat
ISBN: 1-58381-045-5

Vol. CS-215   COSMIC EVOLUTION AND GALAXY FORMATION:
STRUCTURE, INTERACTIONS AND FEEDBACK
The 3rd Guillermo Haro Astrophysics Conference
eds. José Franco, Elena Terlevich, Omar López-Cruz, and Itziar Aretxaga
ISBN: 1-58381-046-3

Vol. CS-216   ASTRONOMICAL DATA ANALYSIS SOFTWARE AND SYSTEMS IX
eds. Nadine Manset, Christian Veillet, and Dennis Crabtree
ISBN: 1-58381-047-1   ISSN: 1080-7926

Vol. CS-217   IMAGING AT RADIO THROUGH SUBMILLIMETER WAVELENGTHS
eds. Jeffrey G. Mangum and Simon J. E. Radford
ISBN: 1-58381-049-8

Vol. CS-218   MAPPING THE HIDDEN UNIVERSE: THE UNIVERSE BEHIND THE MILKYWAY
THE UNIVERSE IN HI
eds. Renée C. Kraan-Korteweg, Patricia A. Henning, and Heinz Andernach
ISBN: 1-58381-050-1

Vol. CS-219   DISKS, PLANETESIMALS, AND PLANETS
eds. F. Garzón, C. Eiroa, D. de Winter, and T. J. Mahoney
ISBN: 1-58381-051-X

Vol. CS-220   AMATEUR - PROFESSIONAL PARTNERSHIPS IN ASTRONOMY:
The 111th Annual Meeting of the ASP
eds. John R. Percy and Joseph B. Wilson
ISBN: 1-58381-052-8

Vol. CS-221   STARS, GAS AND DUST IN GALAXIES: EXPLORING THE LINKS
eds. Danielle Alloin, Knut Olsen, and Gaspar Galaz
ISBN: 1-58381-053-6

PUBLISHED: 2001

Vol. CS-222   THE PHYSICS OF GALAXY FORMATION
eds. M. Umemura and H. Susa
ISBN: 1-58381-054-4

Vol. CS-223   COOL STARS, STELLAR SYSTEMS AND THE SUN:
Eleventh Cambridge Workshop
eds. Ramón J. García López, Rafael Rebolo, and María Zapatero Osorio
ISBN: 1-58381-056-0

Vol. CS-224   PROBING THE PHYSICS OF ACTIVE GALACTIC NUCLEI
BY MULTIWAVELENGTH MONITORING
eds. Bradley M. Peterson, Ronald S. Polidan, and Richard W. Pogge
ISBN: 1-58381-055-2

ASP CONFERENCE SERIES VOLUMES
Published by the Astronomical Society of the Pacific

PUBLISHED: 2001 (* asterisk means OUT OF STOCK)

Vol. CS-225   VIRTUAL OBSERVATORIES OF THE FUTURE
eds. Robert J. Brunner, S. George Djorgovski, and Alex S. Szalay
ISBN: 1-58381-057-9

Vol. CS-226   12th EUROPEAN CONFERENCE ON WHITE DWARFS
eds. J. L. Provencal, H. L. Shipman, J. MacDonald, and S. Goodchild
ISBN: 1-58381-058-7

Vol. CS-227   BLAZAR DEMOGRAPHICS AND PHYSICS
eds. Paolo Padovani and C. Megan Urry
ISBN: 1-58381-059-5

Vol. CS-228   DYNAMICS OF STAR CLUSTERS AND THE MILKY WAY
eds. S. Deiters, B. Fuchs, A. Just, R. Spurzem, and R. Wielen
ISBN: 1-58381-060-9

Vol. CS-229   EVOLUTION OF BINARY AND MULTIPLE STAR SYSTEMS
A Meeting in Celebration of Peter Eggleton's 60th Birthday
eds. Ph. Podsiadlowski, S. Rappaport, A. R. King, F. D'Antona, and L. Burderi
IBSN: 1-58381-061-7

Vol. CS-230   GALAXY DISKS AND DISK GALAXIES
eds. Jose G. Funes, S. J. and Enrico Maria Corsini
ISBN: 1-58381-063-3

Vol. CS-231   TETONS 4: GALACTIC STRUCTURE, STARS, AND
THE INTERSTELLAR MEDIUM
eds. Charles E. Woodward, Michael D. Bicay, and J. Michael Shull
ISBN: 1-58381-064-1

Vol. CS-232   THE NEW ERA OF WIDE FIELD ASTRONOMY
eds. Roger Clowes, Andrew Adamson, and Gordon Bromage
ISBN: 1-58381-065-X

Vol. CS-233   P CYGNI 2000: 400 YEARS OF PROGRESS
eds. Mart de Groot and Christiaan Sterken
ISBN: 1-58381-070-6

Vol. CS-234   X-RAY ASTRONOMY 2000
eds. R. Giacconi, S. Serio, and L. Stella
ISBN: 1-58381-071-4

Vol. CS-235   SCIENCE WITH THE ATACAMA LARGE MILLIMETER ARRAY (ALMA)
ed. Alwyn Wootten
ISBN: 1-58381-072-2

Vol. CS-236   ADVANCED SOLAR POLARIMETRY: THEORY, OBSERVATION, AND
INSTRUMENTATION, The 20th Sacramento Peak Summer Workshop
ed. M. Sigwarth
ISBN: 1-58381-073-0

Vol. CS-237   GRAVITATIONAL LENSING: RECENT PROGRESS AND FUTURE GOALS
eds. Tereasa G. Brainerd and Christopher S. Kochanek
ISBN: 1-58381-074-9

Vol. CS-238   ASTRONOMICAL DATA ANALYSIS SOFTWARE AND SYSTEMS X
eds. F. R. Harnden, Jr., Francis A. Primini, and Harry E. Payne
ISBN: 1-58381-075-7

Vol. CS-239   MICROLENSING 2000: A NEW ERA OF MICROLENSING ASTROPHYSICS
ed. John W. Menzies and Penny D. Sackett
ISBN: 1-58381-076-5

ASP CONFERENCE SERIES VOLUMES
Published by the Astronomical Society of the Pacific

PUBLISHED: 2001 (* asterisk means OUT OF STOCK)

Vol. CS-240   GAS AND GALAXY EVOLUTION,
A Conference in Honor of the 20th Anniversary of the VLA
eds. J. E. Hibbard, M. P. Rupen, and J. H. van Gorkom
ISBN: 1-58381-077-3

Vol. CS-241   CS-241 THE 7TH TAIPEI ASTROPHYSICS WORKSHOP ON
COSMIC RAYS IN THE UNIVERSE
ed. Chung-Ming Ko
ISBN: 1-58381-079-X

Vol. CS-242   ETA CARINAE AND OTHER MYSTERIOUS STARS:
THE HIDDEN OPPORTUNITIES OF EMISSION SPECTROSCOPY
eds. Theodore R. Gull, Sveneric Johannson, and Kris Davidson
ISBN: 1-58381-080-3

Vol. CS-243   FROM DARKNESS TO LIGHT:
ORIGIN AND EVOLUTION OF YOUNG STELLAR CLUSTERS
eds. Thierry Montmerle and Philippe André
ISBN: 1-58381-081-1

Vol. CS-244   YOUNG STARS NEAR EARTH: PROGRESS AND PROSPECTS
eds. Ray Jayawardhana and Thomas P. Greene
ISBN: 1-58381-082-X

Vol. CS-245   ASTROPHYSICAL AGES AND TIME SCALES
eds. Ted von Hippel, Chris Simpson, and Nadine Manset
ISBN: 1-58381-083-8

Vol. CS-246   SMALL TELESCOPE ASTRONOMY ON GLOBAL SCALES, IAU Colloquium 183
eds. Wen-Ping Chen, Claudia Lemme, and Bohdan Paczyński,
ISBN: 1-58381-084-6

LISTINGS OF IAU VOLUMES MAY BE FOUND ON THE NEXT PAGE

INTERNATIONAL ASTRONOMICAL UNION (IAU) VOLUMES
Published by the Astronomical Society of the Pacific

PUBLISHED: 1999

Vol. No. 190   NEW VIEWS OF THE MAGELLANIC CLOUDS
eds. You-Hua Chu, Nicholas B. Suntzeff, James E. Hesser,
and David A. Bohlender
ISBN: 1-58381-021-8

Vol. No. 191   ASYMPTOTIC GIANT BRANCH STARS
eds. T. Le Bertre, A. Lèbre, and C. Waelkens
ISBN: 1-886733-90-2

Vol. No. 192   THE STELLAR CONTENT OF LOCAL GROUP GALAXIES
eds. Patricia Whitelock and Russell Cannon
ISBN: 1-886733-82-1

Vol. No. 193   WOLF-RAYET PHENOMENA IN MASSIVE STARS AND STARBURST GALAXIES
eds. Karel A. van der Hucht, Gloria Koenigsberger, and Philippe R. J. Eenens
ISBN: 1-58381-004-8

Vol. No. 194   ACTIVE GALACTIC NUCLEI AND RELATED PHENOMENA
eds. Yervant Terzian, Daniel Weedman, and Edward Khachikian
ISBN: 1-58381-008-0

PUBLISHED: 2000

Vol. XXIVA   TRANSACTIONS OF THE INTERNATIONAL ASTRONOMICAL UNION
REPORTS ON ASTRONOMY 1996-1999
ed. Johannes Andersen
ISBN: 1-58381-035-8

Vol. No. 195   HIGHLY ENERGETIC PHYSICAL PROCESSES AND MECHANISMS FOR
EMISSION FROM ASTROPHYSICAL PLASMAS
eds. P. C. H. Martens, S. Tsuruta, and M. A. Weber
ISBN: 1-58381-038-2

Vol. No. 197   ASTROCHEMISTRY: FROM MOLECULAR CLOUDS TO PLANETARY SYSTEMS
eds. Y. C. Minh and E. F. van Dishoeck
ISBN: 1-58381-034-X

Vol. No. 198   THE LIGHT ELEMENTS AND THEIR EVOLUTION
eds. L. da Silva, M. Spite, and J. R. de Medeiros
ISBN: 1-58381-048-X

PUBLISHED: 2001

IAU SPS   ASTRONOMY FOR DEVELOPING COUNTRIES
Special Session of the XXIV General Assembly of the IAU
ed. Alan H. Batten
ISBN: 1-58381-067-6

Vol. No. 196   PRESERVING THE ASTRONOMICAL SKY
eds. R. J. Cohen and W. T. Sullivan, III
ISBN: 1-58381-078-1

Vol. No. 200   THE FORMATION OF BINARY STARS
eds. Hans Zinnecker and Robert D. Mathieu
ISBN: 1-58381-068-4

Vol. No. 203   RECENT INSIGHTS INTO THE PHYSICS 0F THE SUN AND HELIOSPHERE:
HIGHLIGHTS FROM SOHO AND OTHER SPACE MISSIONS
eds. Pål Brekke, Bernhard Fleck, and Joseph B. Gurman
ISBN: 1-58381-069-2

INTERNATIONAL ASTRONOMICAL UNION (IAU) VOLUMES
Published by the Astronomical Society of the Pacific

PUBLISHED: 2001

Vol. No. 204 THE EXTRAGALACTIC INFRARED BACKGROUND AND ITS COSMOLOGICAL
IMPLICATIONS
eds. Martin Harwit and Michael G. Hauser
ISBN: 1-58381-062-5

Vol. No. 205 GALAXIES AND THEIR CONSTITUENTS
AT THE HIGHEST ANGULAR RESOLUTIONS
eds. Richard T. Schilizzi, Stuart N. Vogel, Francesco Paresce, and Martin S. Elvis
ISBN: 1-58381-066-8

Vol. XXIVB TRANSACTIONS OF THE INTERNATIONAL ASTRONOMICAL UNION
REPORTS ON ASTRONOMY
ed. Hans Rickman
ISBN: 1-58381-087-0

Complete lists of proceedings of past IAU Meetings are maintained at the
IAU Web site at the URL: http://www.iau.org/publicat.html

Volumes 32 - 189 in the IAU Symposia Series may be ordered from:

Kluwer Academic Publishers
P. O. Box 117
NL 3300 AA Dordrecht
The Netherlands

Kluwer@wKap.com

All book orders or inquiries concerning ASP or IAU volumes listed should be directed to the:

The Astronomical Society of the Pacific Conference Series
390 Ashton Avenue
San Francisco CA 94112-1722 USA

Phone: 415-337-2126
Fax: 415-337-5205

E-mail: catalog@astrosociety.org
Web Site: http://www.astrosociety.org